A Practical Farmer's Journey

A Practical Farmer's Journey

A History of Community, Stewardship, and Resilience

Ronald L. Rosmann

HAMILTON BOOKS

HAMILTON BOOKS

Bloomsbury Publishing Inc, 1359 Broadway, 12th Floor, New York, NY 10018, USA
Bloomsbury Publishing Plc, 50 Bedford Square, London, WC1B 3DP, UK
Bloomsbury Publishing Ireland, 29 Earlsfort Terrace, Dublin 2, D02 AY28, Ireland

BLOOMSBURY and the Diana logo are trademarks of Bloomsbury Publishing Plc

First published in the United States of America 2026

Copyright © Bloomsbury Publishing, 2026

All rights reserved. No part of this publication may be: i) reproduced or transmitted in any form, electronic or mechanical, including photocopying, recording or by means of any information storage or retrieval system without prior permission in writing from the publishers; or ii) used or reproduced in any way for the training, development or operation of artificial intelligence (AI) technologies, including generative AI technologies. The rights holders expressly reserve this publication from the text and data mining exception as per Article 4(3) of the Digital Single Market Directive (EU) 2019/790.

Bloomsbury Publishing Inc does not have any control over, or responsibility for, any third-party websites referred to or in this book. All internet addresses given in this book were correct at the time of going to press. The author and publisher regret any inconvenience caused if addresses have changed or sites have ceased to exist, but can accept no responsibility for any such changes.

Library of Congress Cataloging-in-Publication Data Available

ISBN: HB: 978-0-76189-204-5
PB: 978-0-76188-079-0
ePDF: 978-0-76188-081-3
eBook: 978-0-76188-080-6

Typeset by Deanta Global Publishing Services, Chennai, India

For product safety related questions contact productsafety@bloomsbury.com.

To find out more about our authors and books visit www.bloomsbury.com and sign up for our newsletters.

To my parents, Ray and Ellen Mertens Rosmann

Praise for the Book

Kathleen Delate, Professor of Organic Agriculture and Horticulture, Iowa State University, Ames, Iowa.

Take a fascinating walk with Ron Rosmann on his quest to be a "practical farmer," driven by a deep understanding and respect for the ecology of the farm he stewards. Learn about the strength and stamina Ron and his family demonstrated in charting a new path, away from government machinations that define agriculture today, to a self-sustaining homestead. The book is part historical treatise on complex farm policies, and part journey into farm life from the 1970s to today. Ron easily weaves the importance of participation in political processes with personal farming anecdotes, which often reveal that inimitable, self-deprecating Rosmann sense of humor, that is emblematic of the thoughtful balance in life that the family achieves together.

Having conducted numerous Iowa State University on-farm experiments with Ron, Maria, and their three sons, I can attest to their extreme love of the land through their focus on "crop diversity, water, and soil quality, and other environmental benefits." Our research showed that their farm hosts some of the greatest levels of soil quality in the country, and that this is a direct result of their commitment to long-term soil health through extended crop rotations, integration of livestock, cover-cropping, and compost applications. They serve as role models for anyone interested in transitioning to organic production, and generously share their knowledge across the state, the nation, and the world. It has been a privilege and honor to work with the Rosmanns and reading this book has enhanced my connections to a family so greatly admired by all.

Paul Mugge, Organic Farmer, Sutherland, Iowa

I have enjoyed many experiences similar to those of Ron's growing up on an Iowa farm in the decades following the 1950s, and it brings back a host of memories. It is a great history lesson and so much more. It is also quite thought provoking as Ron shares his thoughts, passions, and a bit of his philosophy and values. He has a wealth of experience as a farmer and policy influencer, and has spent his life helping others and trying to improve the lot of all the myriad of stakeholders in American agriculture. He is advocating for social, economic, environmental, and ecological justice for all Americans.

Ferd Hoefner, Founding Policy Director, National Sustainable Agriculture Coalition, Washington, DC

A Practical Farmer's Journey is indispensable reading for anyone seeking to understand U.S. agriculture, its past, and the new directions we should seek. The book combines an insightful family and community farming history with spot-on observations of U.S. agricultural policy and food system inequities. With bravery and humility, Ron draws the reader into his internal dialogue and openly questions dearly-held assumptions about the role of agriculture in U.S. society and farm policy. Standing on a rock-solid analytical foundation and a lived experience as a practical farmer and a prominent sustainable and organic agriculture advocate, he issues a compelling clarion call for integrating social and ecological justice in farm policy, with a key role for mid-scale family farms in that transformation. Ron's concluding analysis of agriculture's role in climate change and his research-based mitigation strategies are invaluable tools for achieving a more sustainable agriculture and planet.

Ricardo Salvador, Senior Scientist with, Director of, and now Advisor to the Food and Environment Program, Union of Concerned Scientists, Cambridge, Massachusetts

Ron Rosmann has distilled much needed guidance for our era. Although the core of his reflections naturally have to do with his lengthy and successful career as an Iowa farmer, this, to me, boldly addresses the complex technical, socioeconomic, and existential questions of our times with remarkable depth. Rosmann and his wife, Maria, are exactly the type of thoughtful, conscientious, and erudite citizen farmers that Jefferson had in mind when he envisioned an agrarian nation of enlightened agriculturalists. Set aside whether you have an interest in farming or not. If you value sober grappling with the big questions of our time and about humanity's future, read this book to peer into the thinking of a deeply intellectual fellow traveler who describes his understanding of humanity's current condition, his place in it, and most importantly, his assessment of the way forward. Profoundly stimulating and hopeful.

Tom Frantzen, Organic farmer, Alta Vista, Iowa

Ron's work is a comprehensive story of his farm family's journey through decades of change in the Iowa landscape. His research begins with an explanation of what happened to the Native Americans and continues through the social and environmental evolution that brings us to our current times. This is a bold and inspiring book from an exceedingly dedicated farmer who spent his life supporting and defending his rural community.

Vic Madsen, Organic farmer, Audubon, Iowa

Ron has written a fascinating story about the settling of the Iowa prairie where his family farm is located. It explores in a comprehensive manner the inter-connectedness of the history of Native Americans, the Missouri River, and European immigrants settlement to the region. It is both a personal and communal story. The land and its new communities mixed together to shape and form a new country. Its insights into our historical knowledge of the land and rural community, how they have endured, are critical to understanding what to do about our very uncertain future.

Contents

Acknowledgments x

Introduction 1

Part I **Beginning of the Promise** 5

1 Before Iowa Became Iowa 7
2 Embarking upon a Promise: Rosmann's and Westphalia Begin 21
3 Dad and Mom, the Great Depression, and Father Duren 33
4 Grade School at Saint Boniface 47
5 A Working Laboratory 59
6 A Love of Nature, Wildlife, and the Seasons 71
7 The Decade That Defined Who We Would Become 91

Part II **Breaking the Promise** 101

8 Starting to Farm in the 1970s Boom 103
9 Good Times, Hard Times, Getting Married, and Beginning a Family 117

Part III **Restoring the Promise** 127

10 The Real Journey Begins 129
11 Growing People, Not Just Crops 157

12 Matters of Church, Religion, Science, and Technology:
A Layperson's Perspective 167

13 Agriculture and a More Livable Climate 181

14 What Can and Will We Do? 203

Notes 219
Bibliography 234
Index 244
About the Author 249

Acknowledgments

It took over nine years to complete this book because of the realities of organic farming and the responsibilities that come with operating your own business. First of all, I would like to thank my best friend, partner, and teammate, Maria. We have celebrated daily our married life since 1978. She has been a constant source of support and encouragement in this writing endeavor. Without her, I could not have done it. I also want to acknowledge her signature role in helping to financially support our farm. She has done this through the marketing of our beef, pork, popcorn, and eggs, and many other unique grocery items at her on-farm store, "Farm Sweet Farm."

Second, I would like to thank our three sons and their spouses for their guidance and help, too. They had an important role to play in providing suggestions for the book. Two of our sons have now assumed the daily physical and financial obligations of the farm which has enabled me to become more semi-retired.

I owe a great deal of thanks to members of the Practical Farmers of Iowa (PFI) organization that started in 1986. I have known some of these men and women since its beginning. It has to start with Dick and Sharon Thompson, who, along with Larry Kallen, Executive Director of the Iowa Institute for Cooperatives, and Iowa State University soil agronomist, Rick Exner, began to meld their vision into an organization that was both timely and ahead of its time. Through the years, I have learned so much from PFI farmers and have developed lifelong friendships and shared learning experiences in how to become a better steward of the land. How could I name them all? Tom and Irene Frantzen, Vic and Cindy Madsen, Paul and Karen Mugge, Doug Alert and Margaret Smith, and Dan and Lorna Wilson have all been sources of mutual friendship and guidance. I am honored to know them. I am also fortunate to collaborate with both Sally Worley, executive director of Practical Farmers of Iowa and with Teresa Opheim, former executive director of PFI.

There were other visionaries who, early on in the 1980s, saw the need and worked very hard for a more sustainable agriculture both in Iowa and beyond. One was Jerry DeWitt, Iowa State University Extension leader, who championed on-farm research and ensured that the early funding of PFI was made possible. He worked tirelessly to promote the partnership between PFI and agricultural extension in Iowa. Fred Kirschenman is another visionary of brilliant intellect who helped me to form my own ideas about a more sustainable agriculture. He was a 5,000-acre organic farmer from North Dakota who later became director of the Leopold Center for Sustainable Agriculture at Iowa State University. Bob Steffen from Nebraska is another farmer for whom I am grateful for teaching me how to compost manure.

I also wish to acknowledge three individuals who helped in the early reviewing of the manuscript. These are Gary Guthrie and Kamyar Enshayan, as well as Matt Liebman, who gave of his time and expertise to check for scientific accuracy. I would also like to thank Cornelia Butler Flora for her time and expertise in reviewing the manuscript. She serves as distinguished emerita sociology professor at Iowa State University. Her book *Rural Communities: Legacy +Change* is utilized around the world for studying social justice for rural communities. It is in its fifth edition.

There are also other organizations to acknowledge. Three of these are ones that I either was a board member of or worked closely with over the years in agricultural policy, organic farming research, and community and spiritual concerns. These are Ferd Hefner and the National Sustainable Agriculture Coalition, Bob Scowcroft and the Organic Farming Research Foundation, and Jim Ennis and Catholic Rural Life. I also want to acknowledge Kathleen Delate, who leads the Iowa State University Organic Program. She has pioneered organic agricultural science for over twenty-five years with the second oldest organic and conventional research comparison in the nation.

I want to acknowledge my father and mother and my three brothers. Mom opened up the world of nature and its books about it. Dad taught me, by example, how to become a farmer and steward of nature and of the land. My youngest brother Larry, who was born with Down syndrome, taught me how to love and care for one another. My two older brothers, Joe and Mike, provided me with sibling and adult guidance and support, and I grew up wanting to become like them.

Finally, I want to acknowledge the gift of land, our ecology, and our community. I have spent a lifetime living in paradise on this land. I want others to have this opportunity and experience. Our future may depend on it.

Introduction

When you love what you do, work is not a burden. Over time, it can become a vocation filled with great joy and happiness. That is how I feel about the land, farming, and the natural world. It is difficult to separate my occupation from what I believe in. For me, it is trying to do the best possible job in being a good shepherd to the animals we raise and care for, and also trying to be the best possible steward of the land upon which we grow food.

I have been a farmer for over fifty years in western Iowa, though our family has cultivated some of this land for more than 140 years. I started farming in 1973, at the request of my parents, after graduating from Iowa State University with a degree in biology. They were nearing retirement, and my dad's health was in decline. I had not intended to become a farmer; I wanted to attend graduate school. However, the call back to the land was far too strong for me to ignore. Upon returning home, I farmed like most everyone else around me for the first ten years until 1983, when my wife, Maria Vakulskas Rosmann, and I decided to stop the use of all pesticides on our then 320-acre farm. The farm was certified organic in 1994, and 2024 marked the forty-first year of no pesticide use. The farm has grown in size to 700 acres, with two of our three sons farming with us today.

I want to present some of the historical narrative of where I live in Shelby County, fifty miles from the city of Omaha, Nebraska. The small rural community of Westphalia, Iowa, the family farms surrounding it, and the Catholic parish of St. Boniface are where our family calls home. Our town of less than 150 people and our parish celebrated its 150th anniversary in 2022. The farm and town were idyllic places to grow up.

America's story is very short, only about 500 years. In terms of where our family lives, it started about 300 years ago when French fur traders first traveled on the Missouri River, forty-five miles west of our farm. In preparation for this book, I felt the need to learn more about the indigenous peoples who lived here before we did and the importance of the river to them. The fact that we quickly pushed them all aside should still haunt us and remind us to honor and respect them in our actions today. The life-sustaining natural gifts that were theirs for thousands of years were taken from them. The promises made to them by our government were convenient ones at the time, ready to be broken if and when circumstances changed. That seems to be the case for much of our American history.

Much of this book focuses on three of those promises. First is the beginning of a promise in the sense that so much opportunity was created when my ancestors from Slovenia and Germany came to America. There was so much promise in the ability to buy land and to start a farm from the virgin prairie. There was so much promise in starting new businesses, a new town, and a new community held together by the bond of a common faith. Thousands upon thousands of farms and towns did just that all across America. We became a nation of immigrants looking for something better. Many found it.

The second area of focus is the breaking of that promise as time passed and telling the truth about how and why and when it began to happen. Much of my lifetime has been spent living with the reality of the decline of small and medium-sized diversified family farms. Much of my lifetime has been spent living with the decline of rural communities, and the schools, churches, and businesses that embodied them. The promises made and broken have encompassed two competing forces: the exploitation and the abandonment of rural areas.

The third area of focus presents my thoughts and ideas about how to restore that promise. I am offering a plea for learning to live in harmony with our Mother Earth and to do so with a sense of urgency in the face of a rapidly changing environment and climate. Pope Francis said it as well as anyone when he asserted there needs to be a union of social and ecological justice.[1] We cannot keep treating the Earth the way we do. She is crying out. She is angry. She has a right to be, and She will get the last word in this argument.

The word "anger" seems to sum up what so many people around me appear to have these days. There are so many grievances out there; some warranted maybe, some not. Many Americans buy into the notion that bigger and cheaper are always better. But are things really better if they are bigger and cheaper? We still have not factored in the costs of the harm to our own health and the health of our environment in how we live, grow food, and pack both animals and people into small spaces. The injustices and inequities in growing food have always been with us. Government farm policy has a great deal to do with how farming has evolved into what it is today. We could have said "no" to the actions that did harm. We could still do so today.

All life forms, including humans, face many threats. The perils facing basic ecological life-giving cycles and processes that make our planet a place where life of all kinds can thrive may be the most ominous of them. Therefore, it would be wise to revisit the carbon cycle, the nitrogen cycle, the water cycle, photosynthesis, respiration, and so much more. The premise for the book, started in 2015, has changed a bit since then. I thought that much of the latter part of the book would be devoted primarily to our changing climate. Isn't that the great existential crisis the world is facing? It may be, but it is so much more than global warming, rising sea levels, and worsening storms.

With farming, it would be prudent for us to learn and to care about excessive nutrient loads of nitrogen and phosphorus leaving our farm fields. This fertilizer excess is going into the Mississippi River and then on to the Gulf of Mexico. This excess causes massive algal blooms that result in oxygen deprivation for the sea creatures living there. Those that cannot move away from this area are especially hard hit. That includes shrimp, crabs, and oysters. Some years, the Dead Zone has been bigger than the state of New Jersey.

We need to learn and to care about soil erosion and the decreasing quality of soil. We need to learn and to care about monocropping, lack of diversity, and the use of too many pesticides. We need to learn and to care about the competing uses of soil for food versus fuel. Sixty-two percent of Iowa's corn goes to produce ethanol annually. We need to learn and to care about species extinction and the possible loss of entire ecosystems.

Agricultural economists and business analysts must refrain from continuously saying we will need fewer farmers in the future. On the contrary, the world will need many more farmers for many reasons, including viable rural communities and cultures and local and regional food economies, not to mention the satisfaction and human rewards that come from being a farmer. Appropriate future technology needs will continue to require maintaining that reality as its backstop for having more farmers. Technology is not cheap. It has become one of the biggest factors in the demise of smaller and medium-sized farms like ours. Some people have told me that how we farm, with organic practices, is not realistic for others to do. I say hogwash to that! It is untrue. There has not been the will to do it.

A great sense of humility is needed in all of this. None of us will be on Mother Earth for a very long period of time; thus, we are simply planting the seeds for a harvest that will not be ours to reap. That said, the more we know about the growing of food, the history of farming, and the problems that rural people have encountered, the more it will inform us on how to be practical farmers now and in the future.

PART I

Beginning of the Promise

Chapter 1
Before Iowa Became Iowa

When you are in doubt, be still, and wait;
when doubt no longer exists for you,
then go forward with courage.
So long as mists envelop you, be still;
be still until the sunlight pours through and dispels the mists—as it surely will.
Then act with courage.

—CHIEF WHITE EAGLE[1]

Before Iowa became a territory or was a recognized state within the United States, the white man began settling along the Missouri River. This is where my family settled, and it would become our home.

The Missouri River

So the journey begins. Close your eyes for a moment. Imagine a canoe paddling up the Missouri River near present-day Council Bluffs, Iowa. The year is 1714. You may be the first white man on this stretch of the Missouri. What do you see? You see a river so wide and meandering that it may be hard to find the main channel at times. If it is late summer, you see far off in the distance, stands of tall-grass prairie, eight feet high in places. You could not have envisioned that more than 300 years later, people would see another sea of grass, only this time, it would be eight-foot-tall corn plants.

You see a beautifully colored collage of prairie flowers that could match anything in your homeland. You might encounter other humans canoeing on the river, their skin a different color. You will have to be very careful, and your meeting will have to be one of an appearance of both strength and submissiveness. You are on their river. You have to be friendly and confident in your approach. You have already survived previous such instances, or you would not have made it this far. You see and hear many species of birds and mammals along the river. One particular animal catches your attention, if you are lucky enough to see or be startled by the flapping of its paddle-like tail. It is the reason you came here in the first place. Your fellow citizens back in your homeland of France and in much of the rest of Europe think they have to wear its fur on top of their heads. It is the humble beaver.

The history and development of this country owe so much to this lowly rodent. The fur trade drove much of the settlement and economy of the three principal colonizing countries of France, England, and Spain, but especially France and England. This went on for 250 years from the early seventeenth century to the middle of the nineteenth century (1600–1850); the trade and competition for the beaver fur hats worn by Europeans played a role in the great European wars of that period.

> The fur trader is unaware that he is being watched by a small band of warriors of the Omaha tribe. They have silently been following the strange bearded, pale-skinned man for quite some time. They are traveling on foot through the forested hills on the eastern Iowa side of the river. They had heard about the French trader Bourgmont who was living with the Missouri tribe further south on the river. Could it possibly be him? They know a little about the French fur traders. They helped to force the Omaha's to migrate further west from their original home.[2]

The original Omaha tribe began as a larger woodland tribe comprising both the Omaha and Quapaw tribes. They inhabited the area near the Ohio and Wabash rivers until about the year 1700. Not long after, the Omaha nation moved west and built a village of about 400 dwellings and 4,000 people on the Big Sioux River near its confluence with the Missouri near Sioux City, Iowa.[3]

Originally, the name Omaha translated to "those going against the wind or current." The Omaha became the first tribe to master equestrianism on the Great Plains, which temporarily gave them an advantage over their Sioux enemies. Conflict with the Sioux forced a split off of part of the tribe into the Ponca. The Omaha's retreated a little further south on the Missouri to an area in northeast Nebraska in 1775 near present-day Homer, Nebraska. There they developed a village named "Towantongo." It was home to Chief Blackbird and at least another 1,100 people around the year 1795. Blackbird was a notorious and skillful leader who realized early on that it would be to their advantage to trade and become friends with the White Man. He ruled much of the fur trade with the French on the section of the river that borders the two states of Iowa and Nebraska. In 1800, a smallpox epidemic killed Chief Blackbird and at least another 400 residents of the village. He is buried on a high knoll looking over the Missouri River. Lewis and Clark visited the village of Towantongo in 1804.[4]

After the death of Chief Blackbird, the Omaha nation never regained its former prowess. They became more vulnerable to both the good and the bad intentions of white settlers. They became known as a "people without a resting place." They started to be bounced around like a pinball. Because they were a peaceful people who mostly wanted to adapt to the white man's culture, they became victims of ruthless traders who continually took advantage of them. There were any number of well-intentioned Indian agents who wanted to do right by the Omahas. For roughly four decades, from around 1819 to 1856, the well-meaning agent was in constant battle with the traders. While the agent was advising the Indians to give up the chase and settle themselves, the traders were urging them to hunt for more skins. But the skins were becoming harder to find. The traders also introduced the destructive force of alcohol. It both robbed them of their dignity and the vitality of their traditional culture.

They were also vulnerable to the death and destruction wreaked upon them by the much stronger Sioux tribes, who seemed to have a particular disdain for the Omahas. This resulted in their fear and refusal to be forced to settle near the northern part of the Nebraska Territory where the Sioux lived. They attempted to settle around Bellevue, which is south of present-day Omaha, Nebraska. Whenever they asked for protection from the whites against the Sioux, it did not come.

By 1847, their numbers had dwindled to under a 1,000 people. Finally in 1856, they were granted a home in the Blackbird Hills where their beloved chief Blackbird was buried. Their troubles were far from over with Whites and the Sioux, but at least they had been granted a "resting place."

Today, the largest community in the Omaha Nation is Macy, Nebraska, which is around 70 miles northwest of our farm. Their numbers have grown to over 5,000.[5]

The Missouri River is the longest river in North America. Together with the Mississippi, it forms the fourth longest river system and the third largest drainage basin in the world. Some would say that the Mississippi River flows into the Missouri River, not the other way around. However, the Mississippi was explored and named first. If it had been named the Missouri River all the way to the Gulf of Mexico, it would be the longest river in the world.

The word Missouri comes from a Siouan Indian tribe whose Illinois word, *ouemessourita*, and according to some linguists, means "those who have dugout canoes"—although many of us call it the "Big Muddy" because of its ongoing legacy of picking up and carrying so much soil downstream. Fifty or more Native American tribes lived along the Missouri River, traversing it on both sides with a great variety of cultures and histories. One was the Ponca Tribe, whose chief was White Eagle and whose home was where the Niobrara River and the Missouri River meet on the borders of present-day Nebraska and South Dakota. At least twenty-six local tribes lived along the upper stretches of the river in the Dakotas and in Montana.

The history of the Missouri River and its meandering stories began to take shape much earlier than I was taught during my school days. Two early French explorers, Marquette and Joliet, are given credit for first seeing the mouth of the Missouri River near modern-day Alton, Illinois, in 1683.

Mari Sandoz, Nebraska historian and writer, has compelling evidence that it happened about fifty years earlier, between 1630 and 1640. She learned this from her father, Jules Sandoz, who obtained the story in 1884 from some of his Plains Indian friends. Her books about the Plains Indians and the settlement of western Nebraska in the second half of the nineteenth century are both wondrous and disturbing. Our oldest son's spouse is from Scottsbluff, Nebraska. She is a descendant of this author. Her first and perhaps most famous writing is *Old Jules* and is the story of her father.[6] Jules befriended the Native Americans because he repaired their guns and made powder and lead shot available to them. He also spoke French and wore a beard. Both were important because the earliest White Men in the Upper Missouri region were French and wore beards. They were the roaming *courier de bois* (wood-runners; or hunters of the woods), the French fur trappers and traders coming from the east from the Montreal, Canada, region. These unlicensed and very independent fur traders went by a number of different names. I was familiar with the name "voyageurs." To the English of Hudson Bay of that day, fur traders were known as "runners of the woods," and to the Dutch of Albany, New York, they were known as "bush lopers."[7] We do not know for certain, but those early Frenchmen probably did not come as far south as the mid-Missouri River section, where my family lives. There was no good reason to. The beaver was not as plentiful in that region as in the northern regions.

Ettiene de Veniard, sieur de Bourgmont, the first European known to explore the Missouri Valley and who paddled on the part of the Missouri River that lies forty-five miles to the west of my family's land, most likely came from the south, ascending up the Missouri. This was sometime after 1706, probably around 1714. He was living with the Missouri tribe near what is today Brunswick, Missouri, in the northwest region. Bourgmont established the first fort on the Missouri, named Fort Orleans, in 1723 near Brunswick, which is east of present-day Kansas City, Missouri. He is credited as the first European discoverer of the mouth of the Platte River, ten miles south of Omaha, Nebraska. It is unclear how far north Bourgmont traveled on the Missouri. In his journals, he describes the blond-haired Mandan tribes, so it is probable that he traveled as far north as North Dakota where their earthen homes and villages were located.[8] That could have meant that he paddled on that stretch of the "Big Muddy" that I keep describing as being just forty-five miles west of our farm. Why do I think this is important? I would say it is partly because I really never gave it all that much thought until recently. The history of our region did not start with the Lewis and Clark Expedition in 1804. There was a great deal happening long before, and it was happening on and around the river that I am only now beginning to know and to appreciate.

Why is it that the older we get, the closer the years seem to come together? The 1800s do not seem that long ago to me now. I knew many people when I was growing up in the 1950s and 1960s who were born in the last twenty years of the nineteenth century. What they experienced and the lessons to be learned from those experiences are intriguing and inspiring to me now.

Our family came to live on this land and call it their home. Now it has been ours for over eighty years. It is not just any land. It is our land. But is it really our land? In 1876, Chief Joseph of the Nez Perce said it well: "Do not misunderstand me, but understand fully with reference to my affection for the land. I never said the land was mine to do as I choose. The one who has a right to dispose of it is the one who has created it. I claim a right to live on my land. I accord you the privilege to return to yours."[9]

Iowa, Our Home

This piece of land where I grew up and have now farmed for fifty years lies in western Iowa. The farm is located in Lincoln Township in Shelby County. It is nestled about halfway between the county seat of Harlan and my little hometown of Westphalia.

Our area of western Iowa really does not have much significance either historically or in its attractions to entice visitors or inhabitants. There are not many trees; it is not a tourist destination. It does not offer any great natural, cultural, or recreational places. There are no mountains. Iowa may be considered a flyover state, but the area between Des Moines, Iowa, and Omaha, Nebraska, is noticed even less. Western Iowa is at times the forgotten portion if you look at Iowa as a whole. Eastern Iowa was settled first. We sometimes suffer from the complex that we historically have been shortchanged here, politically and economically. The state starts on the Mississippi River and ends at Des Moines in the center of the state. Some of our residents in western Iowa might identify more with Omaha and the Nebraska Cornhuskers sports teams than they do with their own state.

Perhaps, Iowa is not that much different from other states in its formation. Many developed from east to west.

But what happens if you look below the surface, and I mean literally below the surface? You will discover arguably some of the richest soil and food-producing capabilities in the world and realize how that soil could be farmed and, in the process, offer some of the basic solutions to our future well-being as a nation and as a planet. I said "could be farmed" because I believe strongly we still have not figured out how to farm in a way that both land stewardship and now climate change call for. We have allowed so much of our productive soils to be compromised and, even in some cases, destroyed. I also believe that this land does not give me the right to do as I choose on it.

Perhaps the greatest agrarian and rural community writer I've had the honor of meeting is Wendell Berry. Berry has stated that he is "an uneasy believer in the right of private property because I know that this right can be understood as the right to destroy property." He has also stated that "the two great ruiners of privately owned land are ignorance and economic constraints."[10] Economic constraints are defined as external economic factors that affect business and are usually outside of its control. I would argue that it is more often the case that many of these constraints could be controlled if government and economic policies wanted to. Small and medium-sized businesses of all kinds, including most of our farms, have had economic and environmental constraints imposed upon them throughout our nation's history because that is how big business has intended it to be.

The area in which my family now lives was once a vast, fairly shallow ocean that Canadian scientists call the Bearpaw Sea. Perhaps that is why we find so many fossilized remains of small arthropods that appear to be in the form of very ancient trilobites. Trilobites actually are among the very oldest of marine fossils, existing as long as 500 million years ago. Their skeleton resembles that of the horseshoe crab found today, only smaller. We find them in the crushed limestone rock that we purchase annually to stiffen up the muddy farmyards that come with the arrival of the spring thaw and the frost heaving of our soils. This sea existed just sixty-five million years ago, a relatively short time in geological terms. North America consisted of two islands between this narrow sea, which extended all the way down from north to south. The Appalachian Mountains on the eastern island came into existence about 450 million years earlier. The Sierra Nevada mountain range was just forming as a series of jagged volcanoes on the western island. The Rocky Mountains were not yet in existence. It was at the height of the period of the dinosaur, which would abruptly come to an end when an asteroid over ten kilometers wide collided with the Earth in what today is the northern part of the Yucatan Peninsula of Mexico. The impact flattened all of the trees in North America. The heat was estimated to be one thousand times that of the sun. There was more oxygen in the Earth's atmosphere at that time, and vast forests covered most of the two islands of North America. It was a warm period of time. While most life was wiped out in much of the Western Hemisphere, a few species of plants and seeds and mammals and amphibians that lived underground survived, and eventually life again covered our continent. None would ever reach the size of the largest dinosaurs.[11]

The bison that grazed so much of the land where I now live appeared around 130 thousand years ago, when they crossed the land bridge that connected Asia and North America. The first humans crossed that land bridge nearly 14,000 years ago. This was toward the end of the last great ice age, which had reached its pinnacle in North America about twenty thousand years ago. The last ice age to affect Iowa was known as the Wisconsin Glaciation, and it made its final retreat about eleven thousand years ago.

That change brought on the great soil-building ecological system of the tallgrass prairie, which covered much of Iowa. The tallgrass prairies required warm summers and around thirty to thirty-five inches of rainfall for optimum growth of sometimes six to eight feet or more for species such as big bluestem and Indian grass. They also needed fire to prevent the natural succession of trees and forest from gradually taking over. The annual dying back of the grass, both through fire and cold winters, produced large amounts of organic material, leading to the creation of very rich soils high in carbon and organic matter. Where I live, the soils are predominately composed of silt loam. By definition, silt loam soils contain not less than 70 percent silt and clay and not less than 20 percent sand. Silt's mineral origin is feldspar and quartz. Feldspar, a pale-colored crystal consisting of alumina silicates of potassium, sodium, and calcium, is the most abundant mineral on the Earth's crust, making up over 40 percent of the crust. Quartz is the second most abundant mineral and is composed of oxygen and silicon atoms. Loam contains sand, silt, and clay in relatively equal amounts along with good amounts of organic matter and is a great plant-producing medium. Clay consists of silica, alumina, magnesia, and iron, as well as some sodium, potassium, and calcium. It contains very little organic matter and is fine in texture. Water makes it very sticky.

The name, Iowa, has a number of possible origins. Some claim the word means "beautiful land." Some claim it is of Dakota origin and by the French was written as *aiouez* and applied to the branch of the Otoe tribe inhabiting territory west of the Missouri River. In the Dakota language, Iowa means "something to write or paint with." Today, Iowa is often referred to as the "land between two rivers," the Missouri and the Mississippi.

There is little historical evidence of Native American villages in Shelby County. It appears that Shelby County was mostly used for hunting purposes and for passing through to the Missouri River. Perhaps warriors passed through on occasion. This would seem to be validated by the fact that only two arrowheads have been found on our property as far as I know, unless it happened before my days of growing up. The first was back around 2003 when a young man who was working for us found an arrowhead while walking through the cattle yard near the farm house in which my dad was born. The young man found it, so I told him it was his to keep. The Native American period the arrowhead belonged to has not been ascertained. The second arrowhead was found by our son David just a few years back in another one of our farm yards, about a quarter mile from where we live.

None of the western Iowa Indians had villages where our farm is located and, for that matter, in all of Shelby County. Our land has high ridges, long sloping hillsides, and wide valleys. There were very few trees and areas for protection from the elements. A well-defined trail in the northeastern part of the county that ran along the Nishnabotna River appeared on the original government land surveys of Shelby County. The July 1913 issue of the *Iowa Journal of History and Politics* included the following: "These [Ioway] Indians and their Sac and Fox friends probably hunted all over western Iowa, while bands of Sioux descended from the north and Otoes and Omahas crossed the Missouri from the west."[12] Again, the fur trade influenced the region, with the British fur trade penetrating the Des Moines River and also as far west as the Missouri to compete with the French for their share of beaver pelts. This was in the latter part of the 1700s.

A number of treaties existed that affected what tribes in our area of western Iowa could call "home." The first occurred in 1830 at Prairie du Chien, on the Wisconsin side of the Mississippi River. By this treaty, the Sac and Fox, Omaha, Ioway, Otoe, Missouri, Mediwakanton, Wahpekuta, and Sisseton bands of Sioux ceded to the United States all of their claims to western Iowa, amounting to about one-fourth of the state's area. Shortly after, in 1833, the Prairie band of the Pottawattamie tribe of Indians was moved from northern Illinois (Chicago) to about five million acres between the Boyer and Nodaway rivers.[13] This band adapted to the open prairie lifestyle in their successive removals westward. This band is to be distinguished from the woodlands band of Pottawattamie who preferred the woodlands as their home and as integral to their culture. The northern boundary of this parcel ran across the northwest part of Shelby County. About 2,000 of the prairie band of Pottawattamie, led by their leader Sauganash, also known as Billy Caldwell Jr., settled near Lewis in Cass County, which is about thirty-five miles east of present-day Council Bluffs. Sauganash was an influential chief born to a Pottawattamie mother and Scots-Irish father. He was raised as a Catholic and was receptive to having his people be served by a famous Jesuit, Father Pierre Jean DeSmet. DeSmet arrived in 1838 to give missionary service to the Pottawattamies. This did not last long. By 1846, the year Iowa became a state, the prairie band of the Pottawattamie nation ceded all their lands in western Iowa to the United States. They were forced into Kansas and spent many years struggling to maintain their rights and reservation. The reservation is located in northeast Kansas and contains 576,000 acres in the shape of a 30-mile square that spans across four counties.

The word Pottawattamie, as we know it, is named for Pottawatomie County (Council Bluffs). For the Indian nations, it is usually spelled "Potawatomi." It means "people of the place of fire." Their homes were called wigwams or wickiups. They were dome-shaped and covered with tree bark. I cannot begin to imagine or appreciate the suffering and demoralization these people had to endure as they were forced to move time and again after the breaking of one treaty after another.

My maternal grandmother grew up near the west fork of the West Nishnabotna River, south of Defiance and ten miles from our farm. Nishnabotna is taken from an Otoe [Chiwere] word meaning "canoe builders." My grandmother was born in 1896 and lived until 1994. I was intrigued by her stories of seeing Indians on the river when she was a little girl. Perhaps they were remnants of the Potawatomi, one of the last tribes remaining in western Iowa. I imagine one saying:

> It was I, a Potawatomi, whom your grandmother saw, or maybe it was just the ghost of my former self. Your people drove me to the reservation but I would not go there. I hunted along the rivers and streams in western Iowa. One of my favorites was the Nishnabotna. White people left me alone. That is because they never saw me. But your grandmother was dark-skinned for a white person, and she reminded me of my people. So I let her get a glimpse of me until I silently vanished in the passing water. Or so she thought. Maybe it was just a ghost. It doesn't matter now. Your people never wanted us to be here anyway.

Today, no Indian reservations lie on the Iowa side of the Missouri River. However, there are two Indian tribal reservations on the Nebraska side, the Omaha and Winnebago tribes. The Ponca tribe does not have a reservation but is headquartered in northeastern Nebraska in the town of Niobrara.[14] There is virtually no physical evidence left to show that indigenous tribes once thrived with their rich cultures and prudent lifestyles. Their journeys did not end well.

Revisiting Our History

My interest in the history of our country has grown greater as I have grown older and have increasingly come to appreciate and grasp the importance and the enormity of it. Also, lately there has been a good deal of seeking out the more complete truths of our American history. This is especially true with the murder of George Floyd in 2020, which forced us to confront more fully, some of the failures, denials, and partial truths of our past.

During the writing of this book, "Critical Race Theory" came to be one of the terms to describe what the truth of our history is about. It is tragic that those words have become demonized by those who feel threatened by their use or the supposed change of emphasis in the teaching of children about American history.[15] I believe very strongly that the teaching and portrayal of our historical past should be as unbiased and objective as it can possibly be. After all, it is our history. If we feel threatened by the events and truths of it, then perhaps we need to come to grips with not learning from the mistakes of the past and our inclinations to repeat them over and over. First, we need to believe in the dignity of all people regardless of race, color, creed, or sexual identity. It is only then that we can begin to face the historical hard truths and begin to rectify the injustices that have occurred in the past. We can still learn from our mistakes. We can still learn as well from the history of our mistakes as it pertains to our relationship with Mother Earth, nature, and God's creation.

Again, it is Wendell Berry who forces us to examine our history; for example, when he stated, "The past is our definition. We may strive with good reason to escape it, or escape what is bad in it, but we escape it only by adding something better to it."[16]

We either have to relearn or to unlearn some of what our traditional history books have taught us. By this I mean that we have to re-examine some of the most significant events and the reasons for them. Some began to occur over 500 years ago.

My own journey and that of my family's journey to Iowa must wait a little while. I feel compelled to first examine what was happening across the ocean in Europe, as the events that happened there foreshadowed what would happen here.

We know that Christopher Columbus discovered parts of the Americas in 1492. He did so in the name of Spain, which soon thereafter claimed much of the southern portions of North and Central America. The other major European powers at the time, namely England and France, quickly followed suit, with the English initially settling on the eastern shores of North America and the French in the Great Lakes region. The thirst for power, empire, and trade fueled much of the settlement of the Americas. But there were other reasons, too, much more personal in nature. England had been in a seesaw battle for human rights for much of the Middle Ages. Freedom of religion and the desire for property rights gave some people the courage to strike out on a ship to the west.

The development of scientific thought during the period known as the "Age of Enlightenment" helped to lessen the fear of falling off the face of the earth in this quest. It was also the early beginnings of the Industrial Revolution, and it began to provide the sturdier ships needed to reach America and at least some of the tools needed to survive once people arrived.

The Age of Enlightenment is also known as the Age of Reason, and among its many revolutions of thinking were the advances made in science, mathematics, and physics. A number of fundamental principles define what the Age of Enlightenment has meant for mankind:

- Laws govern the order of the natural world.
- Human reason can be used to explain much of what we know.
- Individuals have natural rights, including the right to self-governance.
- Progress should mark improvements in society.

The big problem with the last principle is, how do we define progress? That question has reached a crisis point as we ponder what to do in a world of finite resources and a changing climate. The still-prevailing measurement of progress, yearly growth in gross domestic product (GDP), is not working so well in today's world. We are desperately in need of a new enlightenment for thinking about what progress and quality of life could be. Could the material consumptive and capitalistic growth that followed the first Age of Enlightenment be reaching its limits?

The European intellectual movement of the late seventeenth and eighteenth centuries is considered to be the basis for Western thought. It emphasizes reason and individualism over tradition. It rejects the rule of monarchies and the church over people's lives. Its thought is embedded in the founding documents of our country, embodied in the United States' Declaration of Independence and the Bill of Rights.

It seems, though, that the Enlightenment pertained only to white males. Everyone else, including white women, were left out. Basic human rights did not extend to people with black, brown, or red skin, nor to those white races that had different facial features than white Europeans.

Enlightenment came to be expressed in another term that truly came to be associated with the God-given right of white Europeans to settle and to take over the new world. This term was Manifest Destiny. I could imagine the early settlers may have thought:

> I bear the privilege of Manifest Destiny. It is I who discovered the stars and built the machines to sail by them. I cannot be stopped. I will take whatever I want because I am civilized and better than the savage. God intends this as the way it is to be. Your Native American warrior and friend is a brute and a drunkard. He would just as easily take your scalp. Do not trust him. It is I who will tame and claim the land he said was his people's.

The principle of manifest destiny saw the rural American landscape as a boundless source of natural resources and land waiting to be claimed by a growing population of white people. Since many thought they were superior to all other races, including both natives and those of African descent, at the time of their early encounters, they found it easy to justify what they did both in enslavement of Africans and in driving Native Americans off their lands. Their "progress" could not be stopped. It was their mission. The countryside was so huge, and most of it had few people except for the Native American tribes. Most people in power thought tribes could be easily dealt with. Whites did not fully realize until the Indian Wars of the 1850s to 1870s that many indigenous people would fight to preserve their land and rights with the same fervor that the American colonists did to be free of England's rule.

Early in our nation's history, the US Congress saw the need to create new states out of the territory that was the United States and all its lands at least as far west as the Mississippi River. While the words Manifest Destiny had not yet been coined, the thought process that led to its pursuit and validation was already taking shape. As our population expanded, there would be the need for more land where American citizens could live, build their own livelihoods, and contribute to the growth and status of the country. Why not lay claim whenever the opportunity presented itself to take control of the boundless land to the west of the existing thirteen newly formed states?

Manifest Destiny received its validation by all sorts of moral and intellectual reasoning. First of all, it was justifiable to move into and take over lands inhabited by indigenous Native Americans. After all, white Europeans thought they were intellectually and morally superior to them. They also rationalized that Americans would enlighten the native people and do so by teaching them that, to become enlightened, they needed to adopt and practice tenets of Christianity and to adopt the lifestyles of their white superiors. When Andrew Jackson was president in the 1830s, he began to promote the notion of romantic nationalism and wanted to expand the area of freedom for settlement. "Romantic nationalism" can be defined as the celebration and advancement of an ideal based on a nation's language, culture, and history.[17]

The term Manifest Destiny was created in 1845 by an influential journalist of the time, John L. O'Sullivan. "And that claim is by the right of our manifest destiny to overspread and possess the whole of the continent which Providence has given us for the development of the great experiment of liberty and federated self-government."[18] It became divine providence, national superiority, American exceptionalism, and the wish for enlightenment of native peoples that justified westward expansion. This belief was not universal, however. The Whig Party, started in the late 1830s, resisted this expansionist thought. It became the other major political party along with the

Democratic Party, which started formally in 1828. The first two parties were the Federalists and the Anti-Federalists (from which the Democratic-Republican Party was formed). The Whig Party ended in 1856 and was replaced by the Republican party.

Manifest Destiny also became a political tool used by the South to try to expand slavery into new territories and maintain their power in the federal government. The Missouri Compromise in 1820 allowed the new state of Missouri to be a slave state and Maine a free state. The Wilmot Proviso in 1850 allowed California to be admitted as a free state but did not regulate slavery in the rest of the Mexican lands ceded to the United States. Texas became a slave state in 1845 because enough citizens from neighboring slave states moved there and declared their intent.

There were other more telling reasons why southern farmers wanted to move their plantations to new agricultural lands. Tobacco and cotton soils were beginning to wear out. Tobacco especially was very hard on soil, its organic matter, and its fertility. As long as there were sources of new and fertile soils somewhere else, why not just move to them instead of attempting to correct existing soil quality issues? Their land was being destroyed by ignorance and faulty economic constraints. Most southern farmers did not adequately see the need for crop diversity and soil-building crops, and also having livestock to help restore the land's soil production capacity. And, as they moved, they had to take the slaves with them, who functioned as human tractors and combines. They were unwilling participants in the first industrial version of agriculture, with its emphasis on the soil-depleting monoculture commodity crops of cotton and tobacco.

Economic constraints did not limit or restrict how the land was being farmed in the South and in much of the North, too. Economic constraints called for needing more land for a growing populace. This was used as justification for the inability and the unwillingness to accommodate Native American cultures, traditions, or ways of life. The white American view of unlimited land and resources only for themselves was alien to that of both indentured slaves and indigenous peoples.

What we were not taught in history class in the 1950s was the devastating result of diseases that Europeans brought to the indigenous populations of the Americas. It has been very difficult to reach any agreement as to how many indigenous people lived in the Americas when Christopher Columbus landed. Estimates have ranged from as many as 112 million to as low as eight million people. In 1976, geographer William M. Denevan calculated there could have been a native population of forty-three to sixty-five million living in all of the Americas.[19] The number of indigenous people in North America at that time was estimated to be less than four million by Denevan. By 1800, that number was down to 600,000 according to the US Census Bureau. Not only did the wars exterminate Native Americans, but also the germs and diseases brought by the Europeans eradicated the indigenous peoples of America. Over 90 percent died from diseases that included smallpox, cholera, influenza, chicken pox, bubonic plague, sexually transmitted diseases, and even the common cold. By 1890, the number of Native Americans in the United States reached its lowest point at only 228,000. Native Americans had become an endangered species almost as much as the buffalo that so many relied on to live. In 1869, Lieutenant General Phillip Sheridan, a well-known Civil War hero, reportedly stated, "The only good Indians I ever saw were dead."[20] Sadly, that morphed into an even worse statement prevalent among people in general: "The only good Indian is a dead Indian." America nearly got that tragic wish.

Perhaps the most lasting result that grew out of the Age of Enlightenment was the first Industrial Revolution. It was largely the result of new advancements in science, mathematics, and physics. It is considered to have started in England around 1760, with mechanized spinning of textiles being one of its earliest applications. Perhaps the greatest long-term outcome came from the burning of coal to produce steam power and iron production. This led to the concept of mass production and centralized factories and the burning of fossil fuels, which still dictates the terms of our industrial economy today.

It resulted in some of the first environmental clashes between people living in the city and in the countryside. Many in the countryside reacted in anger against the spoiling of the pristine countryside and its intrinsic beauty. The original cottage industries of textile weaving that were controlled by farmers and small towns began to weaken. Smog and soot from the burning of so much coal began to affect air quality, not just in the growing cities but in rural areas as well. Water pollution in streams and rivers from the growing iron and textile factories also had negative effects. Until industrialization, rural nobility controlled much of what transpired in England. As industrialization began to develop, this control began to be transferred to factory owners in the cities. Many of these factory owners were initially members of the middle class who invested in these new ventures. Nobility tended to invest more in agricultural improvements at the time. Their resistance did not prevail, however, as England began to be successful in international trade, which, in turn, created a new demand for more new inventions and new workers.

Something called the Enclosure Movement, a term I had never heard of until a visit to England in 2018, happened about this time in England.[21] It is now seen as a striking symbol of what the new Industrial and Enlightenment Age was to become. Until the beginnings of the Industrial Revolution in the 1600s and 1700s, much of the land was open, meaning it could be used by anyone, including the peasant population for their own subsistence. Peasants built their own organizations of a sort that coordinated and allowed them to manage the resources of food, fuel, and forest. These resources, both natural and cultural, became known as the "Commons." This did not set well with wealthy noblemen who saw themselves as becoming slaves and the peasants as lords. So they began to enclose the Commons, which allowed for more private ownership and bigger and wealthier farms. A bitter struggle ensued after the enclosure rules were enforced, which some said ended the rights of the common man. Smaller farmers and peasants began to move to the growing cities to find work.

This scenario was influenced by the Enlightenment philosophy of people such as Sir Francis Bacon, who many consider to be the father of modern philosophy. Bacon upended the pagan-agrarian understanding of nature as being a caring mother and promoted the view that nature was something that had to be conquered and exploited solely for man's comfort and accumulation of wealth.[22] The word "pagan" comes from the Latin word *paganus*, which refers to a unit of land in a rural district. Originally, pagan meant a person holding religious beliefs other than those of the main or recognized religions and undoubtedly included being a non-Christian. Non-Christian inhabitants of the New World were not viewed as human beings but as resources to be exploited. Bacon's concept was that, as the sole possessor of mind, man was elevated above and cast in opposition to everything else in nature. Nature's subjugation occurred through industry and man's pursuit of knowledge. That kind of thought about how the world works and for whom it works has been nagging at our conscience for well over 300 years. I ponder:

> Our country's history is still very young if you only consider the discovery and settlement by the old-world countries. I am compelled to start with the discovery of our region by the White Man. I am not proud of what our European ancestors who quickly became our American pioneers did to Native Americans who lived in this area where I now call my home. It is easy to say that I would not have supported or taken an active role in the takeover and almost total destruction of their lands and culture, had I been alive then. *But do I really know that for sure? How complicit would I have been?* I was not to be born for another 150 years.

Most of us still think that the first real exploration and study of this whole region happened after the Louisiana Purchase in 1803. Not true. Some of us may also remember from our school history classes that President Thomas Jefferson in 1804 authorized Meriwether Lewis and William Clark not only to lead an expedition to explore the region by traveling up the Missouri from St. Louis but also to ultimately find the much-hoped-for northwest passage to the Pacific. Numerous earlier explorers, namely the French Canadian *coureur de bois* (the early fur traders who lived with the Indians and took Indian wives and became brothers with some of the tribes of the Great Plains), had traveled at least to where the Missouri River had its start in the Yellowstone area. They knew it was still a long way to that great western sea that so many had hoped to find, but they hoped and even thought that it would be over the next ridge. However, when they reached the first of the Rockies, it became apparent that their destination would be much farther than the earliest maps had indicated.

Many of us do not realize that a Canadian, Alexander McKenzie, had already reached the western sea in the late 1700s, taking the northern Arctic route through Canada. So, while the Lewis and Clark Expedition did reach the Pacific, they were not the first to do so.

One of the reasons for the journey was to obtain the allegiance of the Western Indian chiefs to American control of the region versus the intentions of the British in the Columbia River Basin in the northwest and those of the Spanish in the southwest.

The Lewis and Clark Expedition had its first meeting with some of the Otoe and Missouri chiefs of the region at a place they named "Council on the Bluff."[23] Not the present-day city of Council Bluffs, but an area about fifteen to twenty miles north near present-day Fort Calhoun, Nebraska. It was there at a site some distance to the west of the river that the party held the first conference of what the White Man called a parley, meaning a meeting of two opposing sides.

The meeting proved to be congenial for the most part and paved the way for what would be a friendly, or at least tolerable, relationship with the Native American tribes of this area until a number of decades later when the push of western settlers really began and treaties that allowed the white man and Native Americans to coexist were broken by the government in Washington. The journals of the expedition stated that the parties feasted on elk, venison, beaver tail, and wild fruit. The Indians brought "water millions" (watermelons) that were cool and sweet.

This first meeting site at Council on the Bluff and on the west side of the river would later, in 1817, become the first US military fort west of the Missouri River. It was named Fort Atkinson for Colonel Henry Atkinson, who was attempting another expedition to the Yellowstone to quell the British penetration into the fur trade and to further explore and find a faster and easier route to the Pacific. The ambitious plan called for the use of steamboats to navigate the often wild and treacherous Missouri. Five steamboats tried but did not get very far. Two never even made it onto the Missouri. None made it past the mouth of the Kansas River. Ultimately, the White Man's folly and grandiose plans of steamboat travel up the northern and western parts of the Missouri River failed.

That did not stop them from trying later on, though, especially after gold was discovered at Alder Gulch, Montana, territory in 1863. Only since 1859 had steamboats been traveling up the Missouri River to Fort Benton in the new Montana territory. Reportedly, 289 steamboats sank on the Missouri River between 1819 and 1897.[24]

Winter was fast approaching in 1819, and Atkinson and his 1,100-plus troops started to build winter quarters near the site where Lewis and Clark first met the Indians in 1804. A fort was eventually built in 1820 and lasted until 1827. The purpose of the fort was to protect and provide support for the burgeoning fur trade into the Yellowstone and points beyond. More than 1,000 soldiers lived at the quarters, high on the banks on the western side of the Missouri.

I first visited the replicated fort, which was built in the 1980s and early 1990s, in the winter of 2014 and also the nearby site where the first meeting of Lewis and Clark and the natives was thought to have been held. I knew nothing of what took place in our history hardly fifty miles to the west until I was sixty-four years old! It was fascinating to learn that for most of those years of existence, Colonel Atkinson and his troops spent the largest share of their time improving the quality of their daily existence through agricultural improvements, not by military protection and skirmishes.

There was not much to note militarily from 1804 until toward the closing of the fort in 1827. However, agricultural pursuits were many. A vast amount of food was needed to feed the 1,000-plus troops. They planted corn, potatoes, and turnips and harvested over 10,000 bushels of corn the first year of the fort's existence. Much of that rotted, however, because they did not get the crop planted in time so that it would be dry enough for harvest before the bitter cold and snow of the Missouri River winter was upon them. Over the following years, improvements were made to improve their lot. Radishes, lettuce, and onions were planted to combat the scourge of wintertime scurvy. Beef and dairy herds were started. A hundred head of hogs began to be raised and slaughtered annually.

With no place to go and not that much to do in terms of soldiering the forts, members became more like farmers than anything else. More corn meant the production of more whiskey, which became not only a staple for the soldiers to keep their morale up but also a growing industry for trade with the Indians. Corn was used as a weapon long before today. Agricultural pursuits and activities proved to be the fort's demise, as the United States government decided that a post was not needed there, but instead a number of posts were needed farther up the Missouri to protect the growing fur trading industry and to curtail the growing Indian anger over the White Man's further invasion of their hunting grounds and food sources. The US Inspector General of the Army, Colonel George Croghan, referred to the troops as "awkward plowmen [ploughmen], not soldiors."[25]

Everything to the west of the Missouri River came to be known as the New Frontier. It most certainly was not western Iowa. Iowa may have become a state in 1846, but it was nearly three decades later that western Iowa began to be permanently settled, which brought my ancestors from a very unsettled Europe.

Chapter 2
Embarking upon a Promise
Rosmann's and Westphalia Begin

The bosom of America is open to receive not only the opulent and respected stranger, but the oppressed and persecuted of all Nations and Religions; whom we shall welcome to a participation of all our rights and privileges.
— **PRESIDENT GEORGE WASHINGTON, 1783**[1]

This chapter examines the westward expansion of our country as the nation began a new century. At first western Iowa was passed over but the growing railroad empire changed that. Immigrants, hungry for land and a better life, were pouring into the Midwest. My ancestors were among them.

Building a Nation of Immigrants

A closer look at our nation's history reveals at least part of the answer as to why development in Iowa, and especially western Iowa, took longer than could have been reasonably expected. When the nation was first formed, the land of Iowa and other land even farther west of the Mississippi River seemed far away. After the American War of Independence (1775–83), the United States owned all lands south of the Great Lakes and east of the Mississippi River. The thirteen original colonies grew to seventeen states by 1804, the year of the Lewis and Clark Expedition. Nine more, even though they were east of the Mississippi, did not become states until quite some time after the Louisiana Purchase of 1803.

The Northwest Ordinance was passed in 1787, which laid the framework for new territories to eventually become states. A geographical area with 5,000 residents could apply for territorial status. With 60,000 residents, statehood could be applied for. The future states of Michigan, Ohio, Indiana, Illinois, and Wisconsin were all carved out of the Northwest Territory.

The Mississippi River was a natural barrier to the lands west. All the lands west of the river were owned by Spain until Spain sold them to France in 1800. The three great European powers of France, Spain, and England had been playing a contentious chess game for hundreds of years for who would lay and hold claim to new land in the Americas.

Robert LaSalle, a French explorer, was the first to claim the Louisiana Territory for France in 1682. Eighty years later, France had to cede that land to Spain because of its defeat by the British in the French and Indian War (1754–63). Spain did little to develop the region compared to the French fur traders.

In 1800, Napoleon Bonaparte made a deal with Spain to regain the Louisiana Territory in exchange for six warships and an Italian kingdom that Napoleon had conquered. King Carlos IV of Spain wanted this kingdom for his wife's family. President Thomas Jefferson worried that France would send troops to New Orleans to begin forcing the United States to start paying export duties for goods coming down the Mississippi River to New Orleans. Spain had not demanded a duty because of a treaty it had with the United States. Napoleon was facing impending war with England. In 1803, he offered all the Louisiana Territory to the United States for $15 million dollars, which amounted to just three cents per acre. He demanded the money immediately. It was an offer the United States could not refuse. The United States had to borrow the money from two European banks at 6 percent interest. This loan was not completely paid off until 1823.[2]

In 1812, because of the importance of the Port of New Orleans, Louisiana quickly became the first state to be created out of the new purchase. Since St. Louis on the Mississippi River had long been an important city in the early development of the West, it did not take too long for Missouri to become a state in 1821. Arkansas was next in line in 1836. Then came Iowa in 1846.[3] Iowa had become a territory in 1838 and at the time extended north into what are now parts of Minnesota and Wisconsin.

The early settlers in Iowa settled in the eastern portion of the state, where there were trees and forested land for homes, fuel, fences, and farm buildings. By the time Iowa was granted statehood, log homes were beginning to be replaced by board and batten construction, which made for a much more tight-fitting home than did logs. Sawmills allowed for the cutting of boards that, when nailed to a frame, were covered with batten boards that covered the seams and cracks. Settlers came primarily from surrounding states to the east and south of Iowa, that is, Illinois, Indiana, Ohio, Pennsylvania, Kentucky, Tennessee, and Missouri.

After the forested lands were settled, new settlers resisted the settlement of the mostly treeless prairies farther to the west, which consisted of tallgrass prairie and hundreds of prairie flower species. The feeling of protection that timber provided was a very real phenomenon. Early settlements farther west in the state nearly always started out in timbered areas along a stream.

New inventions and the westward push of the Industrial Revolution played an important role in the settlement of Iowa. By the early 1850s, railroads were being developed in the river cities of Iowa along the Mississippi. These four river cities at the time were Dubuque, Clinton, Davenport, and Burlington. Soon after, discussions started about building a transcontinental railroad from Omaha on the western side of the Missouri to the San Francisco Bay area on the Pacific coast. This talk culminated in the US Congress passing the *Union Pacific Railroad Act of 1862*.[4] It was considered in part as a measure to help protect the Union during the Civil War. To reach Omaha, you first had to reach Council Bluffs, Iowa, on the eastern side of the Missouri. The railroad that first reached Council Bluffs and then Omaha on the other side of the river would be able to reap great financial and economic rewards offered by the connection to the new Union Pacific Railroad. By 1857, five of the ten railroads that linked the Atlantic Coast to the western states had reached the Mississippi River. In fact, the first bridge across the Mississippi was constructed in 1856 and linked Rock Island, Illinois and Davenport, Iowa.

Most railroads originated from Chicago, which were rapidly exerting themselves to be the major links in the West to the Eastern cities. By 1860, 655 miles of track already existed in eastern

Iowa. The race to Council Bluffs was on. By January 1867, the first railroad, the Chicago and Northwestern, was in Council Bluffs. In western Iowa, it ran along the Boyer River valley first explored by Lewis and Clark. The Boyer River runs just north and west of Shelby County, mainly through Crawford and Harrison counties on its journey to the Missouri. Even though the Chicago and Northwestern reached the Missouri River in 1867, it was not until 1873 that the first bridge across the Missouri River was built by the Union Pacific, which was ironic, as the Union Pacific and the Central Pacific railroads met at Promontory Point, Utah, on May 10, 1869. Yet a railroad bridge across the Missouri River to link the western frontier to the east took six more years to build. Futile attempts were made to lay track across the frozen water during the winter months. Shifting ice made that idea impossible. The second railroad to reach Council Bluffs was the Rock Island in 1869.[5]

Further settlement of the state of Iowa was intricately linked to the success of the railroads. They would serve as the primary way of that era to connect the many towns that were springing up across Iowa. This situation was especially true over the thirty-year period from the 1850s to the 1880s. Perhaps even more importantly, it also played a pivotal role in the sale of land to settlers who wanted to establish farms throughout the state.

The M&M (known as the Mississippi and Missouri Railroad) was one of the original railroads in Iowa. It developed plans to build three separate lines across Iowa. It reorganized as the Chicago, Rock Island and Pacific Railroad in 1866 to avoid the imminent threat of foreclosure. Capital was relatively scarce during the years before and after the Civil War. Pressure mounted to create federal and state land grants to help finance the construction of the rails and to allow the railroads to sell or mortgage some of that land to help raise capital for payment of construction costs. More often, however, the sales did not cover the interest on the costs of construction. In 1856, land grants were given to the state of Iowa that gave the railroads the right-of-way of two hundred feet in width and six sections of land for each mile of road. These sections were to be staggered on either side of the road, meaning the railroad owned a section, the Federal Government owned the next, and so on. Some railroads received as many as ten to forty sections per mile and a 400-foot right-of-way.

April 1, 1859, marked the first sale of land recorded for what would eventually become a good portion of the farm I grew up on. This deed was for 240 acres of land in Section 4 of Lincoln Township, Range 79 in Shelby County, Iowa. It was deeded from the United States to David Loring. Section 9, which lies directly south of our land, was originally owned by the Mississippi and Missouri Railroad.

Remember, much of the western half of the state was slow to be developed and populated compared to the eastern half of the state. However, much of the northern part of Iowa was settled even later because wetlands with poor drainage for farming comprised so much of that area. Another reason was that, although land was cheap in the beginning, selling for only $7 to $8 per acre, it was even cheaper in the Dakotas and Nebraska, going for about half that much. Arguably, the biggest enticement to skipping over western Iowa was that lands west of the Missouri River—such as the state of Kansas and the eventual states of Nebraska, South and North Dakota, Wyoming, and Montana—were lands that fell under the *Homestead Act* signed by President Abraham Lincoln in 1862. The legislation allowed any head of a household over the age of twenty-one to claim up to 160 acres of land for free and become a permanent owner of that land if they lived on it for five years with improvements. Southern senators had voted against the idea up until the South seceded from the Union, due to their concern that settlers would fight to keep new states from allowing slavery.[6]

Again, I think the real reason for the slower settlement of western Iowa had to do with what again appeared on the surface. There was no metallic gold in Iowa, only "black gold." Western Iowa did not have the lure of the untamed West that attracted all kinds of people, some good, some not so good. The lawlessness attracted some to the West. The hard economic times and turbulence following the Civil War created perhaps the greatest draw. Some settlers were drawn by the promise of getting rich quickly with the discovery of gold in California in 1849, silver in Nevada in 1859, gold in Colorado in 1859, and then gold in the Black Hills of South Dakota in 1874.

A more detailed and careful look at demographic changes shows that just twenty-four years after the United States was established, the shift from rural to urban was already taking place.

In 1776, about 2.5 million people lived in the newly formed United States of America. By 1800, that figure had risen to over 5 million. The population was roughly doubling every twenty-five years until 1900, when there was a population of over 76 million people. From 1800 through 1860, the US population grew over 30 percent every decade. The explanation given was that in a predominantly agrarian society, families tended to be large, which was apparently the case even though the percentage of rural versus urban dwellers in our country began to decline continually after 1790. In 1790, rural persons accounted for nearly 95 percent of the population. By 1900, that percentage dropped to just over 60 percent. The steepest declines occurred during the American Industrial Revolution over the forty-year period of roughly 1880 to 1920. In 1880, rural Americans still made up 72 percent of the population. By 1920, urban Americans outnumbered rural citizens by 51 percent. Urban areas were considered those that had at least 2,500 residents. In 1790, only twenty-four locations in the United States had more than 2,500 residents.

"How ya gonna keep 'em down on the farm" may have been a playful song intended to shock American soldiers returning to rural life after the First World War in France, but its effects were anything but playful for farms and rural communities. After 1900, it took roughly fifty years for the population to double. In 1950, the year I was born, there were about 152 million people, and the rural population had dropped to 36 percent. By 2000, it reached 282 million, and 19 percent were rural. In 2023, the population was estimated at nearly 340 million, with the rural population being 46 million. Between 2010 and 2020, the rural population loss was minimal at just 289,000. However, it is the first decade-long rural population loss in US history. I am imagining what a hopeful immigrant to the United States might say today:

> The contradictions of immigration to our country are everywhere from the statement of George Washington at the beginning of this chapter, to the inscription on the Statue of Liberty, and finally to the wall trying to keep them out today:
>
> 'Am I an immigrant or am I an exile? I was tired and homeless and huddled in masses and your statue says you welcomed me in 1903. Yet if I was Chinese, you excluded me in 1883. So now if I come again, I am not here at all. I am nowhere, a child who cannot find a parent, a parent stuck in the darkness somewhere.'

Where did America's immigrant population come from? In 1790, approximately 3.9 million people were in the United States. Eighty percent of all white Americans were of British origin. Of these 3.9 million, 757,000 were slaves, mostly from the region of West Africa. By religion, most Americans were of Protestant origin. I am Catholic. Less than 5 percent were Catholic until the 1840s. This percentage began to change dramatically by 1850.

To adequately tell the story of how my Catholic ancestors came to this part of Iowa, it is necessary to delve into some of the history of how Catholics came to this country. I remember learning in grade school, in the St. Boniface parochial school in Westphalia, that the first Catholics settled in 1632 in what came to be the state of Maryland and that Lord Baltimore was the person most instrumental in their doing so. I did not realize that, even though many of America's first settlers came here to escape religious intolerance and persecution, what they encountered here proved to be just as challenging as what had occurred across the Atlantic in England. The rift between Catholicism and the emerging Protestant Reformation started by Martin Luther in Germany in 1517 reached England in 1534 when King Henry VIII declared himself head of the new Church of England. One hundred years later, English Catholics settled in the area they later named Maryland, just twenty-five years after the first successful English colony was established in 1607. The first English colony to escape religious persecution was a minority sect of Puritans in 1620 who landed on Plymouth Rock in what became Massachusetts. They were more strident in their beliefs against Catholics but also broke away from the Church of England because they thought that the Church of England did not separate itself enough from the doctrines of Catholicism. Other Puritans held this view, but most of them did not advocate for their separation from the Church of England. A group of colonists representing the majority Puritan views established the Massachusetts colony in 1629 and several other colonies throughout New England.

Their religious activities eventually evolved into the Congregational Church. These religious differences clashed with one another politically and economically in the new colony of Maryland. However, by 1700, only 2,500 Catholics resided in Maryland out of nearly 30,000 residents. Theoretically, but not always in practice, religious intolerance came to be addressed in the First Amendment to the *US Constitution* in 1802, which called for the separation of church and state.[7]

American Catholic immigration remained stagnant, making up less than 5 percent of the population until the middle 1840s. The potato famine in Ireland changed that dynamic. The poor in Ireland were facing starvation. People were coming in droves to many of the larger industrialized cities in the Northeast, such as Boston, New York, and Philadelphia. The Irish comprised nearly one-half of all immigrants to the United States in the 1840s. Between 1845 and 1855, 1.5 million Irish came to America to try to escape starvation and extreme poverty. The Irish were controlled by the British lords and landowners. Britain's slow response and worse, its intended neglect of the Irish, contributed heavily to the situation. At least one million Irish citizens died from the potato famine.

The rapid influx of poor Irish Catholics to America resulted in a growing backlash. Secret societies formed and eventually organized into a political party known as the American Party. Their members came to be known as "Know-Nothings," because when they were asked about what they were up to, they answered, "I know nothing."[8] Former President Millard Fillmore, who ran for the presidency a second time under the banner of the American Party in 1856, became the most notable Know-Nothing politician. He finished third behind Democrat James Buchanan and Republican John C. Fremont but still garnered over 20 percent of the popular vote. It seems that America's troubled beginnings with intolerance toward non-Protestants and non-English once again reared its head. However, the American Party splintered over the even larger, looming question of slavery and lost its influence by the election of 1860.

As the Irish came to be tolerated and accepted in this country, some of their political leaders exercised the same kind of bigotry toward immigrants of different races, such as the Chinese, when they came to places like New York, where the Irish grew to be dominant in their political influence.

Irish Catholics were not the only Catholics coming to America during the 1840s. German and Eastern European Catholics were coming, too. America more than doubled in size with the Louisiana Purchase. The lure of land, gold, and riches, and the potential for new beginnings with the westward expansion set off a fever that had not been seen in the history of the world. How quickly it happened is staggering. It took hardly forty years to expand from Iowa to the Pacific.

Iowa was clearly where immigration from European countries fueled much of the growth of the state. It was written that Iowa welcomed immigration. In 1870, the state published a ninety-six-page booklet titled *Iowa: The Home of Immigrants*. It was published in English, German, Dutch, Swedish, and Danish.[9] Northern Iowa's first settlers came mostly from Indiana and Ohio in the 1830s and 1840s. Settlers to southern Iowa came from Tennessee, Kentucky, and Missouri. There were 46,000 residents by 1840. By 1860, the number grew to nearly 675,000, and by 1880 to nearly 1.2 million. From the time Iowa became a state in 1846 until the early 1900s, European immigrants were pouring into Iowa. Germany contributed the most, followed by Sweden, Norway, Denmark, the Netherlands, and England. Northern Europeans were hard workers, and most came with families. They were considered good stock.

The Beginning of the Promise

My dad's side of the family came from what was then the Austro-Hungarian Empire. I grew up thinking I was Austrian. It was not until the 1980s that I discovered the Rosmann's came from what is now Slovenia. My mom's side came from Germany.

My paternal great-grandfather, George Rosmann, born in 1849, left Gerdenschlag (in the district of Tschernembl, in the Crownland of Krain, County in Hungary, now known as Presleje, Slovenia) on May 19, 1871. The family has a copy of the passport issued by Franz Joseph I, Emperor of Austria. There was obviously much fluidity about who controlled the land from which the Rosmann's came. He was listed as a soldier in the Second Reserve (National Guard). His occupation was listed as "itinerant merchant, surveyor and also boundary marker." He was twenty-two years old. He traveled to the Dubuque, Iowa, and Galena, Illinois, area on the Mississippi River and declared his intent to become a US citizen in 1872. But he did not actually become a US citizen until 1884, when he arrived in Shelby County. His younger brother, John, joined George in northeast Iowa and northwest Illinois about six months later in 1872. They both became peddlers, selling merchandise from large wooden pack frames carried on their backs. George would have been thirty-one years old at the time he married Anna Ertmer, a nineteen-year-old from Galena, Illinois, in 1880. George and John and their young wives must have heard about the new, small German Catholic community in western Iowa called Westphalia, after the German province of that name. The two families bought land side by side in Douglas Township, which was about five miles from Westphalia and about seven miles from where our farm is located. In 1899, my great-grandfather had around 345 acres of land. He and Anna became parents of fourteen children, none of whom died as infants, which was rare for families during that period. I was able to know at least one half of my great uncles and great aunts when I was a youngster and young man. My grandfather, Joseph Rosmann, was the second oldest and was born in 1883. Unfortunately, I never got to know him, as he died before I was born.

I do not know anything about my maternal great-grandfather, except for one story Mom told me. My great-grandmother, Anna, died in 1907, the year my dad was born. My great-grandfather may have developed a drinking problem after she passed. At least that is what I remember Mom saying. I also remember her saying that he became very lonely after her passing. She went on to say that he died from asphyxiation after falling down the basement steps with a gas lantern in 1910. He was sixty-one years old.

Northwestern Illinois and the towns of Galena and Freeport and the city of Dubuque, Iowa, on the western side of the Mississippi were three common areas of initial settlement in the Midwest for German Catholics after they arrived at Ellis Island. Correspondence and news traveled by both word of mouth and in the early newspapers of those communities. Of course, new immigrants could not speak English, but by the early 1870s, many Germans living in America for a period of years had to have learned English. Westphalia, Michigan, which was established in 1836, continued to draw immigrants as well. In fact, my great-grandfather's sister, Mary Rosmann, came to Westphalia, Michigan, and settled there after her two brothers came to Iowa. To my knowledge, no one in our family knows what became of her.

We know how contentious immigration issues have become today. Perhaps they have always been that way. For immigrants, the need to find people akin to them in race, nationality, culture, and religious identity was paramount for their acceptance and ability to create livelihoods for themselves, whether in farming or as craftsmen. Besides, they had decided it was far better to risk what the future in America might hold versus the desperation that was continuing to grow throughout Europe, especially in Germany and the Eastern European countries. Buying land, finding and belonging to something, and having a sense of place and community were intoxicants far too strong to resist.

Until writing this book, I had never given much thought to how quickly and completely the new citizens of Westphalia apparently abandoned their German roots and heritage. I do not remember ever hearing about any relatives left behind in Europe from our family or from any other families in our community, but surely they all did not leave. Was the desperation to leave Germany and the lands around it so complete that ties to their original homelands were all but forgotten? The German language was spoken and taught for quite some time in America. Dad learned German for half of the school day and English in the second half when he was a youngster. That quickly came to an end with the arrival of the First World War. German heritage celebrations never caught on in Westphalia or any of the other Catholic communities that make up Shelby County. They did bring one thing with them from Europe, the desire to build new religious foundations. St. Boniface Church in Westphalia, Iowa was one of them.

Wendell Berry has been quoted as saying, "What I stand for is what I stand on."[10] My belief is that the interdependent community of life supported by the soil that makes up land is what gives land its richness, and thereby, its contribution to the world. The same should be true of our larger human communities. What Wendell Berry has said about the need for a sense of belonging to a place and a community is a universal necessity for a meaningful life. Berry described belonging as a "connection to the land, to a particular place, to a community." It is a "membership." He talked about qualities that make that membership appealing and enduring, such as concern for neighbors, thrift, care of the land, long-term thinking and planning over several generations, self-sufficiency, and simplicity.[11]

It certainly can be argued that the loss of sense of place and community explains in large part why there is so much hopelessness and despair in this country. We have become further and further removed from the land with each successive generation. So many have had to leave their "place" because of diminished opportunities, competition for land and economic wealth, and a competition that ignores the neighbor both on the farm and in the town square. Long-term planning failed to consider that thrift, self-sufficiency, simplicity, and neighborliness were qualities not necessarily rewarded in the corporate economies already developing when towns like mine were just starting around 1870. It just was going to take longer for agriculture and rural communities to catch up to that mentality. My thoughts about change are that . . .

> Nothing stays the same for very long. We all know that in today's world, change is inevitable. We all know that is the case for small towns. Sometimes the changes are too great for one to be able to adapt and change with it. Maybe you think you are too small to fight negative changes. Maybe you are too small. Maybe you just are not ready for the changes. Worst of all, maybe you just do not care. All these far-off questions were unimaginable for the many small towns springing up all over Iowa in the latter part of the nineteenth century. There was too much excitement and hope. This was the beginning of a new promise! My hometown of Westphalia was one of them.

Westphalia was not the first settlement to spring up in Shelby County. Small settlements appeared as early as 1848, started by Mormons and located near woods and small streams in the northwestern part of the county. Galland's Grove was one such settlement, but it did not last, even though there was a church, a school, a gristmill, and a sawmill. The Mormons had been forced to leave Nauvoo, Illinois, and reached Kanesville, Iowa (now Council Bluffs), around 1844. Some of their members left the church and formed the Reorganized Church of Jesus Christ of Latter-Day Saints. A number of them chose to settle in Shelby County.[12]

The largest town in the county was started in 1858 and was named Harlan after one of Iowa's first Senators, James Harlan. It would become the county seat, which is about four miles from our farm. Located along the largest river in Shelby County, the West Nishnabotna, Harlan's population peaked at 5,357 in 1980. Nonetheless, it declined to 4,766 in 2019, becoming less than 5,000 for the first time since the 1960s.

The first settlers who came to start my hometown of Westphalia arrived in 1872. In March of 1872, the Chicago, Rock Island and Pacific Railroad signed a contract with a businessman named A. H. Kettler to colonize the territory. More people meant more business for the railroads and positive development for the state. The only town of any size in the county was Harlan, with 128 people in 1870. By 1880, that number jumped to over 1,300. Kettler began to advertise for new colonists in American and German newspapers.

On August 31, 1872, the first settler, Emil Flusche, arrived from Westphalia, Michigan, where he had been living. He conceived the desire to establish a colony or colonies for German Catholic immigrants. He saw the advertisement in the Milwaukee *Der Seebote* newspaper. Apparently, Kettler and the railroad had already decided which township would be colonized by the Germans, that is, Sumner and its southern border, only two miles north of the town of Harlan.

The story goes that when Flusche and Kettler rode by horse and buggy around the township, which laid predominately on a high ridge of what looked to be very rich, tall-grass prairie soil, he knew he wanted to start a colony and build a church on a site that stood at the tallest point of that ridge. Flusche bought eighty acres of land on contract for $400 and started to build the first house. According to Flusche's historical account, he bored a well six inches in diameter and twenty feet in depth in one day and found plenty of good water. He quickly took over Kettler's job of being the sole agent for the colony. In less than four years, by 1876, Westphalia had over one hundred families. One of Emil Flusche's brothers, William Flusche, wrote to his cousins and nieces back in

Germany a glowing account of how wonderful and relatively easy life was here in Westphalia, Iowa. He sounded like an *excellent* salesman in excerpts from this account written in 1874:

> One and one-half acres of this beautiful land is $7–10 however, only one-fourth must be paid immediately; the rest can wait until it has been earned by farming. Next year we will sow more than 200 acres of wheat, which is certainly enough for cultivating. The average harvest is 20 bundles for one and one-half acre [sic], one bundle is 60 pounds of wheat, $1 per bundle. This is a fine yield. Even more can be earnt by cattle breeding. It does not cost any money and hardly any work. You can find farmers here who have 80-100 cows and oxen. Cattle soon get fat. . . . Weather always or at least most of the time is fine. Rain only falls for some days in spring and sometimes in summer with thunderstorms. During whole fall and whole winter, it never rains. There is hardly any snow. . . . Horses can be kept extraordinarily easy here. That is why everyone who can call some land his own has at least two. Nobody here walks, except for maybe a German immigrant who has just arrived.[13]

Just a year later, in 1875, another new settler, Adam Schmitz, painted a more realistic picture of the challenges of farming in this new land:

> In 1875 the grasshoppers came so thick that a person drinking water at the pump or well was obliged to hold his hand over the glass to keep the grasshoppers away. These pests do much damage to the grain. I, myself, had about 60 acres in wheat and threshed only 180 bushels owing to the damage done by the grasshoppers. After I had paid the threshers and the harvest hands, there was left only enough seed for next year's crop.[14]

Word must have spread very quickly about the new colony of Westphalia. Settlers began pouring into Westphalia Township in just a few short years, coming from Bavaria, Westphalia, and the Rhine provinces of Germany. Some came from Luxembourg and Austria. Social life centered around the Catholic Church from the very beginning. In 1873, a small wooden church was built. The congregation consisted of five people that first year. However, by 1876, it numbered 396. In 1880, the congregation had reached 603 persons or 112 families. Plans were being made that year to build a magnificent brick church with a steeple nearly 130 feet high. Emil Flusche's promise was being fulfilled.

At the time, ties to Dubuque, Iowa, were strong because of the influence of the Catholic Church in Dubuque and its ties to the new Catholic colony of Westphalia. The architect hired to build the church was from Dubuque, and many of the historical buildings still standing there were designed by him. His name was Fridolin Heer, and he was originally from Switzerland.

The limestone slabs that became the foundation of the church were hauled by train from Dubuque County, over 300 miles away. Northeast Iowa and Dubuque's topography consisted of karst and limestone bedrock sediments, which were absent in western Iowa. The church, constructed in less than two years, was dedicated in 1882. My paternal great-uncle, John, and great-grandfather, George, are listed as two of the many individuals who helped to build the magnificent structure. It was named St. Boniface, which is fitting, for St. Boniface is the patron saint of Germany.

The church remains the focal point both physically and socially for the town of Westphalia, although the social and community functions of the church have diminished as time has passed.

Other Westphalia's

Emil Flusche and his four brothers, who came to Westphalia, Iowa, after he did, were not content to remain in Westphalia. In fact, the name died out a long time ago (only three gravestones bear the last name of Flusche in the St. Boniface Cemetery). Emil wanted to start other Westphalia settlements. One was in Westphalia, Kansas, incorporated in 1880. The last was in Texas in 1887. Since there already was a town named Westphalia in Texas, he and his brother Carl named it Muenster, after the capital of the province of Westphalia in Germany.

You will find other Westphalia communities in this country. All have remained quite small. The largest is in Westphalia, Michigan, where my great-grandfather's sister, Mary Rosmann, settled. It has about 1,000 residents today. There is also a Westphalia, Missouri, and a Westphalia, Indiana.

The Westphalia parish that my family attends had already started diminishing in the 1960s when I was a teenager. In terms of sheer numbers, the Westphalia parish reached its numerical peak as early as 1888, when the settlement had 185 families, numbering about 800 persons. The town had most of the important businesses needed for its economic livelihood at that time, including a post office, general store, blacksmith shop, hardware and tinware store, shoe cobbler, saloon, livery stable, carpentry shop, millinery, clothing, and grocery store.

As German settlers began purchasing land ever farther from Westphalia, the need became apparent for the establishment of other towns. This was spurred primarily by the Chicago, Milwaukee, and St. Paul Railway, which platted the towns of Earling, Defiance, Panama, and Portsmouth, all around 1881 and 1882. All these towns were primarily populated by German Catholics from the original Westphalia parish. These towns became larger in size than Westphalia because there was a railroad going through them. Westphalia has never had a population greater than 160 people.

Everyday life provided many challenges for the new arrivals to Westphalia. An account later in life from Antonia Sasse, one of the girls who came with her parents in 1873, is telling:

> My father and mother were lured from their mother country, Westphalia, Germany, with the purpose of seeking their fortunes in America and to find a safe and wholesome environment for the practice of their religion, particularly for their children. They gained this ideal when they came to this township to seek land for it was admirably suited for this purpose.
>
> There was a certain homesteader situated on a hilly quarter section, who was willing to sell his land. Since winter was drawing near, this land appealed to the new arrivals because it was "improved with a dwelling" which attracted any settler on the prairie.
>
> That home was a building which could not be considered a home today. It was 10' x 15' with one room downstairs and a low attic for the second floor. The men folks went upstairs much in the same manner as Abraham Lincoln and spent the night there on a bed on the floor which served well. A stove had been bought at Avoca which was used for cooking and heating. The walls of this building were made of boarding which were set perpendicular and did not overlap. As a result, the wind found ready entrance and when winter came, the floor was whitened with snow. In order to exclude the snow, Mr. Sasse mixed clay with water to form a putty with which to seal the cracks. This served very well when the weather was cold, as soon as it turned warmer, the putty thawed also and the cracks again were open to the weather.[15]

Miss Sasse also relates how they were forced to pay twice for their land. Coming from Europe, they were ignorant of the fact that homesteaders were obliged to remain on the land for a certain number of years and that they were not allowed to sell it within a specified period of time. When they bought the land, they paid $600, which represented their entire savings. In good faith, they believed the land was theirs. After a short time, however, they were informed by the land agent that since the homesteader had forfeited his right to the land by selling it before the specified number of years had been attained, they were obliged to pay the land agent again for it.

Sounds to me more like they were swindled. That was another challenge for immigrants coming to America.

Chapter 3

Dad and Mom, the Great Depression, and Father Duren

I learned the value of hard work by working harder.

—MARGARET MEAD[1]

I can imagine a daydream I could have had about Grandpa Joe:

> My name is Joe Rosmann. I am the one grandparent you never knew. You never knew me because I died three years before you were born. But you do know me because my spirit lives on in the farm and in the good earth where you now farm. Did you know that the name "Rosmann" means a keeper of horses? I was a keeper of horses because I farmed with them. Your father was a keeper of horses, too. In fact, he was the best in the family at it. That was because he had more patience than I did. I was a very hard worker. In fact that was my reputation in Westphalia. That was your father's, too.

All of the towns that started from the original colony of Westphalia, Iowa, were platted along the larger streams in the county in low-lying ground. Westphalia had neither. What it did have, however, was good soil, some of the best anywhere. My grandfather Joe was able to start farming on some of that soil after he was married in January of 1907 to Rose Hodapp of Westphalia. Dad was born on December 17 of that same year in the house that is still standing, where one of our sons, Daniel, now lives and farms. Rose was the daughter of Wendell Hodapp, who was an early settler of Westphalia, a farmer, landowner, and shoe cobbler. My grandfather was able to start farming on some of his land. In 1911, he was listed as the owner of eighty acres in Section 4 of Lincoln Township. By 1921, he was able to buy another 160-acre farm in Section 11 of Lincoln Township. It was about 1.5 miles from where he lived. That farm would later be lost in order to pay the estate taxes upon his death in 1947. Back in that time, you had to live at least three years after you deeded land over to a family member. The family member was Dad. What a terrible economic blow for our family.

By 1919, Dad was twelve years old and had completed only the sixth grade. There was no high school yet in Westphalia. Being the oldest, he had to assume the role of working as an adult on the farm. Hard economic times were once again setting in after the First World War (1914–18), especially for agriculture and farmers. The Roaring Twenties didn't roar into rural America. I find it hard to admit, but I do not even know for sure if he went to school in Westphalia at the Catholic grade school, but I think he must have because of studying German in the mornings and English in the afternoons. That would not have occurred in the public school system. He did not speak much about his formal education. I do know he started school at the one-room country public school, which stood on the corner of our farm in Section 4 next to the road. It was about one-quarter mile from our home. That school opened in 1876. I remember it being there in my early years. We still have a blackboard in our basement that Dad saved when the school was torn down around 1955. It has a bullet hole through it that some prankster must have been responsible for after the school was closed.

He shared a story about the first day he started school. The school was a little less than a mile away from where he lived, and he took a pony and a small cart and drove himself to school. He said that when he got close to the school, the pony bolted, and he fell off the cart and had to walk the rest of the way. The older students made fun of him, and his first day of school apparently was not a good one. Another story from his schooldays occurred when we were taking down the wooden corncrib that stood on the other side of the road from the school. This was in 1977. After we took the corrugated steel roof off, a section of the old wooden roof sheeting that the wooden shingles had been nailed to appeared charred and blackened. I asked what had happened. Dad grinned and told me about some of the older boys at the country school sneaking up to sit among the corn that was stored there to smoke corn silks. They did not have access to real tobacco, so they had to make do with dried corn silks. (Corn silks refer to the dried, brown, hairlike strands on the tip of an ear of corn.) Apparently, they got a little carried away and managed to set the crib on fire but were able to keep it from burning further.

It has been over forty-four years since the death of my father. He died in May of 1980. I still miss him. I always will. He was a very intelligent, quiet, and gentle man. That is what I remember. My two older brothers reminded me that he was a strict disciplinarian. They must have borne the brunt of that because I don't remember it quite that way. When he talked, you had better be listening because what he had to say was important. He was the oldest in his family, and the responsibilities and hard work that it brought helped to define him his entire life.

When he was just twelve years old, he started his own business in the winter, selling and delivering corncobs for home heating in the town of Tennant, which was seven miles away. He first had to scoop them onto the wagon and then shovel them off at people's homes. Coal was expensive and hard to come by, but farmers had plenty of corncobs. Dad performed this task for a number of years to help keep the family farm afloat. We had a cob-burning furnace in the basement of our house until at least the mid-1950s. We also burned corncobs in the cookstove in the kitchen, even though we had an electric stove by then. I sat by the cookstove to listen to the *Lone Ranger* radio show in the early 1950s. Corncobs were used as a fuel on farms for many decades. They were cheap, plentiful, clean, and easy to light. However, they burned quickly and were very hot. It took a good supply for a cookstove. It took a much larger supply to keep a large furnace going.

Nearly all of the farmwork in the 1920s was still done one of two ways: by hand or by horses pulling implements that were quickly coming on the scene. The first tractor would not come to our farm until sometime in the 1930s. Dad became a very accomplished "horseman." He learned to work with mules and the often problematic "westerns," which may have been nothing more than glorified saddle broncs that were supposed to learn how to pull a wagon or a plow. They also may have been crosses between draft horse breeds such as Belgians or Percherons and smaller western mustangs and saddle horses. I know they were smaller than the pure draft horse breeds.

The Belgian draft horse was the preferred breed for our farm. You must remember that by the time I was beginning to learn about farming in the 1950s, Dad had been farming with horses for over forty years. He broke or trained the teams to work for himself. Today, I tell people that he was an old-timer. But then I suppose I could be considered one, too, as we had a team and sometimes two teams of Belgians until I was nineteen years old and in college. I learned many of the old ways of farming, and I feel lucky I did.

Farming had a rhythm. Dad had adhered to it his entire life. He was used to getting up at 4:00 a.m. every day and doing chores. That meant milking cows by hand, feeding the cows, pigs, and workhorses, and getting the teams harnessed and ready to go so that after breakfast he was ready to go to the field. Of course, in the spring, it meant getting the fields ready for planting oats and corn. The summer necessitated haymaking and cutting weeds, cultivating corn, and threshing oats. Fall, and even early winter, signified harvesting corn by hand. Dad worked steadily, and that is something that he taught me. He did not like to have to work in a panic mode. He knew when to quit because he worked with the sun. He did not like or believe in working late at night. It helped there were no lights on the horses! When 6:00 p.m. came along, it was time to quit. He did not like daylight saving time when it was first introduced in the 1960s. It meant that sometimes you had to work until at least 7:00 p.m. in the evenings if you were putting up hay or something that demanded the heat of the sun to get it dry enough for harvesting. To him, the natural rhythm was being upset. Weekends never existed for him and other farmers of that era. For that matter, it did not exist for any of us children, either. We worked Saturdays like it was any other workday.

Sundays, however, were sacred for my parents, although chores related to the daily milking took higher priority, so he still got up early in the morning. On Sundays, instead of going to the fields early in the morning, he attended Mass at 8:00 a.m., where he sang in the choir. There was the fasting from midnight until receiving communion, which was church canon law until 1955. After the Second Vatican Council in 1962, fasting became a one-hour requirement before receiving communion. Today, very few people, if any, would think seriously about fasting.

The early mass on Sunday mornings was known as "Low Mass," which meant no singing of the Mass parts by the priest or celebrant, though hymns were sung, and no incense or the pomp and circumstance of High Mass. In the history of the Catholic Church, the Low Mass was sometimes referred to as the "Mass of the man in the pew." This meant that the Mass was simpler and shorter and could be done on a much more frequent basis during the week if desired. Some church scholars say that it actually helped to spread the practice of going to mass for Catholics.

After attending the early mass, Dad would come home to eat breakfast and then attend High Mass at 10:00 a.m. He told us that this was just about the only social life he had when he was growing up in his teens and through his twenties. What social life there was centered around church and family. The rest of the day was spent resting or visiting family relatives, except for the afternoon chores that still had to be done no matter what.

In the evenings, during at least part of the year, we went back to church for devotions or Benediction. Devotions consisted of prayers and litanies to the Blessed Virgin, St. Joseph, the Sacred Heart of Jesus, or the Saints. Benediction consisted of the exposition of the Blessed Sacrament (Holy Communion) in a monstrance, which was an ornate vessel used to display the consecrated host, which is the body of Christ in Catholic teaching. The actual service took about one-half hour, but by the time we visited for a little while and drove the three miles back home, we had spent an hour.

The Church did not require that Catholics attend the Sunday evening Devotions and Benediction as it did for attending the Sunday Mass, but our family adhered to it much of the time. This was the routine for many people of that era, even if you lived in town. Not every Catholic attended Mass every Sunday, but you could probably count them on two hands in Westphalia. The strict discipline of abiding by the Church rules was the norm for most people in our parish when Dad was growing up; if you were German Catholic, it was probably further amplified. You did not question it. The practice of one's faith had as much discipline attached to it as the practice of farming. Dad had a steady and quiet discipline and never complained about it, at least not to his children. When he and Mom retired to Harlan in 1975, the practice of going to daily Mass was important to both of them. They made time for formal prayer and meditation.

Grandpa Joe had a reputation for being a relentlessly hard worker. He probably felt he had no choice and likely grew into it naturally. I was once told that Grandpa would go out during the middle of the night to work if he couldn't sleep. Dad told me that Grandpa Joe suffered from stomach ulcers and had a great deal of financial stress that kept him awake. Unfortunately, working hard was no guarantee for success. You also had to work smart, pray for good weather, and hope that pests like grasshoppers wouldn't take some or all of your crop—and also have a good dose of luck. It might have helped, too, if you had a rich uncle or married into money. Most farmers had neither.

Hard Times

The 1920s and 1930s were especially difficult for farming. Dad was also in his early twenties and thirties during that period of time. He let it be known to our family that he did not have a great deal of confidence during those years and felt very isolated. Luckily, he had brothers and sisters to ease that situation. His sister Adeline, born in 1909, was two years younger, and his sister Anna was five years younger, born in 1912. Next came Walter, born in 1914, and finally Leonard, born in 1918. A tragedy happened when Dad was just nineteen years old, when his brother Walter developed a ruptured appendix and died at the age of twelve in 1926. His dad apparently said it was probably just a stomach ache and did not call a doctor until it was too late. Later in life, Uncle Leonard, eight years old at the time, told me he never quite forgave his dad for not seeking a doctor's aid sooner.

Dad did not have the chance to develop many friendships because of his total dedication to the farm and church. Not being able to attend high school did not help his self-esteem either. That changed after he met my mom, Ellen Mertens.

Mom was born on January 9, 1919, on a farm about ten miles from Dad's home. She was the oldest daughter of Henry and Katie (Kirschbaum) Mertens. Mom had been a schoolteacher until she got married. She received her teacher's certificate when she was just sixteen years old and began teaching in one of the Greeley Township one-room country schools. She was born while her dad was still in the US Army in France. The war had ended on November 11, 1918, but he did not get out of the service to return to the farm until the summer of 1919.

Grandpa Mertens was drafted into the Army in March of 1918 and reported to Camp Dodge in Des Moines, Iowa. He was twenty-four years of age. Grandpa was a lucky man in the sense that he missed the deadly flu epidemic at the camp by just a few months. The 1918 Spanish flu epidemic did not hit Camp Dodge until October 1, 1918. By then, Grandpa was serving in the infantry on the frontlines in trenches in France. This was in August of 1918. More than 10,000 of the camp's 32,000 soldiers were stricken with the flu, and more than 700 died. Historians estimate that deaths numbered over 1,000 and that the Army was accused of cover-ups consisting of an unmarked mass grave and experimenting with antiserum compounds given to Black soldiers who later died.[2] Many of us did not know much about the 1918 flu epidemic until we started living through our own Covid-19 pandemic.

I would sometimes ask Dad, when I got a little older, about the 1930s and the Great Depression (1929–39). I grew up hearing about them, often in conversation with adults or relatives in Westphalia. I wish I had listened more closely.

The year 1936 stands out as the worst year for that time. It started out as a year with extreme cold and heavy snow. The summer turned to intense heat and drought. The farmers' corn harvest was worth only eight to ten cents per bushel. It was more prudent to burn corn as fuel than to pay the cost of freight to sell it. The best one could hope for was holding onto the farm. Grandpa Rosmann, along with the help of Dad and his siblings, was able to do that.

Some of the neighbors could not. The family that had lived on the farm where I grew up and still live today was not able to survive. They lost the farm in 1929. I do not know when the family actually moved off the farm, but I do know they were justifiably very upset about losing the farm. While viewing the abstract of land ownership from when this family started farming it in the 1880s, I found the farm was riddled with mortgage after mortgage to both other individuals and insurance companies. That was not unusual, as many other farmers faced similar circumstances.

My grandfather purchased the farm in 1929 and assumed the mortgage, which was $32,000 or $100 per acre. The original mortgage was for $44,000 or $137.50 per acre on the 320-acre farm. Somehow, my family was able to hang on to it, and the mortgage was fully paid off in 1943.

My parents felt bad about the plight of the family who lost the land that he spent most of the rest of his life on. In the 1950s, my parents helped to start two direct descendants of the family in farming. My parents rented two 160-acre farms to them. This went on for many years. Our family could have been farming 640 acres from the 1950s to the 1970s, which would have been a large farm for that period. Mom and Dad's philosophy was that it was more important to have another young farm family in our community. In the twenty-first century, that seldom happens.

I want to emphasize again that extremely hard physical work and mental anguish defined the first three decades of the twentieth century, as people tried to hold onto their land. To this day, I marvel at how Dad and his dad and Uncle Leonard hand-dug, with spades, the clay tile lines to drain the wet areas of our farm. They dug about four feet down to install four-inch-diameter red clay tiles. They did this not only for a few hundred feet but also for about two miles and on both sides of the swampy areas. It is no wonder Germans have historically been known for their stoicism! The times called for this scenario, and the farm and the work often came before anything else, even family. I carry my own burden from that tendency as well.

Sometime after 1939, Mom moved to Westphalia to become the third-grade teacher at St. Boniface Catholic School. This was after first teaching in country schools for a number of years. She must have known Dad at least to some extent. Her uncles Tony Mertens and John Kirschbaum were married to two of Dad's aunts. It was not uncommon in those days for German Catholic boys to marry German Catholic girls from the bigger German Catholic towns of Earling, Defiance, Panama, and Portsmouth that started after the establishment of the "mother" parish and community of Westphalia. My parents were married on October 1, 1941. Dad was thirty-four years old and Mom was twenty-two years old, and they moved to the farm where I grew up. It was the next farm south of where Dad had grown up, the one that was lost during the Great Depression.

By the end of the First World War, Westphalia was already in a slow period of decline. The great influx of immigrants from Germany to rural areas of the country was coming to an end. Discrimination against Germans for their ties to their homeland during the First World War and agricultural economic decline after the war were two primary reasons. What had started out as being so idyllic and promising just a few decades earlier was being challenged by geographical and economic realities. The pace of the Industrial Revolution juggernaut, the rise of capitalism, and the growing power of corporations were pressing on the bucolic rural order that was still trying to become more fully established. The other German Catholic communities that all started from Westphalia were doing better because a railroad line passed through their towns. By 1926, however, the parish still had 130 families and about 160 people in the town.

A savior arrived in town that year by the name of Hubert Duren, the new thirty-four-year-old pastor from Baraboo, Wisconsin. He was an imposing figure, standing six feet, two inches, and weighing 250 pounds. He had a plan. Part of the plan dealt with the parish property, which at the time consisted of the beautiful church completed in 1882, the rectory constructed in 1888, a cemetery, a convent that was home to five sisters, and a three-room wooden schoolhouse. His plan of action took shape over the many years that he was pastor of St. Boniface until his death in 1962. It was called the "Complete Life Program." It coordinated what he thought were the basic tenets for a successful parish and community: Religion, Education, Recreation, Commerce, and Credit. His first concern was the education of Westphalia's children. As there was no high school, he convinced the parish to build a new school that would accommodate all grades.

In 1927, with the help of most parishioners, including Dad and Grandpa, a modern brick school was built. It was large enough for all students from grades one through twelve. The basement served as the town hall and a gymnasium for basketball. In 1933, the parish bought the last remaining tavern in town and moved it to the parish grounds and named it St. Hubert's Clubhouse after Rev. Duren's patron saint and because St. Hubert is the patron saint of sportsmen. For a time, no alcohol was served there, but it did contain a pool and billiards table. Rev. Duren was an accomplished player of both. The club also contained a library and reading room. The clubhouse is still in operation today and carries the enviable distinction of being one of the very few church-owned bars in the country.

The Wood Cutting Story

I recall this story from my father and my own recollection:

> The year was 1934 and the Great Depression was deepening. Farmers in the community had very little money and were growing more depressed by the day. The new school was built and in operation, but there was little cash to heat the church, school, and other parish buildings. Since Father Duren's family in Wisconsin owned and operated a sawmill, Father Duren had an idea. He learned of a farmer about twenty-five miles away who wanted a fifty-acre timberland cleared so it could be farmed. The farmer would give the timber to the parish if they cleared it. All through that winter, volunteers and the high school junior and senior boys, some seventy people in all, worked to clear the timber. Everything was hauled by truck back to Westphalia. The logs good enough to be cut into boards and lumber were stacked east of the church where Father Duren's brother set up a sawmill. The rest of the wood was used to heat the school and parish facilities for three years. This was accomplished by a large boiler in the basement of the school. The sawmill eventually was moved to one of our farming neighbors about a mile from us. I remember going there with Dad to buy lumber when I was little. The project exemplified

the spirit that was growing in the community. It spoke of leadership, willing cooperation, and the sense of pride in completing a very difficult task. Father Duren was an idealist, but also a practical pastor.

Building the "Complete Life"

Father Duren's plan and vision for the community of Westphalia was founded in Pope Leo XIII's encyclical *Rerum Novarum*. It was subtitled "On the Conditions of Labor" and written in 1891.[3] Pope Leo articulated the Catholic Church's response to social conflict that was arising due to abuses of unrestricted capitalism and industrialization. He called for property rights, trade unions, and the right of collective bargaining. These measures would help to ensure class harmony, not class conflict. It strongly stated that the operation of market forces must be tempered by moral considerations. This encyclical is considered to be a foundational text for modern Catholic social teaching.

In 1931, Pope Pius XI wrote an encyclical on the fortieth anniversary of *Rerum Novarum*, titled *Quadragesimo Anno* (In the 40th Year), "Reconstruction of the Social Order."[4] It built upon Pope Leo XIII's message of social justice for the working class and families. It criticized state-sponsored socialism, communism, and also unbridled capitalism. He wrote that human rights and Catholic moral teaching should not be replaced by the state. Parents, families, and communities have both rights and responsibilities. The encyclical also said that, in exceedingly distressing situations where illness, injury, and natural disasters occur, the state has the responsibility to provide public aid for the common good.

Father Duren interpreted this to mean that everyone should have the right to a just and fair living. Cooperative consumer buying power was a component of that thinking. He proposed the idea of the Westphalia Consumer's Cooperative Association. It started out in 1937 in a small, old building. In 1942, the cooperative grocery store was built. The following year, a gas service station was added. Several years later came a cooperative meat locker plant. As time went on, a cooperative farm supply and feed store came into being that served Westphalia until the 1980s. These entities became associated with the "Double Circle" coop that grew into Farmland Industries, headquartered in Kansas City, Missouri.

The last piece of the Complete Life Program was the idea of credit. The Westphalia Community Credit Union was established in 1939. It was one of the first rural credit unions in the state. It remained in Westphalia until the late 1990s. A branch office was opened in 1984 in Harlan and was named the Town and Country Credit Union. It has grown to include another office in the town of Avoca.

There was a great deal going on in our little town. Father Duren was a very gifted man. He was a talented artist and woodcrafter. He was an excellent musician and wrote many songs, both for liturgical use and more intricate compositions for bands. He started and led the community band in Westphalia composed of over forty students from St. Boniface School. It regularly traveled to other communities in Iowa to perform. Some thought of him as a Renaissance Man. Some thought he was a benevolent dictator. Others just saw him as a dictator. To this day, there are still complaints that when the cooperative grocery store started, the private grocery store in town was put out of business. It is probably true that if he wanted to do something, no one in the parish was going to stand in his way. The pastor ruled the parish and that was just the way it was. I never questioned it when I was a youth. It would have to be said that as a youngster, I was afraid of Father Duren. That is mostly because he was a strict disciplinarian. Physical corporal punishment for the older high school boys was a very rare but distinct possibility. This came in the form of a leather belt that Father Duren wore over his black cassock. The threat of its use served as an effective deterrent. Despite his shortcomings, the overwhelming majority of parishioners loved and respected him. He did so much for the parish and community.

During the late 1930s and 1940s, Westphalia, like many places in America began to come out of the Great Depression. However, nearly twenty years of depression in rural areas took its toll. In 1932, farm income reached the lowest point of the period. Many farmers were defaulting on their loans and losing their farms. They were becoming desperate. The new president, Franklin Roosevelt, took desperate measures to rescue farmers first from indebtedness and then by increasing crop prices. In 1933, one in every five workers in the country was a farmer. If farmers' incomes were raised, they would once again be able to buy goods produced by the cities that were also struggling desperately. Roosevelt's programs refinanced farm loans at a lower interest rate and paid farmers to cut production both in livestock and in crops. Iowa was at the epicenter of the farm protests against farm foreclosures starting in 1932, much of which occurred in western Iowa under the leadership of Milo Reno. He helped form the Farmer's Holiday Association that tried to get farmers to withhold their products from the market. He was somewhat successful. Much of the association's work centered in western Iowa and in locations not far from Westphalia. Herbert Hoover, one of Iowa's native sons, was defeated in 1932 in his bid for a second term as president. The following year, Roosevelt declared a banking holiday so that the *Emergency Banking Relief Act of 1933* could be passed, which created the Federal Deposit Insurance Corporation (FDIC).[5] People once again began to have faith in banks and returned to depositing their money in them.

In 1933, 90 percent of America's farms still did not have electricity. That year, Roosevelt started the Tennessee Valley Authority (TVA) to build dams and public power plants on the Tennessee River to help improve the economy of the region. The Rural Electrification Administration (REA) was formed in 1935 to bring electricity to much of the rest of rural areas nationwide. Private electric power companies could not make money bringing electricity to the rural areas. The response by Roosevelt was to begin Rural Electric Cooperatives (REC) to do so. Dad was hired to help string the lines for bringing electricity to our county. We still proudly display the boots that he wore to climb the utility poles. The World Health Organization (WHO) reports that

> worldwide, 22 percent or 1.3 billion people still do not have electricity. About 2.7 billion people still rely on solid fuels like wood and coal for cooking. An estimated 2.8 million deaths occur each year from the fumes of their emissions. About 63 percent of our electricity still is generated by the greenhouse gas-producing fossil fuels of natural gas and coal. Coal is still at around 30 percent of that total.[6]

Father Duren's efforts to get people to work together on projects needed to help the parish, community, and farmers to survive, worked in tandem with the federal efforts to break the stranglehold of the Great Depression. The Second World War (1939–45) appeared ever closer in the late 1930s, and its effects also bolstered the farm economy. Even with the onset of the war, Westphalia kept forging ahead with its local economic endeavors of cooperation.

In 1943, Westphalia was chosen as the most fitting town in America to celebrate the centennial of the International Cooperative Movement. The other site was in Rochdale, England, where the cooperative movement began in 1844. The Rochdale Society of Equitable Pioneers was a group of twenty-eight businesses that banded together to open their own grocery stores selling items that they otherwise could not afford to purchase. Much of the impetus for its beginnings was the mechanization of the Industrial Revolution that was forcing more and more skilled workers into poverty. By 1854, there were over 1,000 cooperatives throughout England.[7]

The cooperative wholesale society played a pivotal role for England in both world wars. It helped to provide a stable supply of food as well as energy requirements for the war efforts. The Rochdale Society of Equitable Pioneers traded independently until 1991, when it joined forces with other cooperatives in England. Today, it is known simply as the "Cooperative Group." It is still one of the largest cooperatives in the world, with nearly six million members. It is also into the funeral, insurance, electrical, and legal businesses in the United Kingdom. It owns over 6,500 supermarkets and convenience stores, making it the sixth largest supermarket chain in England. The branding of its cooperative advantage label has helped to keep it strong. Today, it incorporates a wide variety of socially and environmentally responsible requirements for its products. These include renewable energy sourcing, fair trade, non-genetically modified organism (GMO) ingredients, banning and restrictions on pesticides in food products, responsible fish sourcing, and community and individual member dividends for its brands.

Although the level of cooperatives in the United States has declined somewhat, some very large cooperatives are still present. Cenex Harvest States (CHS), which includes the Cenex brand with energy and Harvest States with supply, food, and grain sectors, is the largest in the United States with over $30 billion in revenue in 2017. Land O' Lakes, Growmark, Ocean Spray, Blue Diamond, Crystal Sugar, and a number of dairy cooperatives are among the top one hundred on the list of cooperatives. The Organic Valley-Organic Prairie cooperative, headquartered in southwest Wisconsin, has had a huge influence on our ability to produce and market beef and pork in our organic operation. We have raised beef for the cooperative for twenty-five years and pork for more than twenty. Organic Valley is primarily a dairy cooperative and, despite the pandemic, had sales of over $1.2 billion in 2020.[8]

Cooperation has been an elusive goal for farmers throughout our rural history. Farmers have not been able to join together or to agree in large enough numbers to have the strength of organization to control the prices they receive for the products they grow. This has generally been the case for farmers throughout all of history. It certainly was true in the latter part of the 1800s. The Agrarian Revolt, from about 1870 to 1900, that occurred after the Civil War was the most widespread in the history of our young country. It had everything to do with the land boom and westward expansion. It had to do with the increasing power of railroads and corporations over the shipment and pricing of food for the large metropolitan areas that were developing on the East Coast but also in growing cities in the Midwest, like Chicago.

The Beginning of Agricultural Economics and Boom and Bust Cycles in Farming

As urbanization and the growth of cities in America spread and as new states and the population began to dramatically increase, new phenomena came into being as well, which have come to be known as agricultural boom and bust cycles. Boom and bust cycles of inflationary land speculation and accompanying high prices for what the farmer produces, then followed quickly by rapidly falling land and production prices, have been the norm in this country since the 1870s.

There is much to blame for this. Some falls on monetary policy, some on government farm and ranch policy, some on corporate control of prices, some on changing technology, some on trade policy, some on the weather, and some on the farmer.

The Civil War and its aftermath evolved into a new era for American agriculture. At first, agricultural production was growing faster than the country's population. Farming was slowly changing from one of mostly subsistence—meaning just feeding one's own immediate family—to that of feeding more and more townspeople and the growing urban centers. As farming slowly became mechanized and as new lands opened up to the west, competition for land increased. The concept of a national agricultural economy began to take hold.

A quick look at these phenomena from after the Civil War up to the present will help to make this more apparent. The *Homestead Act of 1862* helped to usher in this new agricultural economy. This *Act* allowed free land of up to 160 acres to anyone filing a claim and farming that land or living on it for a period of five years. The *Act* applied only to the Great Plains, predominantly the Dakotas, Nebraska, Kansas, Colorado, and Oklahoma. While eighty million acres were settled by this *Act*, more acres were bought through auctions and land speculation.[9] That figure was 108 million acres. An even larger amount of 300 million acres were acquired by states and railroads. During this period of settlement (around 1860 to 1897), agricultural output more than doubled from the previous decades. From 1870 to 1880, farm output rose by 53 percent, but the population of the country increased by only 26 percent. Much of this increase in production was because of the opening up of new agricultural lands and also because of advances in farm mechanization. The mechanical reaper invented by Cyrus McCormick in 1831, the mechanical thresher designed by Hiram and John Pitts in the 1830s, and the steel moldboard plow developed by John Deere in 1837 revolutionized farming by allowing farmers to produce more crops at a lower cost for a rapidly expanding market. The steel moldboard plow was a significant improvement over the cast-iron plow, as the steel's smooth and polished surface allowed the plow to cut through the soil without sticking to the plowshare or cutting edge.

As these new farmlands were being developed, production increased and prices began to fall, even though drought and insect damage affected the Great Plains states. The most important crop economically may have been cotton, but you couldn't eat it. Oats led production in 1862 for food for workhorses and humans. This was followed by wheat, especially in the new western states. Other staple grains included barley, buckwheat, corn, flax, sorghum, peas, and beans. The railroads controlled the prices of shipping these products. Inflation and the subsequent tightening of monetary policy, namely the reliance on the gold standard versus the free coinage of silver as currency that would add to the money supply, led to the Agrarian Revolt that started in the 1870s. This revolt grew in various forms until higher prices and more favorable growing conditions finally returned after 1897. The revolt led to the creation of the first organized farmer groups, such as the Farmers Alliance and the Grange, and some of the first cooperatives in the country. Most would fail, though, as times improved. The first boom and bust cycle, from about 1860 to 1897 was about over.

By the end of the nineteenth century, most of America, including its rural areas, was completely settled. Arizona and New Mexico (in the Desert Southwest region) lagged behind somewhat, and the only frontiers left in the twentieth century were the final two states, Alaska and Hawaii.

The next agricultural boom occurred from approximately 1897 to 1914, followed by a bust from the end of the First World War (1919) through the Great Depression, with the collapse of the stock market in 1929, until roughly the beginning of the Second World War. The bust occurred much sooner for agricultural areas than it did for urban areas. It is no coincidence that boom-and-bust cycles have paralleled the occasion of the great wars. This phenomenon seems to have come to an end with the end of the Second World War.

The historical boom and bust cycles of agriculture are shown below in the average price of land from 1860 to 2023 in Shelby County, Iowa (Iowa State University Extension and Outreach).

Table 3.1 Average Land Prices from 1860 to 2023 in Shelby County, Iowa

Year	Price
1860	$7.00/acre
1870	$20.00/acre (boom)
1880	$20.00/acre (bust)
1890	$39.00/acre
1900	$44.00/acre
1910	$128.00/acre (boom)
1920	$312.00/acre (boom)
1930	$149.00/acre (bust)
1950	$236.00/acre
1960	$423.00/acre
1970	$660.00/acre
1980	$1,984.00/acre
1990	$1,120.00/acre (bust)
2010	$5,506.00/acre (boom)
2013	$9,719.00/acre (boom)
2015	$8,288.00/acre
2020	$7,878.00/acre
2021	$10,237.00/acre (boom)
2022	$12,373.00/acre (boom)
2023	$12,750.00/acre (continued boom)

Source: William J. Petersen, ed., "Iowa Land Values, 1803–1967," in The Palimpsest (Iowa City: State Historical Society of Iowa, 1967), https://farmland.card.iastate.edu/files/inline-files/Murray-1967-Palimpsest-Iowa-Land-Values-1803-1967_0.pdf (accessed February 7, 2025); and Iowa State University, "Historical Iowa Farmland Values Survey by County," Ag Decision Maker, File C2-72, https://www.extension.iastate.edu/agdm/wholefarm/html/c2-72.html (accessed February 7, 2025).

Average Land Prices from 1860 to 2023 in Shelby County, Iowa

The end of the Second World War in 1945 launched an unparalleled period of growth and technological advancement for the country. Farmers in the community of Westphalia shared in those "good times" as well with big gains in yields of crops, especially corn with the new use and development of corn hybridization, chemical fertilizers, and pesticides. Westphalia's influence in Catholic rural social teaching and its efforts to promote social change also were beginning to reach their peak.

In 1947, Westphalia was visited by one of the most influential lay Catholics of the twentieth century, Dorothy Day. Day wanted to see for herself how Westphalia was living out its efforts toward cooperation with one another. Dorothy Day was a complicated and controversial social activist in the Catholic Church. She co-founded the Catholic Worker Movement in the 1930s and became known for her social justice campaigns in defense of the poor, forsaken, hungry, and homeless. She started the first House of Hospitality, a shelter that provided food and housing and advocated for the downtrodden in New York City. There are still 178 Catholic Worker houses in the United States. She was a prolific writer and started the *Catholic Worker* daily newspaper, selling it for one cent per copy. Today, the newspaper is still published seven times each year. While she was accused of being sympathetic to the ideas of communism, seen by some as a solution to the Great Depression, she was actually very much against communism. Day was opposed to its advocacy of class hatred, atheism, violent revolution, and its denial to the right of owning private property. This champion for the poor is now being considered for sainthood by the Catholic Church.

Dorothy Day kept an extensive diary, and her visit to Westphalia was one of her entries in 1947. She quoted Father Duren that day:

> What I would like to see is diversification and decentralization. I'd like to see families on smaller farms. Right now we need a baker, a barber, a shoemaker, a printing press, a feed processing plant, a creamery. And of course we need more houses. I'd like to begin all over again with a little church at a crossroads with a hundred families making a village round about. That is the way America should be. That's the kind of setting which makes it easy for people to be good.[10]

Unfortunately, to me, that is not the historical reality of how America has developed. It has not stayed small and diversified; it has gotten big and more centralized; it has concentrated on making cheaper products in larger numbers—in short, mass production. Westphalia did not have the number of people, either in the town or on the farms, needed to accomplish Father Duren's dream. However, he and our little community did much with what they had.

By the time I was born in 1950, Westphalia was gaining more notoriety in some circles. In the 1950s, Westphalia was featured in my national Catholic geography book as an example of a model rural community. The Complete Life Program was the answer.

Recreation, including sports, was one of the five pillars. Westphalia became known as a baseball town. The first organized team, known as the Westphalia Prussians, started in 1877. As other towns began to spring up, so did baseball. From around 1930 onward, Westphalia and the other towns around it began to take the game even more seriously. The caliber of play continued to improve until it became more like semi-pro ball, even though it was officially amateur in status. Town teams composed of the young men of each of the five Catholic towns and some of the other surrounding towns as well joined together to form the Iowa Western Amateur League in 1947. This league is still in existence today, but there are not enough players in Westphalia to field a team. In its heyday, Westphalia produced one major leaguer who played for the New York Yankees and California Angels. A number of players made it to the minors. Father Duren's nephew, Ryne Duren, was the New York Yankees ace relief pitcher. He came to Westphalia in the 1950s and early 1960s to strut his stuff and participate in some exhibitions. My brothers and I all grew up playing baseball. I was a catcher, although we did not take sports as seriously as some other kids in the community. Our dad needed us at home, especially during baseball season. When the Westphalia Red Sox team still existed, our three sons played with the team for a time.

In a small community with only one church, where everybody in the community was Catholic, at least by name, the church and the town were one and the same. That did not seem peculiar or unique to me when I was growing up. That's just the way it was. That kind of parochialism with a narrow focus only on the local community did not seem detrimental when I was a youth. That feeling changed dramatically when I came back to begin farming in the spring of 1973.

An Idyllic Place

Rev. Duren had a profound effect on me and how I came to view what the world should be like. Yes, he was strong-willed, but it was hard to argue against his ideals and vision for the church, school, and town—in other words, the complete life. I had what I thought was the complete life growing up. It was all divided up into neat little boxes. I had a happy and secure home life. I had the pleasures and experiences of growing up on a farm. I had a school, church, and community life that met my needs for a good early education, for development of my faith, and for feeling like I belonged and fit into the community's social structure. I had everything, even though, like any other kid, I didn't know it. I complained when I had to get up early to milk cows by hand or clean the chicken house on Saturday mornings. Hey, I was a kid.

There are so many stories to tell about my youth. I will start with some of my earliest memories. For some reason, I can "see" myself in a crib when I first became cognizant of some kind of self-awareness. I don't know when that was, of course, but I feel as though that memory is true. There were numerous verifiable memories of when I was very young. For example, I was three years old when I fell on a nail when Grandma Rosmann was watching us. My mom and dad had traveled to Milwaukee to see my aunt, Sister Louis Rosmann, SSSF (School Sisters of St. Francis). I remember running to the house because the nail went in close to my eye but thankfully missed it. I still have a small scar from it. I remember the first night I joined my two older brothers sleeping upstairs. I felt like I had just been accepted into their world as a big kid, no longer a baby. Mom and Dad's bedroom was on the first floor. I had a small bed at the foot of my brother Mike's bed. He was four years older than me. My brother Joe was six years older. I must have been about three or so.

Apparently, it took me longer than most kids to begin talking (which I made up for later in life). An employee who roomed in the bedroom next to ours told me that. He was a Korean War veteran who worked for my parents until he got married and ventured out on his own to begin farming. He would sometimes babysit me. He told me that when it was sweet corn season, I really started to talk, begging for sweet corn. He said that I would say, "Ronnie corn eat," incessantly. Maybe I was always destined to be a corn farmer, but even more so to grow popcorn. Raising popcorn and selling it under our own label has been sort of a hobby for me for many years. Eating it has been nearly a daily ritual for my entire life.

Larry Rosmann—A Life of "Excellent Joy"

There was a life-changing memory when my little brother Larry was born in Children's Hospital in Omaha in 1954. He was born with Down syndrome and also without a rectal opening that required six weeks of hospitalization in Omaha. I remember my brothers and I having to stay in the car while Mom and Dad visited him in the hospital. In today's world, that would not be allowed, but we all learned to take care of ourselves very quickly. I was four years old at that time. Larry's condition and all that was involved were very hard on my parents, but I didn't really appreciate it then.

One thing I do know is that Larry had a profound effect on our family. He helped us feel responsible for the well-being of those less fortunate than us. He formed the social conscience for all of us. He may have had diminished mental capacities, but he taught us all so much about what was really important in life. Since I was the sibling closest in age to him, I spent a great deal of time with him. We hung out together often, and I taught him numerous funny and silly make-believe games that he enjoyed even as an adult. Maria and I became his legal caregivers later in his life. He was able to attain a considerable amount of independence during his adult life, living for more than twenty of his last years in a private care facility in our county for mentally and physically challenged adults. He was always considered to be the life of the party, and his love for music was well known. Larry was allowed to help conduct the Christmas concert put on by the Harlan Community High School concert band when our three sons played in it. During the last years of his life, Larry would throw out a phrase every now and then that had an endearing and wonderful meaning for all of us in the Rosmann family. If he was having a particularly good time at an event or family gathering, he would exclaim: "excellent joy!" Larry died in Maria's arms on May 9, 2008. He was fifty-four years old.

> There are holes that are dug in this life. Some you dig yourself by the stupid things you do. You wish you would not have dug that hole. Some are dug for you that you can't do anything about. They are just there. But what about the hole left by the death of a brother? Especially a brother like Larry who was incapable of ever hurting anyone? You can never quite fill up a hole left by that. The dirt in the hole always washes away, exposing the reminder of my own mortality. That's the only way I can explain it with Larry.

Chapter 4
Grade School at Saint Boniface

Only the educated are free.

—**EPICTETUS**[1]

The importance of having a school in one's small town can never be underestimated. It gives a town meaning and purpose for being both in the present and in the future. People do not appreciate that fact until it is gone. Losing the local school is another one of those holes never to be filled, no matter how well the town may do in the future without it. Having a school gives people hope and a feeling that things are okay and right in the world.

However, the pursuit of education must be cultivated, and that was the case for Westphalia in the early days. The first school in Westphalia was the township school, built in less than two years. It opened in December 1874 with sixteen students present, and classes were held for only about three and a half months. By 1879, the school had eighty-four students, with only one teacher for the large number of eager youth.

The following excerpt is a description of what the school was like in the 1880s. It comes from one of Westphalia's citizens describing the school life of her parents in 1883.

In September 1883, Ann Schwarte was sent to school for the first time at the age of seven. After attending Mass offered by Rev. J. A. Weber, she went to the old school located in the southwest corner of the town. This was Westphalia's first school. Upon entering, she saw she belonged to a class of husky boys and shy girls and that her only book was the German ABC book; along with a slate pencil, which were her writing material. Ann's teacher was Bernard Kaupel.

After Ann had attended school for about two months, she was sent home because she was one of the youngest students and the school was overcrowded for want of accommodations. Preference was given to the older pupils. She returned the next year to find the school even more crowded. The carpenters of the colony were paid to make ordinary, oblong, backless benches, and those were placed against the wall on three sides of the room. The children were forced to sit on these all day long and lay their books beneath them on the floor.

In 1884, the new church was completed, and the girls from the over-crowded previous school were sent into the first old wooden church to be taught by Miss Elizabeth Gollobith.

Usually the last day of school, July 4th, was the test day. On that unforgettable day, the pastor came into school and listened to each child recite. At four o'clock, the dismissal hour, the pastor announced to the teacher the results, and she in turn announced to the pupils who were permitted to take the next higher reader in September.

My father, Ben Blum, was taught during his brief school days by Mr. Bernard Kaupel. He started when he was about nine, but unfortunately, he had to herd cattle for his father and attended school only when he wasn't needed at home. When he returned to school, he could not find his books and at each class to which he reported without his text, he was blessed with a poke from the teacher's rod, or a spanking, now and then.[2]

Experiences of this type were common for both pupil and teacher. Some of the older boys and girls came to school during the winter months only. Teachers were disadvantaged by a large number of very young pupils who demanded much of their attention. When teachers needed to use disciplinary measures, they were dubbed a "wielder of the willow, slippery elm twig, hazel brush, or hitching strap." At a district school board meeting in 1884, William Flusche was empowered to see personally to the regular attendance of pupils and to use means to force regular attendance. (In 1889 in Iowa, 75 percent of all eligible school-age children were enrolled in a school. However, only 47 percent attended on a regular basis.)

In 1884, the wooden parochial school was built to relieve the congestion of the public Westphalia township school. Three School Sisters of St. Francis, specifically a teaching order from Milwaukee, Wisconsin, were invited to teach, and a convent was built for them.

From 1874 until 1927, boys and girls who had completed the eighth grade at St. Boniface and who wished to continue their education in a Catholic school setting were required to leave home and seek this opportunity elsewhere, mostly in academies, which were generally in the form of boarding schools either in Omaha or Des Moines. Secondary education was offered only to those who gave promise of entering the services of the church or some professional field. Secondary education was associated with people of wealth, although that was not the case in Westphalia. Some thought it was a waste of time and money for youth remaining on the farm to have an education. It took time for people in rural areas, just as it took time in urban areas, to recognize the value of secondary education. For example, although secondary education began in Boston, Massachusetts, as early as 1635, it took until 1821 to have secondary education available to all Boston students. That represented a span of 186 years. In the small town of Westphalia, it took only fifty-three years.

When Father Duren came to Westphalia in 1926, he immediately envisioned adding a high school to the parochial grade school. Not everyone was on board. On February 22, 1926, a box social and card party was held as the first event to announce the beginning of the project. A hatchet was given as a door prize. After the prizes for the card party had been awarded, Father Duren took the hatchet and began to tear down the old wooden grade school. An up-to-date building was constructed with considerable aid from volunteers. Dad and Grandpa Rosmann were among the many who did so. The school was modern in all of its appointments, including the architecture, ventilation, lighting, mahogany woodwork, terrazzo floors and stairs, radio wiring, gymnasium, and auditorium. When then Secretary of the Iowa State Board of Education W. H. Gemmill visited the school to judge the merits of its accreditation, he remarked: "This is truly the most beautiful school in the state of Iowa. Not the largest nor the costliest, but the most beautiful."[3]

Happy School Days

Mom read to my brothers and me before we started school in the first grade at St. Boniface in Westphalia. She had a great love for literature as well as having been a teacher prior to her marriage. That gave her first three sons a head start in their formal education. Along with learning to work and help on the farm, formal education became the priority for all of us. The fact that Dad had not been able to attend school beyond the sixth grade helped to make sure that schooling was stressed.

I can honestly say I do not ever remember having more than a few bad days in school until I was in the seventh grade. For that matter, I do not remember many bad days growing up on the farm either, and most of those involved being sick or getting hurt for some reason. I have to admit I was kind of a "runt" when I was little. My ribs stuck out, and I was convinced that I had at least one or more tape worms because I was so skinny. I did not. However, if the truth be told, many kids of my era growing up on a farm and especially those with pigs like ours, did have a case or two of Ascariasis or roundworms. I was no exception. It may have come from walking around the farm barefoot. Of course no one ever admitted it or freely shared that kind of information with anyone if they could possibly help it. What ten-year-old boy in their right mind would tell anyone about that kind of incident? Well, I did tell my mom, and the proper medicine (possibly albendazole) cured me of that very embarrassing occurrence. An estimated four million people have intestinal roundworms at any one time in the country today. It is actually estimated at least fourteen percent of Americans have had a case of a roundworm infection. In today's world most of them come from our dog and cat pets. It was a big shock for me to find that the Center for Disease Control and Prevention (CDC) estimates over 100 million Americans have some kind of parasite in their bodies.[4] That was the case in 2014, and there is no reason to believe it has improved much, if at all.

I loved school, and I had a great deal of fun with my fellow, mostly male, classmates. As a group, we were all pretty well-behaved kids. Or maybe we were just pretty good at hiding any transgressions, but I doubt it. I started first grade in the fall of 1956. Twelve boys but only two girls were in my class, which was just great as far as I was concerned. Who wanted to talk to a girl anyway? Remember, Mom was the only female in our household.

I was taught by the School Sisters of St. Francis through the eighth grade. I never had a lay teacher until I was a freshman in high school. For many years, St. Boniface had as many as nine sisters teaching and living in the convent that stood between the church and school. Nuns were, of course, always a great mystery to my fellow classmates and me. Most were quite strict when it came to discipline. Most were very good teachers. Of course, we had our favorites, and only a few incurred the injustice of nicknames that the older high school boys dished out. The nuns' individual rooms in the basement were off limits. Of course, their layers of clothing were a mystery entirely of their own. The convent remained a mysterious place as well. Rarely was a student allowed in the convent, and then it was on the second floor, where the piano was located, for lessons by the music teacher.

In my estimation, it is noteworthy that very few students fully appreciated or understood just how good a deal they were getting in Westphalia, or in any Catholic school for that matter. The selfless service provided by the multiple teaching orders of sisters generated an excellent education at a low cost. The sisters were paid virtually nothing and were expected to live in extreme austerity and self-denial. They did so for many decades in their service to our community and to others, both rural and urban.

Very early on I learned about venial and mortal sins, the Ten Commandments, and all the wicked things that could send you right to hell if you even so much as thought about doing them. It kept me on my toes and generally out of trouble. I remember a particular incident that happened in second grade. (The first and second grades were in the same room.) One day our teacher had to leave the room for some reason, and she gave strict orders for us to stay at our desks, read, and be quiet. She was away for quite some time, and we all became restless after what seemed like an hour of strict obedience. I do not remember any other teacher coming in to check on us, either. There were many tall windows in our classroom that had spring-loaded shades on the top windows, and they were all pulled down, covering the windows. For some unknown reason, a number of the shades suddenly snapped and rolled up. All of us boys were convinced it was the devil at work. We sat quietly until Sister Romaine returned. I do not remember anyone saying they did not like Sister Romaine. I can imagine her saying the following words about her years at St. Boniface:

> I am Sister Romaine. Yes, I was your first-grade teacher and your second-grade one, too. You didn't know how much I loved all of my school children. Yes, we did not show our emotions very much, but I can assure you that I did love you. I was proud of the work I did during the years I was in Westphalia. I was proud I could give you the first building-blocks for a lifetime of learning. I taught you reading, writing, and arithmetic, respect for one another, and about how much Jesus loved you. Yes, I was strict, but not all of you youngsters behaved like "little angels" either!

Knecht Ruprecht

In the folklore of Germany, *Knecht Ruprecht* translates as farmhand Rupert or servant Rupert. As school children at St. Boniface, we simply knew him as Rupert. Rupert was supposed to be a companion of St. Nicholas who wore a long beard and either a long fur coat or a coat covered in pea straw. He also wore a black mask over his face. In folklore, he was also known as Black Peter because of the soot on his face from going down the chimneys at Christmas time.

The tradition in Westphalia, started by Father Duren, had Rupert accompanying Santa Claus at the annual school and parish Christmas party. I can still picture the scene of Santa Claus and Rupert as they made their way down the steps to the hall where we were all sitting on the edge of our seats. First, there was a loud bang on the wooden door that scared us half to death. Santa was in the building. But so was Rupert. No kid in their right mind would have considered being naughty or bad right then!

Rupert was dressed in a floor-length fur coat and had on black five-buckle overshoes. He looked very menacing to a six- or seven-year-old. While Santa Claus handed out candy, Rupert carried a bag of long sticks he would hand out to some of the high school senior boys. First, he would make them kneel down and ask if they knew their prayers. He would make them recite the "Our Father" or a "Hail Mary." It was all in good fun, and the crowd would be hooting and laughing. Next, he would ask them if they had been good. Even if they answered yes, he would still hand them a stick. As young school children, we were deathly afraid of Rupert. But even then, it was a "fun" scared. I knew he wouldn't be picking on any of us little kids. If he did, I would have jumped out of my skin.

By the time I was ten years old, my male classmates and I were getting quite used to Rupert making his annual appearance at Christmas. It was the highlight of the Christmas season. Our coop grocery store furnished the candy each school student received, regardless of age. I remember it consisted of a mesh stocking filled with all sorts of wonderful goodies of that era. There were Butterfinger and Baby Ruth candy bars, double-stringed suckers, Life Savers, Wrigley's gum, Tootsie Rolls, peppermint sticks, and so much more. After Santa and Rupert made their quick escape out of the school hall, we would run as fast as we could to see where they went. We never did find out. It made us wonder what was really going on. Sadly, the tradition stopped a few years after the death of Father Duren in 1962.

St. Nicholas

Ruprecht was a common name for the devil back in Germany and first appeared in literature in the seventeenth century. St. Nicholas and Rupert had a certain amount of negative social stigma attached to them. To some, St. Nicholas represented country nobility and Rupert, the ignorant peasantry. Sometimes, Rupert went around the village during the Christmas season threatening to hand out sticks or lumps of coal to those deemed not to be good. The goal, of course, was to keep children behaving well so St. Nicholas would come with gifts for them on December 6. This is the feast day of St. Nicholas, patron saint of children. Setting out your shoes the night before to be filled with candy or other treats by St. Nicholas by the following morning became a favorite tradition. This tradition followed through to the Christmas customs celebrated in Westphalia. My family did it while I was growing up, and my wife and I continued the tradition with our three boys, and they are doing the same with their children. You can never have too much Christmas!

Martin Luther and the Protestant Reformation frowned on the celebrations that commemorated the feast days of saints. Luther thought it put too much emphasis on them and not on Christ. As a result, the actual date of gift-giving came to be associated more with the celebration of the birthday of Jesus and was moved a few weeks later to the 24th or 25th of December. In spite of this, the feast of St. Nicholas remained a very popular event for most Protestants in Europe.

The story of St. Nicholas is a beautiful one. He was actually a Greek Christian bishop who lived in the fourth century. He became known for his generous gifts to the poor and especially for providing dowries to three young women given in marriage so that they would not have to become prostitutes in order to survive extreme poverty.

Just Getting There

Sometimes getting to school was not easy during winter. School buses did not pick up Catholic students and drop them off at the Catholic school. That situation did not change until the mid-1960s in Iowa. Four-wheel-drive vehicles had not come along yet. If the snow was too deep for a chained-up car to get through, it meant riding in the back of our farm truck for three miles. The back of the truck was open, so we nestled in between bales of straw to stay warm until we made it to school. On occasion, Dad would get us there only to find out that school had been called off. We were never sad about that happening, of course.

During winter, school recess was a fun time. It meant building elaborate forts with rolled-up snowballs, where we defended ourselves against the grade above. After school meant snowball fights until Dad arrived to pick us up. Dad and Uncle Leonard took weekly turns getting us to and from school. School permits issued to fourteen-year-old students eased that burden for them. That did not help, though, unless you had a second car or could drive the truck. We did not have a second car until I was sixteen, so we had to drive the truck. It was a one-ton GMC with a livestock box on the back for hauling cattle and hogs. Top speed was barely fifty-five miles per hour. It had windshield wipers that worked on a vacuum system, so they moved very slowly when the engine was struggling.

One fateful day in a seventh-grade class, three of my male classmates and I decided it would be a good thing to shoot spitballs with our fingers across the room. Our teacher, Sister Jarento, was working for a month in Montana at an Indian mission school, and the principal was teaching us during that time. Of course, we were caught in this little endeavor, and the punishment was brutal. Each of us was "assigned" to make a thousand spitballs by the next morning and bring them to school in a paper bag. One other classmate and I took the situation seriously. I told my mom and dad, and naturally, I worked all evening trying to get them produced. My cousin Ken, who was in the third grade, even made a hundred of them. Sheepishly, I brought the sack of spit balls to school the next day and presented them to the principal.

The other two classmates must have figured Sister would not enforce that punishment and did not produce any spitballs that morning. They were sorry they did not; however, when they had to make ten thousand of them and glue them on sheets of paper in groups of ten. I never learned how that turned out.

On rare occasions, corporal punishments were doled out by certain nuns. Most of us knew what the metal side of a ruler felt like across the knuckles. We generally took it with a sense of complete acceptance and even a certain amount of pride.

Little corporal punishment was doled out at home, too. If we boys talked back to Mom, Dad would give us a quick but firm slap across the face. That only happened a few times that I recall. Usually, if there was some bad behavior, Dad would *threaten* to pull out the razor strap. This was the leather strap he used to sharpen his razor. I do not remember ever being the recipient of it. I do remember my two brothers finding out about it after all of us were caught playing cards and making a ruckus in the middle of the night. That was the end of our nighttime pitch games.

You had to learn to defend yourself physically against some of the bullies in the grade above you. Sometimes it was with boys two grades above you. Since I had two older brothers, I was kind of used to having to defend myself anyway. They were not bullies; they were just older brothers. I had a few black eyes and bloody noses, and I gave out a few as well during my grade school boxing career. My classmates and I took great pride in the fact that we all got along so well, which meant we all stood up for one another against atrocities committed by boys in the grades above us.

I am not saying everything was perfect in how discipline was handed out and how you were expected to conform to the code of conduct during my growing-up years. The unspoken word carried a great deal of weight in those days. There was not much room for discussion or communication in working out problems. Compliant behavior was a whole lot easier, but that did not mean it always worked that way, nor was it always the most desirable way to resolve conflict. When very bad things happened, I think it was because there was little if any communication between children and parents over the big stuff. Sometimes bullying occurred because family life was not stable at some homes. Sometimes high school students abused alcohol and speed (cars), and tragic life-ending or life-changing events happened. Our community was not immune from that. There were a handful of families in our parish and farming community who were ostracized for reasons that may have involved alcohol, perceived reputations involving economics and not working hard enough, mental illness, and legal problems.

When someone had to be admitted to a hospital for mental illness, it was referred to as suffering from a nervous breakdown. Everyone in the community knew about it, and it created much suffering for the affected people and their families. The stigma of suffering from depression or any other mental breakdown carried negatives in our community, where everyone knew everyone else's business, or so they thought. It has always been one of the biggest negatives for small towns. You could get lost in the anonymity of a city, but not so in a rural area.

> The stigmas associated with mental illness have lessened, but it is still very hard for people to admit they need help and then to seek it. This is especially the case for men who are farmers. There are so many things that farmers have very little control over when it comes to the profitability of their vocation and occupation. The stress associated with economics has historically been one of the principal drivers for farmers undertaking suicide. The farmer always thinks it is his or her fault that the farm is facing financial failure. The persona of being a failure is taken on. This is especially true when farms have been in the family for a long time. It is imperative that our politicians all over the country realize this dynamic and put more resources into rural mental health. Rural America suffers from poorer physical healthcare than urban areas. It is even worse with mental health. There is an extreme shortage of professional counselors, psychologists, and psychiatrists in rural areas.[5]

The Closing of St. Boniface High School

I did not get to graduate from St. Boniface High School in Westphalia, but I did graduate from the eighth grade in 1964. I had a wonderful teacher that year. Her name was Sister Bartholomew Marie; she was from Louisiana. She was quite young and could easily relate to her students. Everyone liked her. She began a civics club in which we learned about state and national governments. She taught Iowa history, and the highlight of the year was the first school field trip that our grade had ever taken. We traveled on a bus to Sioux City, Iowa, a hundred miles away. I remember visiting a Jewish synagogue, the Sioux City Air Base, Sergeant Floyd's monument, and going bowling. Sergeant Charles Floyd was a member of the Lewis and Clark expedition who died in 1804 near Sioux City on the Missouri River, most likely due to peritonitis from a ruptured appendix.

In 1964, St. Boniface High School combined with one of the other small Catholic high schools in the county. The state of Iowa mandated schools to have a minimum enrollment of one hundred students in ninth through twelfth grades. Westphalia had only about sixty students in those four grades. Each of the five German colony towns had their own high schools until that time.

In 1910, Iowa had a total of 406 high schools. Over the next twenty years, through 1930, that total mushroomed to over 900. By 2015, the total had declined to 453. Suburban areas added high schools. Some urban areas saw a slight decline. Rural areas saw a major decline that is still occurring today.

It was decided that St. Boniface would combine with St. Paul in Defiance. The schools in the towns of Panama and Portsmouth combined, as they were geographically close to each other.

Earling's school was big enough to stand on its own. Around 1960, there was serious talk about consolidating and building a central Catholic high school. Father Duren proposed it to be on the outskirts of Westphalia, as that was the most centrally located area. By that time, the Catholic high schools in the five colony towns were fiercely competitive and independent. None of the other towns bought into the idea. Had there been agreement on it, the next fifty years might have been a little different for the community of Westphalia. The three combined Catholic high schools lasted only for three years, and in 1967, they were all closed for good. The grade school in Westphalia was able to remain open for another ten years, through 1976, with grades one through four. Defiance maintained grades five to eight. As of 2023, Shelby County had only one Catholic school, in the town of Harlan, with students from preschool through fifth grade.

St. Boniface High School had graduating seniors from 1929 to 1964. Close to 400 students graduated during that time. Well over 1,000 students passed through St. Boniface, either through grade school, high school, or religious education programs from 1929 to the present. The reality today is that you can count the children attending religious education programs in Westphalia on two hands.

Teachers open the door, but you must enter by yourself.

—CHINESE PROVERB

Not everyone is willing to enter the door, even if it is opened by a good teacher. A good case could be argued that, for St. Boniface School during its first sixty to seventy years of existence, at least some students attended only because they had to. The economic realities of farming and of living in a small, all-Catholic rural community dictated in large part what the bulk of graduating seniors would do with their lives. The first choice for girls until the Second World War was to get married, become a mother, homemaker, and farmer's wife with all the taxing duties that entailed. Not only did she cook the meals, but she also grew much of the food in a large garden, raised chickens, and helped milk the cows, too, depending on the farm's circumstances. The water needed for so many of the necessities of a home first had to be hand-pumped into a bucket and then carried from the well to the house, with the bucket filled to the brim.

Hot running water, inside toilets, electricity, washing machines, and a host of other time- and labor-saving devices did not become commonplace on most farms and in most small towns until well into the 1930s. Many farm homes were heated first by wood and then by corn cobs. The cast-iron cook stoves had to be kept running all day to heat water, cook meals, and wash dishes and clothes. The farm woman's work was literally never done, which is depicted in a picture hanging in our kitchen as a reminder. It shows a homemaker washing dishes with hot water coming out of the faucet. It is titled, "The Farm Woman's Dream," and goes on to say, "Make your dream come true. Consult your County Agent or write to your College of Agriculture for information on water supply systems for farm homes." It came from the University of Missouri Cooperative Extension Service in Agriculture and Home Economics, 1920.

Educational Discrimination

In Westphalia, a second option for high school girls was to enter a convent to become a nun. Westphalia alone had over seventy-five girls become nuns during its first one hundred years. Young girls were encouraged to participate in a very worthy vocation to serve Christ and the Church. Parents took a great deal of pride in one or more of their children becoming a sister or a priest. Two other vocations for girls that grew out of the Great Depression and the Second World War were secretarial school and nursing. Teaching was always an option from the earliest days; however, you had to be single. Once you were married, you could no longer be a teacher in the Catholic school system. This was also the rule in public school systems. In fact, at the beginning of the Second World War, 87 percent of school boards would not hire married women. Seventy percent would not retain single female teachers who married. The practice of discrimination against married female teachers did not formally end until the passage of the *Civil Rights Act* in 1964.[6]

For boys in rural settings, the first choice was farming. In fact, some boys did not finish beyond the eighth grade well into the 1950s because some thought they did not need a formal education to become a farmer. They could obtain all the education they needed right at home on the farm, and it came with its own on-the-job training. Military service was a second common option. Some had no choice about being drafted during the great wars of the twentieth century. In peacetime, many boys were still encouraged to enter a branch of the military for patriotic reasons and as a good way for a young male to grow up and become a man.

Entering seminary to become a priest was another option for young men in the Catholic communities of Shelby County. However, the number of priests coming from Shelby County was less than a third of the number of young women entering the convent.

The idea that a farmer might want to send his son or daughter to college to study farming must have seemed absurd to some farmers. I do not know of anyone from Westphalia who pursued it until the 1960s. Certainly, some young men and women in the public school systems in Shelby County did go to Ames to study agriculture long before I did. However, the idea had very few to open the door and to encourage it in the Catholic high schools of our county.

Land-grant universities with the intent of teaching agricultural and mechanical arts began in 1862 with the signing of the *Morrill Act* by President Abraham Lincoln. It was named after Rep. Justin Morrill of Vermont, who championed its passage.[7] The idea was first promoted in the late 1840s by Johnathon Baldwin Turner, who advocated for the establishment of publicly funded agricultural and technical educational institutions. The law gave every state and territory 30,000 acres for every member of Congress from that state to be used for the establishment of land-grant universities. The money from the sale of those lands would be used to fund the establishment of these institutions of higher learning. Even though agriculture was supposed to be emphasized, more students studied engineering than agriculture well into the twentieth century. As early as 1873, the Grange agricultural organization condemned the twenty-four land-grant institutions at the time for their failure to attract agricultural students.

How the land was made available for these educational land grants has rarely been discussed openly, even as recently as 2019. I certainly was not aware of any of the details. Here, again, it is a case of needing to acknowledge the truth of our historical past and to teach it in our classrooms. The *Morrill Act* redistributed as much as 17 million acres from more than 160 violence-backed secessions of land made by close to 250 tribal nations in our country at the time. Fifty-two universities owe much of their beginnings and successes to this *Act*. For example, here in Iowa, Iowa State University sits on land originally inhabited by the Ioway Nation. The United States obtained the land from the Meskwaki and Sauk Nations in the Treaty of 1842.[8] In the treaty, the Native tribes were referred to as the Sac and the Fox. They ceded land west of the Mississippi and north of the Missouri border. It stipulated they had to completely exit Iowa for settlement in Kansas within three years' time, by 1845.

It is imperative to speak about how the historically Black colleges and universities (HBCUs) came to be a part of the land-grant institutions in this country, although no land was given to them for their creation and ongoing financial well-being. The first *Morrill Act of 1862* denied admission to Black students because of their race. The second *Morrill Act of 1890* attempted to rectify that situation, at least in the sense that they were to be able to establish their own colleges and universities.[9] US Representative Justin Morrill of Vermont became a US Senator in 1867 and served until his death in 1898. He championed the legislative framework for both *Morrill Acts*. While the 1862 land-grant universities annually received full one-to-one matching federal funds, 60 percent of the 1890 institutions (of which there are 19) did not receive full matching state funds. Most of these nineteen colleges and universities are in the South. In spite of these and other inequalities, their track record of graduation rates and expertise has been extraordinary.

It is also important to note how African Americans have been disenfranchised in terms of farming and land ownership since they were freed by the Emancipation Proclamation and the surrender of the Confederate Army in 1865. An examination of our country's Black citizens' opportunities to have meaningful employment in agriculture—and to be able to have livelihoods in farming in the first place—shows that, even after studying agriculture at an 1890 institution, Black citizens had few family farms to go back to. By the end of the Civil War, there were 4 million freed slaves. By June 1865, 40,000 of them had received forty acres of land from their former owners through Special Field Order No. 15 issued by General William Sherman. However, after Vice President Andrew Johnson became president following the assassination of Abraham Lincoln, he oversaw the Reconstruction period, and ownership quickly reverted back to the former slave owners. Today, 45,000 out of 3.4 million farmers identify as Black. Only 1.4 percent of farmers identify as Black, compared to 14 percent one hundred years ago.[10]

There is no doubt that the creation of land-grant universities did much to initiate and promote post-secondary education for ordinary white Americans. It has been referred to as a Bill of Rights of sorts for publicly funded education. The *Morrill Act of 1862* specifically stated: "In order to promote the liberal and practical education of the industrial classes in the several pursuits and professional of life."[11]

However, there is also no doubt that we need to first acknowledge and then begin to figure out how to reconcile another injustice and myth of our Manifest Destiny. We could begin with the federal government's willingness and commitment to recognize Native tribes that still do not have formal affirmation with all of the rights that go along with that designation. Indigenous people still remain largely absent from the student population, staff, faculty, and curriculums in land-grant institutions. At least a dozen or more of the fifty-two state land-grant universities still generate revenues from their unsold lands from this *Act*. As a start toward reconciliation, it is time that some of these institutions' money, as well as specific revenues from the rest of the land grants, be allocated toward indigenous student scholarships, faculty, staff recruitment, and curriculum development.

It was not until after the end of the Second World War that new educational options began to filter down into rural areas. The *GI Bill of Rights* (*Servicemen's Readjustment Act*), passed in 1944, gave veterans the opportunity either to finish degrees that were interrupted by military service or to pursue another degree. An unprecedented number of engineers (450,000) came out of that program along with 238,000 teachers, 91,000 scientists, 67,000 doctors, and 22,000 dentists.[12] The *GI Bill* helped to pave the way for the new educational focus on the sciences, teaching, and the goal of improving public health.

Education can change the world for the good, but ignorance may be the greatest threshold to poverty. For me, living on and growing up on a farm did not lend itself to becoming an ignorant person. There was always so much to do, observe, and learn. It became my "working laboratory."

Children must be taught how to think not what to think.

— **MARGARET MEAD**[13]

Chapter 5
A Working Laboratory

I'd rather be on my farm, than be emperor of the world.

—GEORGE WASHINGTON[1]

This chapter provides a picture of the evolution of handwork on a farm to mechanization—from horses to tractors—and the dwindling of the number of farms as the old ways give way to the new. It also describes the social and physical value of having neighbors. It ends with a description of the Blizzard of 1975!

Learning the Old and New Ways of Farming

Growing up on a farm meant there was always something to do, whether it was work or play. Much of the time, the work was play to me. My dad even thought so, at least in the sense that he would say that work was his play. But that was because he never took much time for play, so he made the most of it.

Since we had work (draft) horses, sometimes Dad handed me the reins when I was riding in the wagon with him on such jobs as fixing fences or hauling hay in the winter. An early memory of mine was riding on the bundle wagon when we were threshing oats. A large wagon was used to pick up the bundles of oats that had been stacked so that the small stacks would shed most of the water when it rained. The sheaves of oats also dried out more quickly because the bundles were stacked vertically. This was done during the month of August. It was fun for everyone except maybe for Dad because he had to make sure everything was working correctly and manage the operation.

Harvesting began in July with the oats being cut and tied into bundles. Our John Deere B tractor had a hand clutch, so I was able to drive it when I was eight years old. It was too hard for me to press down on our other tractor's foot clutch until I got a little older and my legs were long enough to reach it. I remember my brother Joe riding on the seat of the binder, barking out instructions for me to do a better job of driving! Did I listen? I tried as best I could for an eight-year-old!

The binder automatically tied the clumps of oats with sisal twine and dumped them out the side of the platform onto the ground. Sisal twine is made from a type of agave plant, the same plant used to produce aloe and tequila. It is native to southern Mexico. Brazil and Tanzania are today's biggest producers. Sisal twine was used by nearly all farmers back in the day before plastic twine started to replace it.

The history of the binder must begin with the history of the reaper. In 1831, Cyrus McCormick invented the first mechanical grain reaper that cut the grain and dropped it on the ground behind the reaper. The farmer still had to rake up the grain. It was then stored in barns to be fed to livestock. The next major improvement happened in 1872 when a reaper was invented that bound or tied the grain in a bundle and dropped it on the ground. The first binder used wire, which could be dangerous for livestock if a farmer failed to cut the wire off entirely from the bundle of grain, and livestock happened to eat it. In 1878, the binder that used sisal twine to tie the armfuls of grain was invented by John Appleby from Wisconsin. This type of binder did not change much until after the Second World War.

However, there were few efficient ways to thresh the grain, dehull it, and separate the straw and chaff from the grain. The first threshing machine was invented in England around 1786 by the Scottish engineer Andrew Meikle. Early threshers were hand-fed and horse-powered. Throughout the 1800s, improvements were made to threshing machines. They became much larger and did a better job of separating the grain from the plant. They became powered by steam engines and finally by tractors.

The mechanization of agriculture caused social unrest when machines replaced human labor for producing food. Threshing machines became the target of agricultural workers in the largest movement of social unrest in nineteenth-century England, known as the "Swing Riots."[2] Threshing machines were destroyed by angry farm workers who were being replaced by the machines. Low wages, growing poverty and hunger, and the required 10 percent tithing requirement of the harvest going to the church helped to drive the unrest. Workhouses, where workers lived and were fed, and tithe barns, where the grain was stored for the tithing requirement, were destroyed. Nineteen people were executed, 505 transported to Australia, and 644 imprisoned. The riots were historically linked to the Enclosure Movement as discussed in Chapter 1.[3]

By the 1940s, we had our own threshing machine, and we would neighbor back and forth with Uncle Leonard and sometimes another neighbor to separate the oats from the straw. Stationary threshing machines were the first mechanical combines. Our machine was driven by our 1952 model International Harvester Farmall Super M tractor, which we still have. It still runs, too! A long leather belt ran off a drive pulley on the side of the tractor to the large pulley on the threshing machine to run the thresher. The separated oats poured out of an elevator on the side of the machine into a wooden wagon. It became my job when I was little to level off the oats. My older brothers were big enough to pitch oat bundles with a fork to hand-feed the grain-eating monster. The straw was blown out a long metal tube onto the ground to be stacked later. It was quite the operation to behold, especially for a little kid. I remember it as being absolutely a fun time. The best part was when my mom would fix a large dinner at noon for the men and then make sandwiches, Kool-Aid, and cookies for a lunch break in the afternoon.

Back before more mobile tractors, big, slow-moving steam-powered machines went from farm to farm to do this work. I never experienced that. Those heavy steam tractors were expensive for their time, costing between $2,000 and $5,000, which is why each community had only a few machines owned collectively by a number of farmers or by one farmer who went around the community to thresh oats. I do not know when Grandpa Joe or my dad purchased their threshing machine. The serial number on the McCormick Deering threshing machine that we used when I was growing up indicates it was built around 1929.

> The impersonal hand of government can never replace the helping hand of a neighbor.
> —HUBERT H. HUMPHREY[4]

Neighboring back and forth for certain farm jobs was one of the best parts of growing up on a farm. It was done for such labor-intensive jobs as baling small square hay bales or putting up loose hay and for shelling corn. Small square bales were tied with sisal twine. Loose hay was not tied. Because the corn was picked in the ear before the advent of combines that shelled the corn off the cob, a custom-hired machine that did this came to the farm for this job. We neighbored back and forth with Uncle Leonard and two of our other neighbors for over forty years. Our farm was located on the southern boundary of our Westphalia parish. The two other farm families we neighbored with were Protestant. That was a good thing. It broadened our horizons even as children.

One of the first topics I wrote about when I came back to the farm in 1973 was about neighboring. I was twenty-three years old. I thought it so pertinent that I decided to put most of it into this book. The last remnants of performing some farming jobs on a neighbor or communal basis ended with the turn of this millennium.

Neighbors and Haying[5]

In the early days of farming in our area, a farmer depended upon his neighbors for a variety of things, sometimes, his own economic survival. If he couldn't get along with his neighbor, his life was bound to be "poorer and harder" because of it. Religious differences were transcended and tolerated for the sake of being good neighbors. It wasn't until I was growing up in the 50s and 60s that this spirit and dependence upon one's neighbors began to visibly be shaken.

Neighboring was the norm in my parents and grandparents' day. Such jobs as butchering a hog or beef, building a barn, making hay, shelling corn, threshing oats, picking corn, and chopping corn silage were generally shared on a neighborhood basis. These were jobs that could not be done so easily alone or even by a single family. For instance, threshing was done on a communal basis because it was not practical or economically feasible for every farmer to own a steam engine or a threshing machine.

Most of these jobs lost their neighboring requirements on our farm during my youth except for three tasks. They were making hay, shelling corn, and chopping corn silage. I still "neighbor" back and forth with the neighbors that I grew up with for these three jobs. Putting up hay was done during the summer so I always got in on that. Shelling corn might be done on Saturdays and in the summer but chopping corn silage was done in the early fall after school had started. Still, we helped after school. It was and still is fun to go to one of the neighbors as a member of the haying crew or for them to come to your place. My very first experience with driving a tractor was when I was eight years old. I began to learn to pull a hay rack with the hay loader attached across the windrows of hay. The hay loader picked up the loose hay and dumped it onto the rack. Usually, two men with pitch forks would be on the rack to stack the load. At least two racks would be kept in operation that way; while one was being unloaded, the other was being loaded. Unloading the hay at the barn was done by means of a "harpoon" fork which had four tines or prongs in it that would be pushed by hand into the hay in the shape of a square about five feet in diameter. The fork was attached to a thick rope and pulleys pulled it up to the top of the barn and then back across the entire span of the barn and then came back around and out the barn again where the end of the rope was tied to our John Deere B tractor. It became my job to back the tractor up and lift the hay up through the large door and into the barn. The fork full of hay would follow the metal track until the men stacking it would holler "woah." The man on the rack would then pull the trip rope. When pulled, the trip rope released a small catch on the fork and the hay would drop. I would then pull ahead with the tractor and the process would be repeated until the hay rack was empty. Six or seven men would be required for making hay in this manner. On a hot and humid day in July, working in the barn is the best way of weight reduction I know of. There would be very little air moving in the hay loft and one's clothes would be drenched with sweat. It actually did help to keep you cool. It was a good feeling to use one's muscles and any tensions either physical or emotional were at least temporarily erased by the hard work. Our two barns full of sweet cured alfalfa and red clover was something our cattle would cherish during the up-coming winter.

The mechanical hay baler changed the process of making hay. Dad and Uncle Leonard bought a New Holland hay baler in 1960. The first balers used wire to tie the bales. Sisal twine later replaced wire as the tying material. I am still using this baler and it has paid for itself many times over. Baling the hay speeded up the process but it was still hard work. Now one had to stack 40–50 lb. bales and more depending on how dry the hay was. About 100 bales of hay could be stacked on a typical hay rack. Usually three racks would be operated by the crew with two men in the barn. I started running a rack when I was 16. To be able to stack a neat and square load of bales was a challenge for me as I weighed only about 140 lbs. I helped to put up about 20,000 small square bales that summer and was quite proud of my newly acquired strength. I was most proud of passing my rite of passage into the adulthood jobs of farming.

Haying was done during my high school days in a variety of ways. By the early 60's, nearly everyone stopped putting up loose hay and purchased balers producing either small square or small round bales. A few people preferred to stack their hay loose out in the field in large piles by using what was called a booster buck. It was a large wooden fork made from oak that was mounted on a tractor which gathered the hay and lifted it up to form a stack where men with forks would stack the hay evenly. We had one of these machines. Stacking the hay so it would not tip over was another skill you had to learn.

Raymond Errett, who was our neighbor to the south, was both a phenomenal mechanic and farmer. He could fix just about anything and always had an ingenious way of solving any farming problem. I remember the time he had a very large and very long stack of loose hay out in one of his fields made by the booster buck machine. He didn't get it fed up in time so it eventually spoiled and became as hard as a rock. The dynamite he stuck into the stack at various places took care of that little problem. Watching things blow up when you were a young farm kid had a lot going for it.

Reading what I had written over forty-seven years ago has allowed me to think about an observation my dad made about the appeal and value of being a farmer, owning one's own business, and being able to work outside. When I came back to the farm in 1973, my dad talked to me about the value of working to forget one's troubles or at least make them seem more tolerable. If I was feeling down about something, he always suggested, "Go outside and do some work. There is always something that needs to be done," or something like, "You'll never run out of work." How true those words have been for me, as I know they were for him. Working outside and performing the task at hand is generally good therapy.

What I remember the most about neighboring is one of our neighbor's hired hands of Danish ancestry. His name was Otto. He was a short and stocky man and was as strong as an ox. I recall many experiences working with this man during my youth. My early experiences usually involved me making some mistake, such as backing the tractor up too far on the hayfork so that he could not trip the rope where he wanted to when the guys in the barn hollered. Another mistake happened in the field, where I drove too far from a hay bale where a person stood on a wooden skid resembling an oversized sled lifting the bales off the ground. Both generally resulted in a good stream of unflattering words. One time, Uncle Leonard's hired hand for the summer, who was his high-school-aged nephew, had been driving on the skid all afternoon for Otto and could not do anything right for him. To get a little revenge, on the last skid of hay for the day, the hired hand put the tractor in "road gear," popped the clutch, and left Otto sprawling on the ground. I think the cows even perked up their ears that afternoon when Otto screamed his litany. Lunchtime always proved to be the funniest. When Otto told his stories, he talked fast and threw in a few expletives every now and then. Sometimes, you had to leave the table because of laughing so hard. Even my dad and Uncle Leonard could not stop from laughing. The young guys and I would be holding our stomachs and rolling on the ground. I do not believe I ever laughed so hard in my life as I did some of those times. It did not make Otto feel badly, either. He just kept right on going. I suppose you had to be there to appreciate it. He was one of a kind, but was a person of good character and was one of the last of his kind in farming, the "hired man." Haying just did not quite have the same flavor after he died in the 1980s. Haying with the neighbors was a communal way of helping, supporting, and getting to know one another. I miss it. I muse to myself:

> I was one of your neighbors. Do you remember when we would stop along the fence, get off our tractors and spend five or ten minutes just visiting? Of course, it always started out with the weather and then was followed by how the ground was working and when we were going to start planting or something along those lines. It was said "good fences made for good neighbors." The fences are mostly all gone now, and there is no time to stop and talk either.

Baling hay today is mostly done by one person making the large round bales or large square bales urbanites see as they drive down the interstates. The hay is covered with plastic net wrap, and most is left outside, not under a roof. There is at least some spoilage in baling hay that way. Yes, it is much faster and easier and requires less labor. It also is not as much fun, and the sense of rural culture due to neighboring is mostly gone.

I started milking cows by hand when I was eight years old, too. Joe and Mike always said I got off easy because they started when they were at least two years younger. My reprieve was due to helping care for our brother Larry. We generally had three to six cows that we milked in the morning before school and again after we got home from school in the afternoons. They were the Holstein breed. Because we only sold the cream, we also had to separate the cream from the skim milk by running a machine that did it. That was done in the basement of the house. We did not drink raw milk. My mom would pasteurize the whole milk for drinking. The skim milk was fed to the hogs. The milking chores usually took us an hour in the morning and an hour in the afternoon. Once my two brothers were in college and I was a freshman in high school, I took over most of the milking chores. Dad had enough to do already.

Milking cows by hand had both its own set of hazards and its fun memories. We first had to get the cows into their individual stanchions, which were metal brackets that fit on either side of their necks. To get them in, we first opened the stanchion brackets so the cows could get their heads through easily. Then we fed them ground corn to entice them to enter and locked the stanchion so they could not back out. After a time, they became so used to the stanchion that the chore was not very difficult. We sat on wooden stools that were simply a four-inch by four-inch block about a foot long nailed to a two-inch by six-inch board on top that was also about a foot in length. The first fun part was squirting milk at the cats, who were just waiting for it to happen so they could lick the milk off their paws and coats. The second was squirting one another if things needed lightening up a bit. The not-so-fun parts happened if a cow stuck her foot into the bucket of milk, which meant you had to throw the milk away, or if you were kicked by a cow. We used what were called hobbles on cows that had a habit of kicking. They were metal cups attached to chains that fit over the cows' hind legs. They could only kick a short distance until the chains tightened up.

I have to admit that milking cows by hand was not our favorite part of living on the farm. On the one hand, we had to get up at 6 a.m. every day and get the milking done in time to leave for school, and then after school do it all over again. On the other hand, it was a ritual for us, and we accepted it for what it was. When my mom made ice cream, the pain lessened.

Far back, far back in our dark soul the horse prances.

—D. H. LAWRENCE[6]

Because horses, both draft and saddle, were such an integral part of our growing up, we also got to learn how to be cowboys as well as farmers. By the time I was fourteen years old, I could harness and hitch a team of Belgians by myself. Even though we had three tractors, my dad was never going to give up the workhorses. He loved them too much. He broke them himself, and all three of us boys helped as best we could, depending on our age. We used our workhorses only for certain jobs when I was growing up. Tractors were doing nearly all of the fieldwork that our draft horses used to do. We used them for hauling manure by pulling a ground-driven manure spreader, hauling hay with a bob sled in the wintertime to feed cattle out in the fields, pulling grain wagons when we ground feed for the hogs, digging potatoes, pulling logs and branches after cutting down trees, and pulling a horse-drawn mower for cutting weeds.

Getting a young colt to accept wearing a harness was probably the hardest part of breaking a thirteen-hundred-pound yearling. Below is how it worked, for instance, with a yearling female.

The colt would become accustomed to walking alongside her mother and was tied to her with a rope and halter. The second horse of the team was generally a gelding (neutered). By the time the colt was a yearling, she was also used to standing next to her mother in the stall. When it came time to put the harness on the yearling for the first time, Dad would stand next to the mother and throw the harness over the mother on top of the yearling's back, and eventually he would be able to reach under and around the well-behaved mother to get the harness connected. At the same time, one of us boys would have the rope twitch (simply a stick with a circular rope) attached to the yearling's upper lip so that the twisted-tight pull on the lip would keep the horse's attention on it and not on the harness going on its back. The science-based theory is that the tight lock on the lip released beta-endorphins that calmed the horse. This sometimes took a while, as the yearling fought having the harness placed on it. The next part was to get the horse out the barn door with the harness attached to both the trainee and her mother. This also took quite some time, as the beginner would sometimes lay down and roll over to get the harness off of its back. When and if that happened, my dad's patience would begin to wear thin. Eventually, though, he always won and managed to get the team hooked up to a wagon that had a heavy load of corn or oats in it. By that time, the yearling was getting used to the harness.

Now, it was time to learn how to pull. Once the yearling had the harness in front hooked up to the singletree of the wagon tongue and the back tugs or chains hooked to the double-tree of the tongue, it was more difficult for the yearling to bolt. A singletree is a round wooden piece that holds up the front of the wagon tongue and connects to the harness below the chest of a horse. It also allows the horse to back up a wagon or implement. A doubletree attaches the harness tugs on each side of the horse to the back of the wagon tongue so that it can be pulled forward. Dad would then get the mother to start pulling, and the yearling would have to start walking, too, because otherwise the front of the wagon would hit its back legs. Because the yearling was hooked up both in the front and in the back and sideways, too, it learned to keep rhythm with the other horse and to pull in cadence with it. After about a mile or two of pulling the heavy load, the yearling was essentially "broke" to work. It would not be long before the yearling began to enjoy and want to pull and work. Horses that are used to working every day are hard to hold back. They just want to work and to pull whatever they are given.

We have not had a horse on our farm for nearly twenty-five years. The last workhorses left the farm in the winter of 1969 after I started college at Iowa State University. My dad was in his early sixties, and the years were gaining on him. It was a very sad day when the horses were loaded onto a truck and taken up to Canada, where they were still used in places in the logging business. A big part of his life went with them as well.

Some scenarios might call again for the use of workhorses, or possibly a mixed scenario where horses could be used for some farming jobs and tractors for others. Most of you are probably thinking this old man has really lost it, but hear me out on the possibilities. It is not entirely out of the question that energy disruptions or other unknown future factors involving fossil fuels could create shortages for agriculture. Even if such scenarios do not happen, horses have some advantages that tractors do not. For starters, they are cheaper to buy than tractors. A team of good horses may cost around $4,000 to $5,000. A new tractor with just 75 HP could cost more than $70,000, and a decent used one could be over $35,000. Perhaps you could farm somewhere between 80 and 160 acres with one fairly small tractor. You could farm about the same number of acres with an excellent team of horses. With two teams, you could increase that to maybe 200 to 300 acres.

We are currently are using eight different tractors on our 700-acre farm. Three of them were purchased new, and five were purchased used. Four of them are front-wheel assist, and four are two-wheel drive. The eight tractors have a combined total of over 72,000 plus hours of use, or more than the average of 9,000 hours each. That is a lot of hours! Four of the eight tractors are of 1970s vintage, and the other four were purchased after the year 2000. Over a farming career of fifty years, the tractors have been driven about 54,000 hours, or over 1,000 total hours every year, equaling about twenty hours of every week spent driving a tractor. Of course, forty-hour workweeks have never existed on our farm. It is more like sixty-plus hours per week.

You are going to say that tractors have so much more power and can get the work done hundreds of times faster. You would be right. However, in terms of energy needed to get the work done, the scales are tipped in favor of old-fashioned horsepower. How could that be?

Low-Tech Magazine is a solar-powered website that looks at the potential of past and present technologies and how they inform sustainable energy practices. It was founded in 2007 and began running its solar-powered server in 2017. In an article published in 2008, *Low-Tech Magazine* looked at historical research with regard to using horses in agricultural settings.[7] A study appearing in 2001, in the *American Journal of Alternative Agriculture*, did just that. Its author was M. H Benda. He calculated that it would take at least 23 million horses to cultivate the present farmland in North America.[8] However, when taking everything into account, including the feed to maintain the horses and the positive manure fertilizer tradeoff, powering agriculture with tractors requires almost 2.5 times more energy than it would take with horses.

I think in some farming scenarios, it would work to farm with horses. The Amish and Mennonite farmers have been able to make it work on their small farms and still be profitable. A 2020 study by Jennifer E. Ifft and Youwei Yang for the Agricultural and Applied Economics Association annual meeting at Cornell University in New York found that New York's Amish farmers continued to grow and were relatively profitable and bid for farmland at a level similar to conventional farmers.[9]

Some Mennonite and Amish farmers use their horses to pull newer farm implements that are quite heavy by using modern technologies and obtaining high yields. It does not have to mean going back to the Middle Ages.

There is a problem, however, according to *Low-Tech Magazine*'s editor, Kris De Decker, Barcelona, Spain: "Encouraging people to watch a horse's rear end instead of a computer screen might prove difficult."[10]

We had a riding horse on our farm until 1999. We only had three riding horses from the early 1950s until then. The first one was named Trixie. I was afraid of her. She was a large quarter horse, and for a youngster like me, that was a bit much. Joe and Mike rode her often. I remember she developed a bad habit of running up as fast as she could to the watering tank and abruptly stopping. Sometimes the rider got pitched into the tank. She did not want to work all that hard, so when she got close enough to home that is what she would do. Maybe it was because none of us could handle her as well as we should have. My dad sold her and then bought a horse that was only about two-thirds the size of a quarter horse. I was eleven years old by then. We named her Star. I quickly claimed ownership of her, and that is when I began to learn how to ride in earnest.

In the 1950s, my dad began to buy feeder cattle from some of the western states in the fall, usually western Nebraska, Wyoming, or South Dakota. They were normally newly weaned calves. He annually purchased at least 100 to 150 head. In November, after the fall harvest of picking corn was done, the cattle could roam over the entire 320 acres we farmed. It became our job to round them all up after school and bring them home so they could be started on a corn silage and grain ration, which we usually did well into December or when the snow prevented the cattle from finding enough to eat out in the fields. That was probably the best chore of all after school, partly because whoever had the job of bringing the cattle home did not have to milk the cows. On occasion, it took two riders to get the cattle in if they were a little on the wild side. Then one of us would ride one of the workhorses bareback. It was a trick to stay on their wide backs and not fall off. Star was a great horse, but was a little too small for the big job of herding many cattle.

When I was fourteen years old, Dad bought a yearling. She was an Arabian/quarter horse who lived thirty-six years on our farm. We named her Samantha, Sam for short. She was a very intelligent animal and was a natural cutting horse. Some people said she was the best horse in Shelby County. The story of how my dad trained her is amazing. Dad wanted me to be the first one to get on her in August of 1964. We got her saddled without too much difficulty, and I got on her. She immediately reared up so high that she fell over backward. I could not get my feet out of the stirrups in time. She landed on my leg, but luckily I did not get hurt. Dad said that was enough of that. He led her out to one of our cornfields and got on her. She started along the tall corn rows as fast as she could but did not try to run in between them. She was probably terrified. When she came to the end of the field, Dad was able to get her turned around and headed back down the corn rows in the opposite direction. This went on for maybe a couple of hours. Dad came back to the barn smiling. She was forever trained to neck rein, which meant she turned when you touched the reins to either side of her neck to turn left or right. She became like a member of the family. She may have saved my life during the terrible blizzard of 1975 (more about that later). When she died a peaceful death in 1999, she was buried on the north side of our grove of trees.

> The horse knows. If you know, he knows. He also knows if you don't know.
> —RAY HUNT, LEGENDARY AMERICAN NATURAL HORSEMANSHIP TRAINER.[11]

I always sensed that Sam and I had that rare and beautiful bond between a horse and a human. We knew each other, and we knew what to expect from each other. She always gave it her all, and she knew she had a job to do, herding and sorting cattle. She loved it, and so did I. She made it easy most of the time, unless there were cattle that thought they were smarter than she. She would never back down or give up. She always won in the end. She always did what you asked her to do. We never rode much for pleasure. It was all about the work. Work was her play, too.

I really should get some horses back on the farm, like a quarter horse and an older team of draft horses. Our three sons never really learned to ride horses like I did. It would be good for the grandchildren, too. The rigors of organic farming and all the other activities in our busy lives have prevented me from doing so, but it might still happen. Still, it takes a great deal of time to work with horses. You cannot just ignore them and expect them to behave and work if you do not spend time with them.

The Blizzard of 1975

The snowstorm in January 1975 was the worst I have ever encountered during my life as a farmer. It imprinted my memory to such an extent that I can still clearly see it over fifty years later. It increased my respect for the force of nature. I cherish the memory of our horse Sam, who helped me through it.

The morning the blizzard started was not unlike any other January morning on the farm. I had done the chores and had taken supplementary hay out to the cows that were gleaning what the cornstalks had left to offer. It was only about half a mile from the farmyard. It was quite warm for a January morning in 1975, around thirty degrees. It was starting to snow. The snow struck me as being a little peculiar, as it was falling in large clumps of flakes. As the morning chores were done by 9:30 a.m., I decided to spend the rest of the morning working on income tax preparation. Farmers have to file their state and federal tax returns by March 1. By 11:00 a.m., a storm was beginning to rage. This time it was small flakes driven by a very strong northwest wind. It was getting much colder, too.

Dad had been working all morning on their retirement home in the town of Harlan, and I was starting to become concerned as he had not come for lunch yet. He had a heart condition that I worried about even more. Mom and I were both relieved when their heavy four-door Chrysler sedan pulled into the yard.

I was starting to become concerned also about the forty-five stock cows, of which fourteen were fall-born cow-calf pairs that I had purchased earlier that fall. I did not think too much about it, though, because we had already had a couple of smaller snow storms that winter and never made sure the cows were home. They were generally better off out in the fields anyway, as long as they could get out of the wind. By 1:00 p.m., the intensity was increasing, and visibility was down to nearly zero. I saddled up Sam anyway and was going to head out to herd the cattle home. I wondered why they did not come home by themselves like they usually did before a very bad storm. They had over 300 acres to roam, and I was not sure where they would be. The pelting, whirling snow was freezing on my face and on Sam's, too. I decided it was far too risky to try to find the cattle. They would probably be home before it got dark anyway. But by the end of afternoon chores, they still had not come home.

All night long the storm raged. The news said the winds were gusting over sixty-five miles per hour. Morning proved to be little different. The wind was still blowing over forty miles per hour, but it appeared to have stopped snowing. It was hard to tell. The temperature was down to zero, but the wind made it feel like forty-five degrees below zero. Once in a while, when the wind subsided for a bit, we saw a patch of blue sky. By 11:00 a.m., the winds were dying down somewhat, so I thought it was time to find the cows and calves. Drifts of snow six to eight feet high covered the farmyard. When I reached the barnyard, I saw that some of the cows and calves had made it home during the night. Twenty-three of them were huddled in the yard. You could not tell they were cattle. Their bodies were frozen masses of ice and snow. They were all hunched up to preserve their body heat. Their eyelids were frozen shut. But they were alive! Where were the others? I started out to the field riding Sam but soon discovered I could not go the usual way. The snow was just too deep. I came back to the yard and picked up a fence pincher to cut wires. The only way out to the high flat was through the south field. By that time, I could at least see some of the time. Sam found her way through the snow mostly on her own as she followed the places where it was not so deep. On reaching the flat, I could see a group of cows huddled up against the fence along the dirt road, or so I thought. They had actually been blown across the road and the two fences on either side. Solitary cows and calves were scattered all over the big field. I did not know how I was going to get them home, but they were alive. What saved the day were the two cows that followed Sam and me into the field because their calves were still out there. After about an hour of fence cutting and pushing and shoving and shouting, the two cows and their calves began to head home. The rest followed. They tried to go home the usual way through the lane, but the snow was four to five feet deep. I had to cut more wires so they could find another way. By the time I got home, my face, hands, and feet were all numb. I thawed out in the house for a time, my face burning like fire. Then I went back out to try to get the stranded and isolated animals home. By about 4:30 p.m., I got them all home except for one cow. She had marooned herself in a deep snowbank and was too tired to move any longer. I carried some hay out to her and dug around her with a scoop shovel so she could at least be a little more comfortable. The next morning, to my surprise, she had made it another fifty feet. I dug a path for another fifty feet or so, and she then walked home on her own.

What a miracle that all of us survived the blizzard. It was estimated that 100,000 cattle and 15,000 hogs died in the storm. Northwest Iowa fared worse than we did. The blizzard was now being called the worst since 1888, if not the worst ever. Within the next eight days, another eleven inches of snow fell. Our roads were closed until the Iowa National Guard came out with trucks and payloaders to create a one-lane road. This one-lane road lasted for most of the winter. The snowbanks on the sides of the road were ten to fifteen feet high in places.

I had some frostbite on my hands, but especially on my face, where most of the skin peeled off. But I was okay. I was young and strong, and so was Sam. She was a remarkable animal. A few of the cows actually lost their tails due to the cold temperatures and the storm. The blizzard marked one of the few times I was afraid of what Nature could do. I learned much from it. It was a hard winter. We did not have the big equipment we now have to better deal with a storm of that magnitude.

We had a series of snow and ice storms over the 2009 Christmas break that were just about as bad. Miraculously, we never lost our electricity from the storm in 1975. In 2009, we lost it for a number of days over the holidays, and our generator quit working as well. Being married and having sons around to help get the farm and family through it made all the difference in the world. In 1975, that seemed an eternity away.

Chapter 6
A Love of Nature, Wildlife, and the Seasons

Everybody needs beauty as well as bread, places to play in and pray in, where Nature may heal and cheer and give strength to body and soul alike.

—JOHN MUIR[1]

While Dad represented at least some control imposed on the land, Mom instilled a sense of adventure and new discovery. She did not do it through physical activities such as hiking, bird watching, camping, or fishing. She was not that much of an outdoors person except for her chickens, gardening, and flowers. Her adventures in the natural world were lived through reading about it. Mom loved to read all sorts of books about history, politics, nature, and wildlife, and adventurous places to travel and explore. She loved to travel to historical places, especially in Iowa. She read many dog, horse, and wildlife books, and she encouraged her children to do the same. When it came to reading, Mom found relaxation, comfort, fulfillment, and entertainment through books. I married someone just like her.

Conversely, I loved to read about such things and then actually experience them. Mom was responsible for helping to create in me a love of nature, wildlife, and the seasons. Dad represented it out in the wild, too. That helped me to realize at quite a young age that I could have both a farm and a wildlife preserve of sorts. I could be both a farmer and a naturalist. I could potentially create a landscape and a farm where there were symbiotic relationships among crops, livestock, forests, and wildlife. I could potentially enhance the ecosystem services of nature and of wild things to increase the productivity of a farm and to improve the soil, too. But I did not think about it in that way when I was growing up on a farm. That would take a college degree in biology and ten years of experience in farming more conventionally with chemicals, followed by forty years of organic farming to begin to both understand and slowly begin to implement ecosystem services.

One could say I was a tree lover from the get-go. I made forestry one of my projects in 4-H, as well as showing pigs and cattle. I especially loved to learn how to identify trees and make displays of their wood characteristics, leaves, and taxonomy. Most importantly, I liked to plant them and to plant many different kinds. Much time was devoted to daydreaming about our farm being a lake surrounded by a forest.

Since our farm was once tallgrass prairie with big bluestem, Indian grass, and switchgrass being some of the more dominant species, trees were sparse. Some cottonwoods and willows were in the wet areas, and that is still the case. We had fewer bur oak forests or oak and walnut savannas than other parts of our county. The high, flat ridges; wide, long valleys; and gradually sloped hillsides in western Iowa lent themselves to fire, not to trees. However, bur oaks have exceptionally tough bark that can survive prairie fires.

I have planted trees on our farm my whole adult life. I have planted thousands. They stand out when compared to the treeless fields of farms in our area. I have planted trees on our terraces, headlands, field borders, the farmyard, and in a small forest that was created from farm ground. My goal was to plant an average of one hundred trees and shrubs annually. Just as is the case with organic farming, I wish I had started sooner and planted more trees and shrubs when I was younger. I wish I had planted more hardwoods early on versus my emphasis on conifers. Native hardwoods capture more carbon. The tallest trees on our terraces are ponderosa pine and blue spruce that were planted in 1983. They are over forty feet tall now and run for over half a mile on a particular grass-backed sloped terrace. I have planted a diverse number of other conifers and deciduous hardwoods. The evergreens include concolor fir, white spruce, Norway spruce, Douglas fir, and white pine; and the deciduous trees include bur oak, red oak, black cherry, black walnut, maples, northern pecan, chinkapin oak, swamp white oak, chestnut, hackberry, pear, apple, plum, peach trees, and hazelnuts, along with a variety of wildlife-loving shrubs. More species are added every year. The act of planting a tree never gets tiring, just one's arms. Tree planting is a satisfying and therapeutic activity. We need to be planting billions of them for reforestation and carbon sequestration.

The result of planting trees and shrubs and habitat for wildlife over many areas of the farm resulted in more wildlife species being able to thrive on our now organic farm. I can count well over sixty species that I may see on any given day in the summer. I am quite sure that there are more than that; I just do not see them all. The fact that we have 630 contiguous acres should help to give some species and their families and offspring enough area to live out their lives, too. I would like to have more knowledge about habitat needs and ranges. In fact, I think all of us need to understand more about that. It would help us to know more precisely how to create areas where wildlife can thrive. The diversity of crops, livestock, perennial pasture and cover crops, trees, and shrubs should help to define and reinforce habitat requirements.

It had always been a goal of ours to build a small pond, and in 2020, we constructed a 1.5-acre pond. It was aptly named "Jacob's Pond" after David and Becky's son, Jacob Raymond. Baby Jacob, full term, was stillborn on May 4, 2021. I cannot wait to see how many new species of waterfowl and other animals will inhabit or visit the pond as time goes on.

The Seasons—Then and Now

Living in Iowa gives you the chance to experience the four seasons. They were more distinct and well-defined when I was younger. Now, with climate change, the lines are more blurred. Now it sometimes seems as if you can witness all four seasons in a period of a couple days. Still, we have seasons because of where we are located on the planet, and I am very grateful for that.

There are two ongoing myths about why we have seasons. First, the closeness of Earth to the Sun causes our seasons. When it is summer in the Northern Hemisphere, Earth may be closer to the Sun than during winter, but negates the fact that the Southern Hemisphere experiences winter, too, even when it is closer to the Sun. This is because it is the angle of the Sun's rays hitting Earth that determines the seasons, not the closeness to the Sun.

The second myth is similar. Yes, it is true that Earth is tilted 23.5 degrees on its axis and that it may be closer to the Sun in summer; however, because Earth is so small compared to the Sun, the distance is inconsequential. Again, the angle of the Sun is more important and causes the four seasons.

When I came back to the farm in 1973, I read a book that inspired me to start writing. It was *From the Land and Back*, by Curtis K. Stadtfeld, an Eastern Michigan University English professor. It was about what life was like growing up on a family farm in central Michigan during World War II and how technology changed farming.[2]

In 2019, I pulled out some essays that I had written back in 1974. They were about the seasons. I had not looked at them for over forty-five years and had forgotten that I had even written these particular ones. Back then, I was still getting used to coming back to farming, and the winter evenings were long. I tried to describe the seasons and what the land was experiencing during each season. My experience with wildlife accompanied each one. It disturbs me to see in these essays' descriptions of birds and mammals that I have not seen for many years. I have done some editing on these essays because of that. It is now a look at the seasons as they struck me as a youth, a young man, and now as a senior citizen farmer.

As a young boy growing up on a farm, my family was very aware of the seasons and how much we depended on nature for our livelihood. There were many signs by which we could identify each season and the transition to the next. Some were obvious, some were not. An obvious sign was in the work we did during each season. Nature held the best signposts. There was the budding of the silver maples in early March to show that winter was on the wane (now it occurs in February). Later, there was the return of the Barn Swallow to show that spring had arrived to stay. The smell of the first cutting of hay in June held the first promise of summer, and the fire-red of the sumac in early fall, the mystique of autumn. The thickened coats of our cows warned us that winter was coming fast. The crunch of hard-packed snow under our boots in January signaled the grip of winter.

A farmer's life is so entwined with nature that it is difficult to separate one from the other. The farmer is only taking on a role that nature is already performing every spring. I consider it to be a sacred role, that of being a co-creator with nature. It is natural for seeds to sprout in the spring and to bear fruit in the fall. The farmer is in the unique position of controlling, to a certain extent, what the farmer wants nature to do. A farm is, in a sense, its own closed ecosystem where the farmer decides the function of the land, what crops will be planted in what fields, and what plants are not wanted, known universally as weeds. In another sense, the farmer is trying to go against what nature wants to do and that is to revert back to the diverse prairie that it was before the farmer arrived. It may seem like the farmer is trying to take on the very role of nature, but in reality the best he or she can do, is to try to become a naturalist, which is to be a good steward where the land is given the best chance to thrive and be healthy while providing food for people to eat. The first principle of stewardship is to recognize that the farmer does not own the land, but is only "renting" it for a time from nature. If the land is cared for properly, nature will be a generous landlord and the two will get along beautifully.

Spring

It seems logical to start with the season of spring for it is nature's way to continually begin anew, to bring new life and change to the land. Spring is the time to begin committing to the process of endless change and an uncertain future.

Placing an exact date on when spring will arrive each year on the farm is hard. With the unpredictability of Iowa weather, it can feel like spring one day and be back to the dead of winter the next. That is why spring is usually associated with the first day of getting out to the fields to disk cornstalks so that the oats could be planted, which was usually around the first week in April (now it is more like the end of March). Back before climate change and a warming planet were combined words, there was the beginning of the spring thaw. This was usually in early March. The landscape took on a different shape. It became a landscape of miniature rivers as the melting snow began its trek to the ditches and waterways of our fields and some ultimately to the Gulf of Mexico. As a youngster, it was fun to play engineer with the small streams that developed from the melting snow. I could dam them up, form new channels to redirect their course, and have a good time playing with my imagination.

Birds

The most delightful indicator that spring is near is associated with the birds that begin their yearly journey back from down south. In early March (now in February), a brave robin or two may arrive just in time for at least a couple more snowstorms to welcome their return. The western meadowlark is a most welcome and early returnee to our fields. This beautiful-sounding and hardy bird feeds on the previous year's seeds until the insect world becomes alive later on. Quite a few meadowlarks disobey their migratory instincts and brave the entire winter, which is especially the case now.

After these two early spring arrivals comes the real northern migration, including majestic flocks of snow geese heading for the far northern reaches of the Arctic tundra in Canada to breed and hatch their young. When I was young, vast and random patterns of black birds and red-winged blackbirds darkened the skies.

When I was a teenager, I began to notice other arrivals while doing fieldwork in April. Among the most interesting of birds was the Franklin's Gull. Every spring, while plowing the rotational hayfields of alfalfa and red clover, dozens of these birds would suddenly appear out of nowhere. They would follow and circle the tractor closely, unafraid as they scooped up the worms and grubs exposed by the moldboard of the plow. They were only around for a few days. Apparently, they actually followed the plow northward into the Dakotas and Minnesota, as they nest in the marshy prairie lakes there as well as in Canada, later in May.[3]

Another bird that seemed to have strayed from its true home by the water is the Killdeer. But unlike the Franklin's Gull, it would spend the summer in our fields close to accessible water. It is a member of the plover family and gets its name from the sharp cry of "killdeer." It also likes the tilled fields in the spring where there are plenty of grubs and worms. Later, its diet includes mosquitoes, click beetles, and wireworms, doing the farmer a big favor in the process. One of the most unique aspects of this bird is the way in which the mother protects her nest. Many times I would get off the tractor just to watch her decoy me away from her ground nest. It was really not a nest at all but simply a small depression in the soil. She would limp along the ground, dragging a wing as if it was broken, scream harshly, or even fall over, all to draw the intruder away from the nest.

A very inconspicuous and mysterious bird that I was lucky enough to see on occasion was the Nighthawk. It seemed like the only time I would see it was in the spring on an overcast dreary day. Much of the Nighthawk's coloration is a sooty-gray, making it very hard to see, whether on the ground or sitting on the side of the trunk of a tree. Its erratic flight always scared me when I disturbed it. The Nighthawk is not a hawk at all. It is a flycatcher and is in the order of Goat Suckers. The name, Goat Sucker, comes from an old European superstition that the bird subsisted by sucking on a goat. The Nighthawk is active only at night or during the twilight hours and then spends the day resting on the ground. While apparently there are still plenty of Nighthawks in Iowa, I have not seen one on the farm for a long time, which is disappointing.

While the Nighthawk really is not a hawk, there are plenty of real hawks in Iowa and around our farm. They are given the name Raptor, which means a bird of prey. Five hawk species nest in Iowa. The two most familiar to me and the easiest to identify are the Cooper's Hawk and the Red-Tailed Hawk. Some of the others that nest in Iowa, such as the Broad-Winged Hawk and the Red-Shouldered Hawk, are harder for me to identify. The fifth hawk that nests in Iowa is known as the Northern Harrier or Marsh Hawk. I am happy to say that I see at least one regularly around our farm today. They fly close to the ground looking for voles, ground squirrels, and field mice. Their quick, swooping, and erratic flight patterns are fascinating to watch. They have earned the nickname "the gray ghost" because of their low-flying flight pattern over open pastures, hayfields, and prairies. They have a range of only a square mile, so it seems we may have our own "gray ghost" watching over the farm.

In the spring of 2020, while using an undercutting plow that does not invert the soil like a moldboard plow does, at least twenty-five or more hawks of various species showed up to snatch up grubs and worms left exposed in the thirty-five-acre field. To date, I have never seen that many hawks congregate in one area. Since about the year 2000, we have frequently seen Bald Eagles as well on our farm. We do not have any nesting ones, however. They do have nests close to us on the Nishnabotna River. Today they are quite common and are actually once again flourishing. Iowa did not have a nesting Bald Eagle for seventy years, until 1977. They are majestic birds, and I am still awestruck when I see one.

Mammals

Birdlife is not the only sign that life is stirring out in the fields. Early in the spring, while the frost is still coming out of the ground, the plains Pocket Gopher starts working out in the pastures and hayfields in search of succulent new root growth of various grasses, but especially legumes like alfalfa. Gophers actually do not hibernate but are most active in the spring and fall. Gophers do more good than I want to give them credit for, as they aerate the soil and allow for better water drainage in compacted soils. Unfortunately, they now can be a factor in how long we are able to keep a hayfield in rotation before it must be torn up and planted in corn, which can be after only one year if there are a large number of gophers. They leave the ground very rough, and we try to run a harrow over our hayfields at least twice a year to smooth them out. The greatest annoyance is that the wet dirt mounds plug up the cutting mechanisms of any machine you are using to cut hay. The worst plugging occurs with a sickle mower, and the most acceptable option is the quick-spinning disc mower or a spinning cupped rotary scythe.

When I was a youngster, my brothers and I would trap gophers. There were a lot more acres in hay and pasture back then, so all Iowa counties considered them a pest, and a small bounty was paid on them. That practice has been discontinued with the monocropping of corn and soybean fields where the fences are mostly all gone with little grass anywhere except on field terraces. The bounty was ten cents for a pair of front claws. We would put the claws in a coffee jar filled with salt stored in the shop. You had to take them to the county courthouse to turn them in. I remember Joe and Mike redeeming over a hundred pairs one year. Compared to them, I was a mediocre trapper.

The best natural predator of the Pocket Gopher seems to be the badger. Badgers dig the gophers out and eat them. This usually occurs after dark when the badger can be away from humans. The only problem is they create holes and mounds even worse than those of the gopher. Foxes, coyotes, and owls are mostly nighttime predators, as well. In California, farmers have been successful in putting up Barn Owl boxes in fields to help control gophers. I have not seen a Barn Owl for many years, although there were some around when I was growing up. They have been considered an endangered species in Iowa since 1977. According to the Iowa Department of Natural Resources, their numbers have slightly increased over the past five years in Iowa. There are two reasons for this. One is that our winters are getting warmer because of changes to our climate. The second is that wildlife experts have found that erecting seven- to eight-foot poles near wooded areas helps keep predators away and thus save more nests. Barn Owls live for only three to four years, whereas a Great Horned Owl can live for twenty. Today we have numerous Great Horned Owls and Barred Owls, much to our delight. On a quiet evening, we can hear them in our home. I have even observed a few Burrowing Owls some years, although they have just been visitors from farther west in Nebraska. Screech Owls used to be very common when I was a youth. I rarely see them today. I saw my first and only Long-Eared Owl while combining corn in the fall a few years back. It was just visiting.

By the time May arrives, most of the spring arrival of the many species of birds is over. The last to arrive is the Barn Swallow, usually around the end of the first week in May. We then know that warmer weather is here to stay. Barn Swallows are very adaptable to living in close proximity to humans, and they range over large parts of the world. They can be aggressive at times in protecting their nests. I remember a specific swallow some years back whizzing just past my ear every time I approached the barn. Barn Swallows only eat insects. When I am cutting hay, watching maybe twenty to thirty of them dive in and out of the area at once to snatch flying insects is quite a delight to see.

While Barn Swallows generally live to be about four years old, their nests can survive twenty years or more. They are made of mud pellets, grass, and straw packed together and thickly lined with feathers.

Spring Rush—Not Spring Break

The spring rush of fieldwork to get the manure (now composted manure) hauled and the crops planted has always meant very long days of working, sometimes fourteen to sixteen hours. Still, there has always been time to observe the wildlife and all that is happening. One had time to think, watch, and notice while performing whatever the task. Before radios were on tractors, I spent more time meditating, singing to myself, and daydreaming. Now with all that is happening every day, National Public Radio (NPR) and old rock stations have largely replaced those activities.

For the decade from 2010 to 2020, we endured cold, wet, and quite late springs. Did this occur because of climate change? Most likely. It does fit the observation of greater weather extremes with a changing climate. These changes made it more difficult to get the spring fieldwork done and sometimes resulted in delayed planting of corn and soybeans, especially for organic farmers. Organic farmers generally wait longer to plant corn and beans because of the requirement to use seeds that are not treated with pesticides to prevent insect damage and seed degradation due to cold and wet conditions. It is a good reason for heavier plant populations, and timing has become more critical than ever. Long-range weather forecasting has taken on more significance.

The three springs of 2021, 2022, and 2023 have meant intensifying drought, with 2023 being the driest year since at least 1988 for where we live.[4]

Once spring arrives, there is no turning back. Daylight Saving Time means another hour of sunshine to get more work done. Spring does not wait for anyone. It seems to know that summer is near, with only a short time left to lay eggs and hatch new life, sprout seeds and form tall stalks of corn, and generally replenish Mother Earth. Spring has one huge advantage over the other seasons: the Sun. Day length reaches its longest hours, and sunlight spurs the new growth and quickens the already feverish pace. It is summer before you know it.

Summer

Summer is such a promising and hopeful time. It seems to know no bounds. It can be a season of seemingly unlimited growth if the rains come when they should. Summer is a time for productivity for the farmer and for nature. For me, it is a time for cultivation and care for the prospects and hopes of a good crop. For nature, it is a time of purpose and a time of some contentment. There is contentment in her long, sunny days and warm nights.

Summer is the season that most every farm youngster for generations has looked forward to more than any other. It is a time for baseball, swimming, camping, fishing, hiking, and exploring. Summer is picnics, fairs, and parks. I participated in all those activities. My youth was not just work. Summer still is some of those things for me. However, as a boy and now as an older adult, there is more to it. I have a different perspective because I have a different lifestyle. It is a way of life that sank its roots early and deep. It is a different way of living because of a different goal. For me, that goal has been a oneness with nature.

Fighting Weeds

As a youth, summer started immediately after the school year ended. That was around May 20, not on June 20, technically the first day of summer. Generally speaking, I liked school a great deal. However, I was always ready for the school year to end. I remember that one of the first farm-related jobs assigned to me was digging Sour Dock and Wild Mustard out of our oat fields. These were two weeds that my dad thought had to be dealt with. There were many more. Canada Thistle and Button Weeds, also known as Velvet Leaf and Butter-Print, topped the list.

Like just about every farmer then and now, Dad believed that a farmer was judged by how weed-free the fields were and by how neat and well-kept the farmyard and buildings were. Dad was an early adopter when it came to some new agricultural technologies. This included anhydrous ammonia as a nitrogen source and 2,4-D herbicide for spot spraying thistles. There was no blanket spraying of cornfields in those early days of herbicides on many farms. It was not until the advent of a new crop, soybeans, that farmers began to apply more herbicides. Dad died in 1980. That was three years before I decided to end the use of all pesticides. The farm crisis years were just beginning. New knowledge about some of the harmful effects of pesticides on both human and ecological health was only beginning to emerge. While Dad thought that every weed had to be virtually eliminated, my goal became one of managed weed control. However, it took me the first ten years of farming to even begin to figure that out. I do not want to make it appear that my dad was a conventional farmer. He used pesticides and fertilizers judiciously and never ended the crop rotations of oats, pasture, alfalfa, and red clover hay. With the legumes of alfalfa and red clover as nitrogen sources, he did not apply additional synthetic sources of nitrogen on those fields of corn. He never ended the raising of some cattle, hogs, and chickens. I think he, too, would have changed his mind about pesticides and would have encouraged and applauded the farm becoming certified organic.

Walking Corn

Have you ever walked corn to cut or pull weeds in August when the corn is six to eight feet tall? I have to admit, in my youth, if I hated any job on the farm, it was walking corn, which was done from the first to the middle part of August when it was really hot and humid. We wore long-sleeved shirts to keep our arms from getting scratched, wide-brimmed hats to keep the leaves from scratching our eyes and face, and bandanas around our neck and face to keep the pollen from scratching our neck and getting into our mouth. We carried a three-inch-wide and two-foot-long corn knife, also known as a machete, which could cut us as much as the weeds if we were not careful. We also had to be wary of running into fierce-looking black and yellow garden spiders. When I was about ten years old, Mike was bitten by one, and I was afraid of them for a long time after that. Dad would carry a small hand sprayer filled with 2,4-D to spray Canada thistle patches when we ran into them. That gave us boys a chance to rest, or rather time to spend throwing clods of dirt at each other or playing sword fighting with a long, thick weed, like giant ragweed or water hemp. By the time we spent seven to eight hours a day for about a week walking fields of corn in August, we wondered if town life might have some merit after all.

Walking corn in August was done in addition to cultivating corn and beans twice in June with mechanical weed cultivators mounted on the front sides of our tractors. I do not think Dad used any tank mixes of pre-emergent herbicides until I left for college in the late fall of 1968. That was when soybeans were first grown in the area. A tank mix refers to blanket spraying an entire field with herbicide before a crop is planted to stop weeds from germinating. We walked our corn for weeds at least until 1970.

By walking corn, I learned to identify all of the bad weeds, or what are called noxious weeds. Those are the ones that the USDA says are harmful to crops. There is a list of over thirty for Iowa. Some that we battled were Velvet Leaf, Cocklebur, Sour Dock, thistles of many kinds, Wild Mustard, Quack Grass, Sunflower, and Horse Nettle. Some are controversial because they are a nectar source for pollinators of various kinds. It is ironic that today's herbicide-resistant weeds are not considered noxious weeds by the Iowa Department of Agriculture and Land Stewardship (IDALS). The weed species of Marestail, Giant Ragweed, Pigweed, and Water Hemp are some of the primary ones that have developed resistance to herbicides.

Hunting

My brothers and I were crazy about hunting when we were young. If you could eat it, it might have been fair game. We hunted cottontail rabbit, squirrel, pheasant, pigeon, red fox, jackrabbit, and also duck and geese on occasion. We ate much of what we hunted, but we had to clean the game ourselves. Mom would not do that part, but she would cook it, and it was usually very good unless you got a tough old duck or goose. I first shot a 12-gauge shotgun was when I was ten years old. I bagged two pigeons in one shot. Luck!

I hunt very little now, and then only for rabbits that eat our trees or raccoons that eat our sweet corn or damage our buildings and grains. I should have learned to hunt deer out of necessity because they eat so many small trees, but I did not. I would rather watch and observe most of our wild animals. Some wild animals that were hunted are no longer here. I feel guilty about that, but hunting was and still is a big part of our rural culture. The animals that may have been hunted then are also less prevalent today because of less crop diversity and habitat. I am not against the sport of hunting if it is managed and regulated properly.

A mammal that we hunted in the 1950s and 1960s was the White-Tailed Jackrabbit. It is not a rabbit. It is a hare and is quite large, around seven to eight pounds. It can run up to 35 mph and leap long distances of over fifteen feet. It lived in the fields. They were numerous when there was more hay, pasture, and small grains such as oats in Iowa. They are now considered a "species of concern" in Iowa. I have not seen any since the 1980s. They were a challenge to hunt. We would run our legs off after them in late May and early June after the first cutting of hay and before the corn and oats were tall enough that you could not spot them. My brothers and I devised a strategic plan where one of us would drive the animal over a hill to where one of us might be located. We were not always very successful, so their numbers were not depleted by our doing, that is for certain.

Red Foxes were hunted then as well because they were plentiful. I would not think of it now. There were more fences and hedge rows and habitat conducive to fox populations then. There were fewer row crops of corn and soy. We had a den of five young fox kittens in our hay shed a few springs back. They became fairly tame until they left the den permanently in late May. We are starting to see more Red Fox on our farm. Because we have significantly increased our tree and shrub plantings out in our fields and have kept our fences, we will probably see more of them as time goes on. That is a good thing. We do not have the Gray Fox on our farm. They live in forested and some urban areas of the state and are declining at even a faster rate statewide than the larger Red Fox.

I am not sure why; coyotes were rarely, if ever, seen in our area of the state back when I was a youth. Today, they are plentiful. One explanation may be due to the increase in their adaptability to live alongside people even in urban areas.

Field Birds

Summer was a magnificent time for the birds of the fields and marshes. There were and still are Bobolinks, Dickcissels, and Upland Sandpipers, but their numbers are declining rapidly. They are some of the most imperiled species of the northern tallgrass prairie region. The birds of this region have had a more widespread and consistent decline than any other group of birds in North America. We still have some because we have large areas of the farm in small grains such as oats, rye, and barley and because we have mixed hayfields and permanent pastures. We have some tallgrass prairie headlands and field borders. The corn and soybean belt is moving northward into Minnesota and the Dakotas because of warming temperatures and longer growing seasons, which will be even more devastating for the birds.

More than 150 species of nesting birds exist in tallgrass prairie ecosystems. Iowa and Minnesota were once covered by over 25 million acres of tallgrass prairie. Today, there are only 300,000 acres, and most are in such small remnants that they are not effective areas for maintaining the healthy breeding populations needed to expand their range. One way to help that is to try to join smaller remnants together in strategic locations. The Loess Hills is considered one of the last possible areas with enough tallgrass prairie for grasslands to be increased. There was an effort to try to do that in the 1990s when a national park was proposed for the Loess Hills. It never came to fruition. That idea perhaps needs to be revisited.

We observe so many other birds in the fields. These are the ones that were plentiful while I was growing up and still are quite common today, including Cowbirds and Horned Larks; Red-Winged Blackbirds and Western Meadowlarks; common Yellow Throats, Chipping Sparrows, and Northern Bobwhite; Purple Grackles and Eastern Kingbirds. There are more, of course, but we either do not see them or cannot identify them. That has changed dramatically for the better since we learned about the smartphone app called Merlin, from the Cornell University Lab of Ornithology. It has allowed us to listen to and identify birds. It is being used around the world, and well over 10,000 species have been identified so far. It could be a game-changer for helping to maintain endangered and threatened species.

A small, common mammal is the Thirteen-Lined Ground Squirrel. This pesky little creature likes to dig up the young corn shoots after they have just emerged from the ground. A large number of them could do quite a bit of damage to a cornfield. It was a popular Sunday activity to go out to a cornfield to shoot ground squirrels with a .22-caliber rifle. The trick was to sit on the ground and stay perfectly still. Before long, the curious squirrel would come out of his hole and stand up on his hind legs to survey the area. This often led to its downfall. During the spring plowing, their holes would be covered sometimes with the moldboards of the plow. Once in a while, I could outrun them as they headed down the furrow left by the plow. That was then.

Wild strawberries used to be found on the north side of our grove when I was young. Some were also in our road banks and ditches. Wild black raspberries have become very common since we stopped the use of pesticides in 1983. That also goes for wild grapes and elderberry. Elderberries have become popular again, and we are often asked by others if they can harvest them. Hippocrates called elderberry the "medicine chest" of all herbs.

Church Picnic

The season of summer was not just associated with all things "nature." The Fourth of July meant our annual church picnic in Westphalia. When I was a little kid, up until I was about ten or twelve years old, the picnic meant receiving three dollars from my godmother to spend at the picnic. I usually went for the ten-cent bottles of sixteen-ounce Mason's root beer or Bubble Up first and the five-cent Butterfingers second. The big bucks were spent on buying firecrackers from other kids at the picnic. My parents seldom bought fireworks. Mom figured we would blow our fingers off. There was always some kid who managed to get hold of some of the big stuff like the firecrackers called M-80s and cherry bombs. They could blow up a beer can or something worse. An M-80 might go for a dollar. You could get a cherry bomb for fifty cents. The purchases were made out in the baseball field, and then the fun began. My friends and I were ultra-careful, and no one ever got hurt.

The Heat of Summer

August meant it was getting time to think about going back to school. The sound that accompanied that dreadful thought was the cicada locust. The term " locust" is inaccurate because that name applies to migrating swarms of grasshoppers. The cicada that I am thinking of is the common large black and green harvest fly, which matures in two years. Their shrill, ear-piercing sound on August evenings has long been their trademark. The most famous cicada is the Seventeen-Year Locust, which has a life cycle of seventeen years of sleep in the ground followed by five weeks of feverish life. I have found the sound of the cicadas to be quite soothing.

Some of the happiest memories of my childhood occurred during the warm and sometimes downright hot nights of July and August. These involved catching fireflies and playing hide and seek and "Star-light, Star Bright," either with family or with friends, neighbors, and relatives. We did not have air conditioning until later in the 1960s, so it could be really hot sometimes in a north-facing upstairs bedroom. When that happened, we could either sleep in the basement or, if no one was staying in the south upstairs bedroom, we would shove the bed up tight against the open windows and sleep with our heads on a pillow on the windowsill. Looking at the stars and talking about exciting adventures and dreams kept us occupied until we finally fell asleep. I did not know it at the time, but I now realize that I was beginning to sink roots into the farm and the good earth that it encompassed. It may be what nostalgia is all about, but it really was a fun and beautiful time. Everyone likes to say they were simpler times. I am not so sure. Summer was a hectic time for our family because we were involved in many activities besides farmwork. The most important was 4-H.

4-H

Our family was into 4-H.[5] Mom and Dad thought it was important. For me, 4-H probably had more to do with being introduced to science than did school. Back then, Westphalia had a 4-H club. Its name, the Westphalia Cooperators, was fitting for the times. We had the standard livestock projects of raising and showing both cattle and pigs. That is because just about every farmer raised cattle and pigs. Quite a few showed dairy cattle. The purebred breeders in both cattle and hogs had at least somewhat of an advantage in winning the purple ribbons. We did not raise purebred cattle or pigs. My brothers and I would pick out what we thought were the best calves from the ones that Dad purchased in the fall. These calves came from one of the western states, such as Wyoming, Colorado, or Nebraska. We did not have a stock cow herd of our own at that time from which calves could be chosen. Our pigs were crossbreeds that we bred ourselves. Other projects included woodworking, entomology, climatology, leadership training, citizenship exercise, crop, livestock judging, and, for me, forestry. 4-H broadened our horizons beyond our little hometown of Westphalia. We made many friends from other areas of the county. It whet my appetite for studying science and agriculture.

In a Pinch

At the age of ten years old, I had one of those lifelong memorable experiences after I had just started 4-H. Because my dad was busy trying to get all the farmwork done in the summer as well as his volunteer organizational work in Westphalia and in the county at large, too, the 4-H cattle did not get what was called "broke to lead" until just before the county fair, which occurred in early August. That meant a ten-year-old trying to hang onto an £1,100 steer with a halter and a lead rope. Leading an animal that large is obviously much different from leading a dog on a leash. For cattle, the rope halter that goes around the head is held very close to the animal's head so that there is not a lot of slack for it to move back and forth. That is, if the animal is tame enough and trained to do so. I learned a hard lesson, as I got my finger pinched between the rope and the 6 × 6-inch wooden upright in the barn while trying to hold onto my calf. It nearly took off the tip of my middle-left finger. I was rushed to the hospital where it was heavily sewn and stitched, and it healed fine. It became the first finger to always get numb from the cold for the rest of my life. I became more careful after that with all the possible ways one could be hurt around animals and machinery on a farm. The worst farming injury for me, up to now, has been getting my ACL torn by a cow that fell on top of me. That was when I was fifty-three years old. I was lucky, as I could have been crushed to death by a whole group of cows that had me pinned against the barn. I had a blood clot in my knee after that.

Farming is one of the most dangerous occupations. Just about every kid growing up on a working farm has had farming accidents of some kind, either when they were young or as they became older. Older farmers run the greatest risk because their reaction time has slowed down so much, and they just can't get out of the way fast enough sometimes. Age causes one to get complacent and think that just because they might have gone through the same routine step or movement thousands of times, they will be immune from anything bad happening. Poor judgment also becomes a problem as well from such things as being alone, tipping over a tractor or skid-steer, or suffocation in grain bins.

> All three of our sons and I have had serious run-ins with mother cows. It seems to go with the territory of raising livestock. The worst happened in late May of 2021. Our middle son, Daniel, was doing the usual routine of tagging a new-born calf with a numbered identification tag that matched the mother's so that we could always keep track of the pair when going to different pastures. A mother cow that had previously been tame enough attacked him before he could even get close to the calf. She drove him into the ground and could have killed him. Nine of his ribs were broken and misplaced. He suffered a blood clot and a partially collapsed lung. He now has five titanium plates holding some of his ribs in place. God was with him that day, that is for certain. He has healed almost entirely, but it is something that will affect him for the rest of his life. People usually never come away later in life totally unscathed from those close brushes with possible death. We learned something from this incident. During the fall of 2021, we incorporated some new gates and a different way to tag calves that hopefully will be much safer.

Who Needs a Vacation?

Anything that could be called a family summer vacation happened on a very limited basis when I was growing up. We went to Arkansas the summer I was ten to visit relatives. When I was fourteen years old, there was a trip to Colorado after I had gone through an emergency appendectomy a couple of weeks earlier. It was not as if we were very much different from anyone else. Farmers usually did not have the time or the money to take off for more than a few days. Weekends were not considered weekends like they are now, of course. I never felt like I was really missing anything by not being able to go on a vacation. I feel much differently about that now. Getting away from the pressures and stress of farming, volunteer work, and meetings is absolutely necessary to maintain good mental and emotional health. It can be just a night of camping and fishing for me or a night's stay in a cabin somewhere with cooking, eating, reading, and sleeping as the only requirements for a great and relaxing break from the pace of life on a farm. Maria prefers a comfortable bed and a hotel, and that is great, too. The Covid-19 pandemic reminded me a little of what life was like growing up when I did, in the sense that with both, I mostly stayed home. I find it quite easy to be a hermit.

The guarantee about summer in my youth was that it went too fast, and it still does. There was some trepidation about the coming of fall because it signaled going back to school, and that meant being in another grade. That was okay in grade school, but by the time high school came around, the thought of the decisions I would have to make about my future weighed heavily. After I reached adulthood, fall became perhaps my favorite season. Even when I was growing up, the coming harvest and the fall hunting of squirrels and Ring-Necked Pheasants made the thought of school more tolerable.

Fall

I wrote the following poem in 1975.

> There is a hint of fall in the air tonight
> Ever so slight, you can hardly tell it is there
> Somehow you get the feeling that summer has turned the corner.
> How can that be?
> Wasn't it just yesterday you looked forward to summer?
> Maybe it's the way the air feels, still and heavy, but a little cool
> A misty gray blanket covering the fields of corn and beans
> The blanket gets heavy, drops of rain patter on corn leaves like machine guns in war.
> There is a hint of Fall in your thoughts
> You start looking for the months that lie ahead.
> What will they be like? What will happen to you?
> There will be the harvest and then far away into the distant future. . . . Winter
> Suddenly a cricket snaps you out of your spell;
> Its friendly chirping breaks the lonely silence.
> It's trying to tell you something if you listen closely,
> Something about the coming of fall.

Corn Picking

During my years of growing up, fall meant primarily one thing: the harvest and picking corn. Because so many of the kids in school in Westphalia lived on farms, school was dismissed for two full weeks when I was in grade school. In the fall of 1964, when I started high school in the town of Defiance at the combined Defiance-Westphalia Catholic High School, corn-picking vacation lasted only a week. That went on for three years until the high school closed in 1967 and all the Catholic high school students in Shelby County descended upon the public high school in Harlan.

From the 1940s until the invention and use of self-propelled combines, which was about twenty-five years later, most of the nation's corn was harvested by a machine mounted on a tractor. It was called a corn picker. It could be taken on and off the tractor so the tractor could be used for other farm jobs the rest of the year. It was an ingenious invention. It took about a halfday to get it mounted on the tractor. It could harvest two rows of corn at a time. The ears of corn would be gathered by moving chains into the snapping rolls where the ears would be stripped off the stalks of corn. The ears of corn would then travel up an elevator on each side to be taken to what was called the husking bed where most of the husks surrounding the ear were taken off. Then it would be transferred to another, larger elevator where it was taken up and then dropped into the wagon being pulled behind the tractor and mounted corn picker.

Indian Summer

An *Old Farmer's Almanac* type of phrase for a portion of fall has historically been called Indian Summer in North America, as its origins and usage were steeped in Native American folklore. It appeared in literature for the first time in the late 1700s (see https://www.almanac.com/). One early description of Indian Summer was a distinct period of warm and hazy weather in the fall that was good for Native American hunters. This was true for the Plains Indians and buffalo hunting and for Native Americans on both coasts as well. The term also may have originated as being a time in the fall after the first killing frost where an area of high pressure settled in for a prolonged period of time. It also is referred to in some European countries and countries of South America, albeit by different names, with the timing based on a country's geographic location. It was a period of time when the fall weather seemed it could not be any more perfect. It meant clear blue skies, crisp mornings with maybe a little frost, and warm sunny afternoons.

In 2020, the American Meteorological Society decided to discourage the term "Indian Summer," announcing it is a relic of the past and disrespectful to Native American people. It recommended that the term to describe this phenomena should be "Second Summer."

It seemed like a farmer could bank on the Second Summer occurring just about every year when I was growing up. It usually occurred around the tenth of October and sometimes lasted for up to two weeks.

Most of the time, this two-week period allowed farmers to get at least a good portion or even the bulk of the harvest done when students (girls and boys) were on "corn picking vacations," which occurred only in the parochial schools. I doubt that the students were idle while at home. Just about every boy who could drive a tractor brought loads of corn into the farmyard and unloaded them into the wooden storage structure called a corn crib. The wagon was raised either by a hoist or by a hydraulic cylinder mounted on the wagon so that the bulk of the corn would fall out by the force of gravity. Hydraulic cylinders did not come along until about 1960. Bigger and more modern farms had elevators inside the corn crib, which were metal buckets attached by chains that revolved in a square not a circle from top to bottom of the crib. The ears of corn in the metal buckets would be dropped into the two sides of the crib, either to the right or the left. Large cribs could hold five thousand total bushels of ear corn. We had that size crib with the revolving cup elevator. We still use it today. We have one of the very few working crib elevators left in the area that I am aware of. We no longer use it for ear corn but use the overhead wooden storage bins for oats and other small grains. It was considered the Cadillac of grain storage in its day.

I loved picking corn. By the time I was in high school, Dad occasionally gave me the chance to run the corn picker. That was a big responsibility, but I welcomed it. I sat in the seat with all of those mechanical parts within arm's reach. That meant being ultra-careful about where I stuck my hands. If the corn picker plugged up with stalks or ears of corn for any reason, I always shut the tractor off before unplugging it. That procedure was mandatory and never compromised. I automatically shut the equipment off and then unplugged it by hand. Many farmers have lost fingers, hands, arms, and even their lives by sticking their hands too close to the moving parts when trying to dislodge a stalk or ear of corn.

Sometimes early snows made picking corn much harder. The corn picker was heavy, but when it was attached to a tractor, we could get through the snow if it was not too deep and keep on harvesting. We could stay warm, too, because the tractor engine threw considerable heat back to us.

First Fruits

Remember Father Duren from earlier in the book? He started something in the fall that was very symbolic and also practical for the economic health of the parish of St. Boniface. The parish built its own corn crib, and Father Duren called it First Fruits. Farmer parishioners were asked to bring a wagonload or more of ear corn that would later be shelled and sold the following spring, with the proceeds going to help fund the parish. It was based on two Bible verses from the Old Testament. The first was Leviticus 23:22: "When you reap the harvest of your land, do not reap the very edges of your field or gather the gleanings of your harvest. Leave them for the poor and the foreigners residing among you." The second was Exodus 23:16: "Celebrate the Festival of the Harvest with the first fruits of the crop you sow in your field." I generally brought the last load of corn from our own harvest to help the cause. The bringing of ear corn to fill the parish crib went on until the 1980s. Very few farmers used corn pickers by that time, but we still did. Our parish still promotes First Fruits as a fundraiser, accepting checks in lieu of actual corn.

Picking corn in the ear started coming to an end in the 1960s. Combines that shelled the corn instead of leaving it on the ear began to replace corn pickers. Steel grain bins began to replace wooden corn cribs. We picked some corn into the early 2000s. We did this for a number of reasons. First, you did not have to dry the ear corn with fans or supplemental LP gas heat to dry the corn down to where it could be safely stored. When picked in the ear, the corn kernels would dry naturally on the cob. Second, it made for exceptional feed for growing cattle as well. The high amounts of fiber in the cob and the shucks become part of a healthy feed ration for beef cattle.

Fall Hunting and Gratitude

I particularly enjoyed hunting red Fox Squirrels and Ring-Necked Pheasants in the fall. Squirrels could be very hard to see in our tall maple and walnut trees. They could be hidden by the foliage. They were extremely quiet and never moved. You had to be very quiet and patient as well. I spent many peaceful and beautiful autumn Sunday afternoons hunting squirrels. Hunting Ring-Necked Pheasants was not always so calming. The sudden clamorous sound of a startled Ring-Necked

Pheasant could make your heart jump for an instant. I got to be a pretty good shot while hunting pheasants.

Autumn probably has more spiritual significance attached to it than any other season. It is the time of gathering the hoped-for bountiful harvest. Every season has its very real bonds and connections to humans and our own cycles of birth, growth, maturing, and dying. This has to go back to the earliest of human activities of the harvesting of wild game to prepare for the long, cold winters that were awaiting them. Not only did many of the annual plants die, but also, if you and your community were not adequately prepared for winter, you could face the prospect of death yourself.

The feast and observance of Thanksgiving had a great deal of significance attached to it. It still does. Farmers hoped every year that they would be done with the fall harvest before Thanksgiving Day arrived. If not, on occasion, after eating a huge Thanksgiving meal at noon, farmers would spend the afternoon working on getting the harvest done before winter arrived to stay.

Winter

I wrote the following poem in 1974.

>Winter, the Lonely Season
>Unrelenting, unmerciful, uncompromising in her actions
>She spreads her cloak across the land
>Covering the cold earth
>
>Winter, the barren season
>She takes her wrath out on us for her desolation
>Picket fences try to slow it down but it's no use
>Dead corn stalks stand like sentinels above her
>A meager diet for hungry cows dreaming of summer pastures
>
>Winter, the hard season
>Except for the hardy
>Like the swoop of a great horned owl she silently strikes down
>Some wisely choose to escape her by sleep, like the ground-hog
>Some insist on challenging her
>Some have no choice
>Some move away entirely, waiting for a sign of spring to return,
>Look out, Winter will fool them, they are the fool-hardy
>
>Winter, the tired season
>A time to rest for the tired land
>Short days and long nights for sleep
>Nature slows down, people slow down
>Time to re-evaluate their lives

Winter, the enchanting season
Frost shimmers on her snowy blanket
Softly and tenderly covering us as a mother would her new-born babe
There is beauty in her wind-swept drifts like waves upon the sea
Ice crystals hang like chandeliers
Evergreens warm in their new white overcoats
Sparrows gossip noisily, feeding on the oats and corn left from grinding
Softly glowing red sunsets, It's Santa Claus making candy

Winter, the prophetic season
There is hope to be found in her hopelessness
She will give way to the new again before long
By the end of February, it's not long until Spring

Our Own Christmas Traditions

Christmas on the farm when I was a youngster was wonderful. We did not have Midnight Mass as the typical time for celebrating the birth of Christ until I was in high school. So traditions were started on the farm. Some we still hold dear today. The tradition of feeding the livestock more grain or hay or the horses more oats on Christmas Eve was started and is still carried out today. Santa Claus would come about 7:30 in the evening. Mom and Dad would make us go upstairs to our bedroom to say the Rosary until Santa Claus arrived. It was the longest fifteen minutes of our young lives. We would kneel at the windows hoping to catch a glimpse of Santa and his sleigh. One time we actually thought we saw him. We did not get a million presents like most kids seem to get today, but the ones we did get were meaningful and usually something that we had asked for. We attended Mass on Christmas morning.

Winter Is So Beautiful When You Are On the Inside Looking Out

Winter took on a whole different meaning when I became older and especially when I came back to the farm. Yes, it could still be thought of as a Currier and Ives romantic scene sometimes, but that depended mostly on whether you had livestock to care for. Yes, winter still was a time where you could slow down some. Part of that was just because the daylight hours were so much shorter. It was dark by 5:00 p.m. For farmers with livestock, winter was and still is usually quite challenging. If the weather was cold and the ground was covered with a good deal of snow, it could mean spending whole days just getting the chores done. By that, I mean just making sure the animals were fed and cared for and had plenty of water available. Keeping waterers free of ice and open for the animals to drink was an ongoing challenge. Until we had better electrical water heaters, some of the watering tanks were heated by the burning of corn cobs. If it was really cold, such as below zero degrees Fahrenheit, we would have to use corn cobs dipped in kerosene on a long stick to thaw out the pipes so the water could keep flowing. The best thing was to have enough hogs and cattle on a waterer so that sheer numbers and greater consumption of water would help to keep the lines open. We always had at least 150 to 200 head of cattle, and we still have at least that many, if not more, at any given time. We raise fewer hogs than we used to but still have around 350 head on any given day. For egg layers, it is around 130 hens.

If we had to start tractors to haul or to grind feed for the livestock, that could be another tough assignment. It meant first being able to get the tractor started. On the one hand, most tractors were powered by gasoline in the 1940s, 1950s, and 1960s, so that helped. Diesel tractors, on the other hand, are tougher to start in the winter and sometimes need to be plugged into a block heater to get them going. They also need a lighter blend of what is called number one diesel to get them started when it is −10 degrees below zero or more. That is because regular diesel fuel can gel up in extreme cold and not flow through the fuel pump into the engine. Tractors, at that time, were not four-wheel drive and had narrow front ends that could easily ball up with both mud and snow. We had to put heavy chains on the back tires to get around and get the work done, which was a good reason my dad had Belgian draft horses to do the daily hay-feeding chores. They would easily "start" in the morning even if it was below zero, nor did we have to start a tractor every day.

Flexibility and planning are key requirements during the winter. A backup plan is a must if it is too cold, the snow is too deep, or the electricity is interrupted because of ice or wind. We have a tractor-driven power take-off (PTO) generator we use when there is a loss of electricity. If the equipment will not start or if we cannot get feed to the cattle, having small square bales of hay that we can carry ourselves can work until the weather improves. Having water sources out in the fields can help, too. We have a tile line that never stops running with water. (That is, until the drought of the last three years!) The tile line has a wide cement pad underneath it so that cattle can drink from it year-round in emergencies. If there is enough slope and water pressure in water tile drainage systems, a farmer can get the underground water to come up so the cattle can drink and then have the water go back down into the tile line. We prefer to keep our spring calving cow herd out in the fields all winter if possible. Why haul more manure to be composted when the cattle can spread it themselves? We also have a fall-calving herd, so the calves are nursing on the mothers all winter. They are also fed out in the fields. Cattle are hardy and can stand extreme cold. What they have to be protected from is the wind, which is another reason we have so many trees out in our fields. Cattle require a great deal of feed just to stay warm and meet their maintenance requirements in extreme cold and windchill. When it is below zero, a mature cow can consume forty to fifty pounds or more of good quality hay just to get by. That is on top of being able to find some cornstalks or other feed sources still left over from the preceding crop-growing season.

As I have become older, I must admit I kind of grew to hate snow. It mostly depends on whether you have to work in it every day. It is hard to keep your hands warm and dry with snow. Yes, I know and realize snow is very necessary for keeping the ground covered, keeping the frost from going in too deep, and for keeping the winter annuals like rye alive. We certainly need snow because of what is happening with our changing climate, but it does not make life easy.

During much of the 1980s, winter seemed to be particularly stressful for Maria and me and our young family. We had three young sons, born in 1981, 1983, and 1986. It also happened that during some of those years, the first consequences of our changing climate appeared to be beginning. There were temperature, wind, and snow extremes. I vividly remember some of the worst storms during those years as being ones where it was not uncommon to have wind chills in the -50 to -60 below zero range. We had only one tractor that had a cab on it to keep the worst of the cold out. Our tractor with a loader and bucket on it for moving snow and manure had no cab. Our loader tractor did not become a cab tractor until 2007. We just toughed it out. You can get by doing that when you are young.

My dad did not have a large, heated shop to fix machinery, do woodworking, or do any of the many other winter tasks in which something needed fixing. Our shop was only about sixteen feet wide and twenty-four feet long. It was heated with a wood-burning stove. In 1996, I built a forty-feet wide by sixty-feet long insulated and gas-heated shop. It should have been done long before.

Not My Winter of Discontent

In spite of its rigors, I wish winter would last longer. I always think there is so much time in early December to get all of the fixing and other projects finally finished. It never works that way. In winter, it is easy to be enticed into ordering all kinds of new and exciting kinds of garden seeds for the upcoming spring. Dreaming of sweet corn, green beans, beets, and tomatoes always wins that round! Gardening is easy until you have to plant one. The same holds true for our larger farm fields and their seeds. During the winter, it is always fun to think of new research projects and new varieties and even new crops to try. I always think there will be much time to get all of those new ideas and projects implemented the next crop-growing season. Hope springs eternal. Maybe the phrase *should be*: Spring hopes eternal. The crop-growing season of 2024 was the fifty-first chance for me to help plant, grow, and harvest a crop. In the grand scheme of things, I find that pretty exciting.

Chapter 7
The Decade That Defined Who We Would Become

Knowing yourself is the beginning of all wisdom.

— ARISTOTLE

This chapter reflects on my life and that of Westphalia, Iowa, during the 1960s and early 1970s, at a time of war in Vietnam, civil conflicts in the United States, and my decisions about college and my life's work.

Fertile Fields for Church Service

A large number of Westphalia's young men and women became priests and sisters during its 150-year history, nearly all during the first one hundred years. After 1970, only a very few young people made that choice. There have been at least twenty priests and over seventy sisters from our little parish. That is a remarkable number given our small size. What are the reasons for this? While this issue was examined to some extent in a previous chapter, another look at it is warranted. The 1960s ushered in a new era in which young people did not want to be told how they would spend their lives. This was true whether it was to be drafted and sent off to Vietnam or discriminated against and beaten down for the color of their skin or denied equal rights if they were women.

Westphalia owed its beginning to those people who wanted to establish a Catholic colony in this particular area of the state of Iowa, which was done in concert with the railroad magnates who wanted to encourage settlement in this sparsely populated but large geographical area. This was the town's reason for being centered around the church and new farming families. These Catholic families tended to be large. Some families had at least one child who died at birth or shortly thereafter. Families needed to be large, as the success of a farm depended in part on the number of people who could contribute to the livelihood of the farm. Many of the farms were not extensive in size. One hundred and sixty acres of land was the norm. The vocational opportunities of that time were limited for young men and women who were members of the first generation of native-born Americans. Young men were encouraged to become farmers and either try to expand the size of the farm or wait to take over the existing family farm. As farm mechanization increased, so did the size of the farms, and for a time, the number of farmers did too, until there was no new land left to farm in our county or any other, for that matter. There was a history of absentee landlords, and competition among farmers to rent their land has always been an issue.

Just about every family in our parish had at least one member from its extended ranks become a sister. If you were lucky enough to have a priest in your family, that was an even greater honor and accomplishment. Our family was no exception. I had two great-aunts who were sisters and one great-uncle who was a priest. Dad's sister Anna joined the Milwaukee Franciscans and served in hospital care her whole life. So it was natural for someone in my family to be encouraged to become a priest, as were both my older brothers. On the one hand, my parents, our pastor, and the school sisters all had high hopes and expectations of the priesthood for my two older brothers.

On the other hand, I think my parents always expected me to be the farmer in the family. I developed that expectation as well. I think it began when Joe and Mike both were in the seminary. I was fourteen years old. That is when I began to take helping my dad and working on the farm seriously.

We were fortunate in our family to have parents who cared about local and national politics, agricultural policy, and social justice issues. We had a pastor who cared deeply about social and economic justice for all people. Mom was a teacher, a voracious reader, and a political junkie before the word ever became popular. She was the oldest in her family. Dad was the oldest in his family. He was a self-made man. He did not have a formal education beyond the sixth grade, but he became a respected leader in our community and in the greater Shelby County. I think the reality of having my brother Larry, who had Down Syndrome and was somewhat physically challenged, contributed a great deal to the attitudes that we acquired and displayed. Mom started the first chapter of what was, in the early 1960s, called the Shelby County Association for Retarded Children (ARC). Larry was in the first special education class. There were three students. Their first names were Larry, Barry, and Terry. The ARC served a great need, but I am pleased that the word "retarded" is no longer deemed as appropriate.

Our family meals were probably a little different from most people's. Politics and the news were discussed at every meal. What struck me the most, though, was that my parents always insisted on treating everyone with respect and dignity. We were taught to be open-minded. Even though "all things local" were important in family conversations, we also cared and talked about the state and national issues of the day. We subscribed to both the *Des Moines Register* and the *Omaha World Herald* newspapers.

Two of the biggest issues of the early 1960s were the fight for racial justice and the war in Vietnam. My oldest brother, Joe, entered the seminary in Conception, Missouri, in 1962. Mike entered in 1964 at the same location. By 1964, Joe was spending the summer in Mississippi helping to build Head Start centers for Black children and participating in the struggle for civil rights. After graduating from Conception Seminary College in 1966 with an undergraduate degree, Joe went on to earn a graduate degree in sociology from the University of Iowa and worked to combat institutional racism in the city of Chicago for a number of years. Mike left the college in 1967 to pursue a career as a clinical psychologist. He graduated with a PhD from the University of Utah.

My parents struggled with accepting the fact that their two oldest sons would not continue their studies to become priests. Just as their sons were struggling to find their direction in life, Mom and Dad were being asked to accept their lack of control over the direction Joe and Mike were taking. Bob Dylan's song about how the times were "a changing" seemed very appropriate for all of our family members.[1]

After Mike left for the seminary in 1964, the perfect little world of my childhood was beginning to come to an end. Not only were the times changing, but I was also changing. Adolescence and its accompanying baggage were confusing and troubling. At least that was true for about two years until I began to figure it all out. I was not ready for it. No one had helped me to prepare for it. I certainly was not going to hear anything about how my body was changing or how to handle that from anyone who wore a habit or a collar around their neck.

The Catholic Church has struggled with clerical sexual abuse for a long time. The record of the clergy's past sexual abuse of children, which in some cases is still happening, points to the critical need for appropriate education and training in self-awareness, sexual orientation, and appropriate behavior. I was not abused.

After some thought and investigation into the history of priesthood in the Roman Catholic Church, I think that the Catholic Church should allow priests the option of marriage. I also think the Catholic Church should begin to allow the ordination of women. These are bold statements from a Catholic layperson with absolutely no theological credibility in church matters. Still, I am entitled to my opinion. Some of the apostles were married. As the early church grew and developed, some of the early leaders, like St. Paul, encouraged celibacy. He and others began to emphasize that celibacy is a special gift of God by which sacred ministers can adhere more easily to Christ with an undivided heart and are able to dedicate themselves more freely to the service of God and humanity. Some Catholic historical scholars have maintained there is no evidence of a general tradition or practice, much less an obligation of priestly celibacy, before the beginning of the fourth century. Even though celibacy came to be seen as a requirement for priests to adhere to, the first written law making the priesthood a diriment impediment for marriage for the Roman Catholic Church appeared as a result of the Second Council of the Lateran in 1139. Strict adherence to this rule did not begin until the Protestant Reformation began when the Catholic Church clarified its rules regarding celibacy and marriage. This was over the period of 1545 to 1563.

Today, Pope Francis has made it known he would consider the possible marriage of priests in areas of the world where there are not enough men entering the priesthood. One such area is in the Amazon of South America, where there is a severe shortage of people entering the priesthood.

My parents never broached the subject of sexuality, either. Frank discussions about sexuality were off the table. In my mind, that was true for most parents of that generation. It was possibly even more so with those of us who grew up in a small, strict, and tight-lipped German Catholic community.

The sheltered and secure school life in Westphalia changed in 1964 as well. St. Boniface High School closed. We had to combine with the Catholic high school in Defiance because it was the bigger town, even though there were considerably more kids from Westphalia in my class. This meant we had to travel every day to Defiance on a school bus, a thirty-mile round trip. It was a difficult adjustment for most, me included. But it, too, had its plus sides, such as playing pitch, five hundred, or hearts on the long bus rides to and from school.

Trying to Leave Childhood Behind

After both of my older brothers were gone from home, my life changed on the farm as well. I had to assume much more responsibility. I welcomed it, though, for the most part. My high school years were ones where I began to really learn what farming was all about, and I started to develop the skills and the work ethic that went with it. Dad came to depend on me for help. I wanted to help him, as I was beginning to slowly see the very first signs that he was physically changing, too. He was fifty-seven years old. I took it upon myself to try to do as much as I could to make the physical labor that he shouldered a little lighter. I was genuinely interested in farming and decided early on that I would attend Iowa State University (ISU), in Ames, Iowa, to study agriculture.

There were other interests that began to gnaw on me as well, however. I was an idealist. That was easy to be when you were fifteen years old. My brother Joe's work with civil rights spurred my curiosity, and I began to read many of the books associated with the injustices that Black people had to endure. There were the ones that I had to read in high school English classes, written by white authors, such as Harriet Beecher Stowe's 1852 *Uncle Tom's Cabin* and Harper Lee's 1960 *To Kill a Mockingbird*. After that, I took it upon myself to read more disturbing and mind- and conscience-shaping books by Black authors, such as Ralph Ellison, James Baldwin, and the autobiographies of Malcolm X and Nat Turner. *Black Like Me* was published in 1961 by a white person, John Howard Griffin, who temporarily darkened his skin so that he could explore the experiences of Black people. *Black Like Me* was one of the first books in which I read about how the color of people's skin defined their entire being. That book shook me.

Many other books were written in the 1950s and 1960s that helped to shape my consciousness. These came from authors like Kurt Vonnegut, Joseph Heller, Ken Kesey, Philip Roth, George Orwell, J. D. Salinger, Ray Bradbury, John Updike, Henry Miller, and John Steinbeck.

When I was a junior in high school in Defiance, I was lucky to have an English teacher who challenged everyone, including me, to read books that questioned total allegiance to the growing power of the military, the wealthy and powerful, and the growing influence of corporations and popular culture on society. The teacher was a former seminarian. He encouraged us to look critically at our government's long history of racial discrimination in the workplace, schools, and businesses. Institutional racism became a term very familiar to me. He encouraged us to think deeply about the war in Vietnam that in 1966 and 1967 was beginning to escalate. I did not have to read books to learn about the war in Vietnam. I just had to turn on the TV at night and watch the evening news to see the growing futility and atrocities of that war. I did try to think more deeply about the war in Vietnam. I wrote against our country's involvement in the war for our school newspaper. I felt like both an outsider and a minority for taking that stand. However, I believed that many other students agreed with me, even though they were afraid to go against their parents or the patriotic stance of both the government and society at large at that time. The opposition to the war was growing and would soon bubble over into both nonviolent and violent protests. By the time 1968 came, the assassinations of Martin Luther King Jr. and Robert F. Kennedy, were seared into my conscience and memory.

All three Catholic high schools closed in 1967 due in large part to growing financial problems. More lay faculty had to be hired, and they required much higher and more livable salaries than the teaching orders of sisters of that era, whose numbers were quickly diminishing as fewer young women were entering convents by the early 1960s. It costs money to run buses, hire bus drivers, and have school lunch programs. None of those things were necessary at St. Boniface in Westphalia. Everyone was responsible for getting to school themselves, and everyone packed a school lunch. I finished my high school education at the public school in Harlan. It was a good year. I already knew many of the Harlan kids because of 4-H.

Was I ready to leave home in the fall of 1968 for college? The answer was emphatically, "No!" I was scared to death of moving from home and going to a large university. Ames was hardly an urban environment, with only around 40,000 people and half of them students. That did not matter. It might as well have been New York City.

The four-year curriculum in the College of Agriculture that I had signed up for was then known as the Farm Operations curriculum, which was meant for students who intended to return to the farm. At the time, Iowa State University was on a four-quarter system, not a semester system. That meant I could start college the second quarter, which was after Thanksgiving. It was easy for me to make the excuse that Dad needed me to help with the harvest. That was true, but it was more because I was afraid to leave home. I suppose it had much to do with my parents' expectations and my own expectations of my duty to care for them and the farm. Their first two sons had not chosen paths that my parents had hoped for, and my impression was they were not about to let that happen with me.

Getting mostly A's all through grade school and high school was not that difficult for me. Both of my parents expected me to, and especially my mom, who had been an elementary teacher. I was fairly bright, but not in math-related subjects. I never understood algebra and why it was necessary to even learn about it. I blamed some of it on my teachers in the parochial school system. The sisters were much better teachers of English and social studies than they were of math and science. I thought I had to get mostly A's in college, too, but the fear of not being able to do so in mathematics requirements scared me a lot.

I entered Iowa State with many intellectual and emotional conflicts, the war in Vietnam and the varieties of turmoil in our country—and my personal ones as well, that is, conflict between my ideals of making a difference in the problems confronting our country regarding race, politics, and the war; and upholding my duties toward the farm and the study of agriculture.

Dad dropped me off at the dorm known as Friley Hall around the first of December in 1968. I started classes a week late because I had attended the National 4-H Congress in Chicago the week after Thanksgiving as a result of a state 4-H achievement award I had won. So, I did not exactly start off on the right foot. Nevertheless, I had no problems fitting into dorm life. I was a sociable person, so that part was okay. I bucklod down to my studies immediately and ended up getting mostly all A's the first year. I should have. Besides hanging out with the guys in my dorm house, about all I did was study. Many of the guys were agricultural and engineering students. I played some basketball on our house's intramural team. Still, I did not feel like I really belonged there. I was already thinking about not studying agriculture and majoring in sociology or political science at some other university.

Even though I felt like I did not really want to be there, I dutifully went home that summer to help Dad on the farm. But I felt like I no longer belonged there, nor did I think I belonged in Ames at ISU. I had the chance to travel to New York City in August of 1969 to both Harlem and the Bronx, where my oldest brother had worked previously with poor Black children at a camp. By the time I went back to Ames that fall to begin my second year of college, I began to feel very conflicted. I was unhappy and did not want to be there. I started looking into transferring to some other college. The University of Michigan was one place I was looking into. Besides my agricultural courses, I began to take more sociology courses, and my adviser started to wonder what was going on. I never said anything. I still had room for plenty of elective courses at that time. I was still getting A's because I thought that is what I had to be getting. I was taking many credit hours, and by the end of the first quarter before Thanksgiving of 1969, I was starting to become a little afraid. Of what, I did not know for sure. I started hinting to my folks over Christmas that I might transfer to somewhere else.

Cutting the Cord

In late January (which can always be a depressing time), there was much snow on the ground. It was cold and gloomy. I was very unhappy. I woke up with a start on one of those kinds of days feeling absolutely terrified. I had never felt that way before. What was happening? I was scared. I put up a good front with my roommate and other dormmates, but inside a storm was raging. I was very lucky in that there was one person I could turn to, one of my professors whom I had had the opportunity to get to know a little the previous fall. He was teaching Biology 101, and I immediately found him to be the kind of caring and loving person I could confide in. I had to talk to somebody. I tried to eat some breakfast and remember going to see him just as soon as I possibly could. Luckily, he was in his office. I told him I did not know what was happening to me but that I needed help. He recommended a psychologist in student health and called him. I do not remember if I was able to see the psychologist that day or if I had to wait a day or two. At any rate, I was walking around in a total depressed daze. I met with him probably once a week from the end of January until May. We all know what "total makeovers" are in today's world. I was going through one. It was like I was becoming a completely different person, even though I was not sure I recognized that person. I did not realize that I had never really been my own person. I had not broken the umbilical cord with my parents, and especially with my mom. I needed to grow up, assert my independence, and figure out who I was to become. That scared me half to death to say the least. In my heart, I did not want to go back to the farm that next summer but instead get a job working with disadvantaged kids. However, also in my heart, I did not want to hurt my parents, either, especially my dad, by not going home to help on the farm. I knew I would be letting them down. What was I to do? Over that early period of about four months, my advisor helped me to begin to understand what I was going through and what I had to do to make things better. One of the first things that I began to realize was that I had not really tried to adjust to life at the university or as a student living in the dorm. I had just been going through the motions, waiting for the day I would leave, not even sure where I would be headed. How could it be better in a place that likely would be farther away from what little comfort zone I had?

Working up the courage to tell my parents I was not coming home that summer was by far the hardest thing I had ever done at that point in my life. There were times when I thought I could not do anything that would help my situation. During those times, it seemed like the easiest thing to do (even though it would have been the biggest cop-out) was to quit school, go back home to the farm, act like nothing had happened, and just fit right back into my mostly sheltered and secure life. However, I began to realize that if I never became an independent person who at least tried other jobs and educational pursuits, I would never be really happy. I would have shut the door on exploring and discovering my own talents. The hypocrisy of claiming the importance of open-mindedness and then not being able to do that for myself, might be more than I could bear.

Meanwhile, throughout all of my own troubles, the war in Vietnam raged on and was getting worse. There was also the very real possibility of being called to serve in the war in Vietnam. My lottery number, 108, was picked on December 1, 1969. In 1970, people with lottery numbers up to 195 had to report to service unless they had a deferment of some kind. I had a student deferment of 2-S. If I dropped out of college, I was certain to be drafted into a war that I strongly opposed. Things could have been worse. Many of my classmates would have no choice but to go to Vietnam unless they had a "get out of jail free card" for staying in school.

Because of this growing discontent with how the Selective Service System operated by drafting young men who were either poor or members of a minority, President Richard Nixon ordered that college deferments be limited in 1971. However, by then, there was growing criticism of the war, and the military was calling up fewer draftees.[2] By 1971, polls indicated over 70 percent of the country thought the United States had made a mistake by waging war against the North Vietnamese. Fifty-eight percent thought the war was immoral. Already in 1968, the fourth year of the war, a majority of Americans thought the war was a mistake.[3]

When spring arrived in 1970, I told my agriculture studies advisor I was dropping out of the College of Agriculture. He was very disappointed and voiced his dismay. I always felt he could not understand my decision. As far as what I wanted to major in, I did not know. Luckily, there was a program at Iowa State for people like me. It was called Distributive Studies. You could major in three different areas of study. I decided on zoology, psychology, and sociology. It proved to be a very good decision.

After the spring quarter ended at Iowa State in 1970, I applied for a summer job at a camp for White and Black Catholic children from Omaha, Nebraska. It was in the hills surrounding the Missouri River about sixty miles from the farm. It was run by an organization in Omaha called the Christ Child Society. I knew about it because my brother Mike had worked there one summer. I was hired to be a senior counselor in charge of about fifteen boys in my unit, with new kids every week for ten weeks of the summer. There was also a one-week camp for troubled teenage girls from a group home in Omaha. I was put in charge of the nature program at the camp. This meant taking children exploring through the woods, doing woodworking and wild animal care, and conducting actual classes on the ecology of the area. I caught snakes, frogs, lizards, interesting insects, and butterflies, and had the children learn about them and take care of them. We even had two coyote pups in a cage behind the nature lodge. I fed them dog food and also trapped mice to feed them. We had hamsters and a very sassy parrot from a pet shop in Omaha that were loaned to us. I had taught myself how to play the guitar the year before, so I led the campfire sing-alongs.

That whole experience of working with children and with a subject matter that I already knew something about was great for my self-concept and general well-being. I finally started to relax a little and began to enjoy life again. It had been a long time. Working with other college students and adults for the summer boosted my self-confidence. I also met a girl who turned out to be my first real girlfriend. I was beginning to slowly grow into the adult person I would become.

I went back to Ames that fall with a sense of real purpose and excitement. I started taking courses of every kind that dealt with botany, zoology, microbiology, ecology, genetics, biochemistry, and the whole gamut of classes that one would take for pre-medicine. I still took some psychology and sociology courses. I started to not only enjoy learning but also to begin to understand conceptual learning. I had not experienced that in the College of Agriculture, which at that time was much more about technical and agricultural topics. It was in the study of the biological sciences that I began to understand how the natural world functions and works. I could not get enough of those kinds of courses. I quit worrying about having to get A's to feel good about myself. That was a good thing because some of those courses were tough. I had to learn how to use a slide rule for physics classes, which readers of my age know was not for sissies. I looked into the possibility of going to medical school but was told I would have to have a higher grade point average to even be considered. I graduated with a 3.0 grade point average. I also had no one to advocate for me to even consider that option. Besides, I did not know what I wanted to do for sure anyway.

My parents were getting used to the idea that I was not coming back for the summers to help. Dad had begun to hire high school students from Westphalia to help him during the summers.

He reduced the livestock numbers so that the rest of the year was not quite as difficult as it had been. I still helped as much as I could when my summer jobs were over and before I went back to Ames. I also helped during holidays and quarter breaks. They had become more accustomed to the possibility I would not come back to the farm after college graduation.

My last summer job before my last year of college started to pull me in another direction. I was hired to be a summer school teacher at a center for disturbed children in Ames in 1972. I continued that job on a part-time basis for the remainder of my last year of studies. I started to take more psychology courses. I had enough credits already for a Bachelor of Science degree in Biology and a double minor in Psychology and Sociology. I also was not very far away from a degree in psychology. More importantly, though, I thought I had received a well-rounded liberal education. But now what? Graduate school interested me but not by continuing in the field of biology. The field was too general, and I did not have enough of a focus on any one part of it to know if I wanted to study it further. I could have become a science teacher, but my attention turned to the field of psychiatric social work and obtaining a master's in social work. I had presented a paper at the Iowa Association of Social Workers that was well received, and I thought I could pursue more training in that field if I so chose. But again the question loomed. Where? It would have required me to take another big step. Even though I lived off campus in an apartment with other students, I was not earning the kind of income that it would take for me to move to a new school and location without financial help of some kind.

In the meantime, my parents had not completely given up on me returning to take over the farm. Over Christmas of 1972, Dad asked if I would consider trying farming for a year, and if it did not work out, I could walk away. I thought that was a very fair deal. It just did not sit right with me that the farm would have to be rented out to someone not in our family if I did not make the decision to come back. My two older brothers were deep into their own respective careers, so they were not a part of the decision. When asked if I thought I could be happy farming back home from which I had worked so hard to cut the umbilical cord from, I said yes. I considered myself an adult by then, and I got to make an adult decision. I made a choice, and I have never regretted that choice.

It has been over fifty years since I graduated with a Bachelor of Science degree in Biology. I am very grateful that I have had many courses in plant, animal, human anatomy, and physiology. I am grateful for the core courses in organic chemistry, biochemistry, microbiology, plant taxonomy, genetics, ecology, and meteorology. The laboratory work associated with the lectures in each of those courses was rigorous. It has provided me a lifetime of understanding the basics of how life functions on Earth. I think it has helped me to become a better farmer. Today, the exciting new discoveries in soil quality require new discipline in understanding and learning how to maintain and enhance it. The existential challenges of climate change require even more rigorous thought and research and discovery. It will require all the wisdom, understanding, and knowledge we can muster, and then some.

However, fifty years ago in 1973, my "graduate school" became the world of agriculture and rural community and the surroundings into which I settled.

I made it through the 1960s and early 1970s. Many young people did not. It was a tumultuous time for so many of us growing up during that troubled era. I wrote a poem about it in 1976.

Dreamers of the Sixties
We were a generation meant for the sixties or so it seemed
A teen-ager seven of those years, I had dreamed
of changing the world with one easy blow

We fought racism and poverty
Institutions and the military
We fought politics and presidents
They say we even made one decide not to run again
We had John, Bobby, and Martin
But they died too just like the sixties
Has anybody here seen them?
We demonstrated for what we thought would make us more free
But the times were telling us it just wasn't meant to be
That simple or that easy, what would be our destiny?
People reacting to life in so many ways
Drugs, expanding your mind, a new consciousness
Flower children, let it all hang out, a new responsiveness
But polarities were formed
Between Parents and Children
Right and Left
Liberal and Conservative
Black and White
White and White
Purple and Purple
Purple and Green! Psychedelics! The picture is getting hazier
My God, we must all be getting crazier!
Survival
This is what they're saying now at best
This generation lost in a haze, they've lost the rest
Energy-Survival
Food-Survival
Money-Survival
Pollution-Survival
Population-Survival
Survival-Death
Death-Death
Is there no alternative?
If not, this meeting is adjourned until 1984
We'll see you all then you dreamers of the sixties
Dreaming dreams never more

Again, over fifty years ago, my graduate school became the world of agriculture and rural community and the surroundings into which I settled. It took some time, but there came to be a feeling of security and wonder for what the future would be.

PART II

Breaking the Promise

Chapter 8
Starting to Farm in the 1970s Boom

The industrialization of agriculture is said to have achieved two goals: to "free" Americans from farming so they could join the labor force in offices and factories, and to make food and farming cheaper so Americans could afford to buy the products offered by new industries.
            —JOHNS HOPKINS CENTER FOR A LIVABLE FUTURE, 2014[1]

This chapter examines some of the US government's intervention into agriculture and the crops gown; the variation of land, crop, and livestock prices; the use of pesticides and the value of crop rotation; and organizations founded and focused on rural farming, and specifically on the corporate influence on the hog and pork business.

A Changing Landscape and a Changed Person

I started farming in the spring of 1973. I came back a much different person than I was when I left in the fall of 1968. What I soon discovered was that, although I had changed a great deal, farming and rural culture were changing even more. The social fabric of rural America, consisting of security, safety, and isolation, was beginning to unravel and was quite evident in my little home community of Westphalia, Iowa.

When I first returned to the farm that spring, one of the locals asked me why I would want to come back. After all, I had a college degree. What was there for me, she wondered. She said I could do better in the city. Unfortunately, I found out later that most people who have grown up in rural areas have chosen or found it necessary to leave. Still, I have spent much of my life fighting that putdown, which so many rural people place upon themselves. It must go back, in part, to a saying that I sometimes heard when I was a youngster. "If you could not do anything else, you could always be a farmer."

It did not take me long to realize I was tired of being a student in a formal educational setting. I had been one for nearly seventeen years, and it felt good not to have to study for the next test. I started to enjoy my freedom. That did not mean I wanted to stop learning. My study of the biological and social sciences had just begun to whet my growing appetite to keep learning more.

I noticed an immediate difference when I came back home. I was treated as an adult by the rest of the community. In a small rural community and county, I knew it was possible to put my extra time and freedom to constructive use, make a difference, become a leader, and not just be a number, which was more often the case in urban environments. In small communities, people think they have to know everything others are up to and try to pigeonhole their every move. That did not bother me. I made up my mind early on that I would try to "get along but not necessarily go along." My life experiences had already begun to teach me the critical value of that. If I thought I was doing the right thing, then I did not worry so much what others had to say about it.

One area that began to draw my attention was politics, and especially the politics that affected farming and rural communities. I became involved in the local Democratic Party. The Watergate hearings began in the spring of 1973, my first year of farming. However, I was most concerned about what was happening in rural America. When I left for Iowa State University (ISU) in 1968, farming and rural life had a sense of stability. By 1973, it was changing to one of increasing volatility and controversy.

> It had come to be accepted that the pigs should decide all the questions of farm policy.
> —ORWELL, *ANIMAL FARM*[2]

Industrial Agriculture Arrives in Rural America

Earl Butz became US secretary of agriculture in the fall of 1971. Clifford Hardin, who had been chancellor of the University of Nebraska before becoming President Richard Nixon's first US secretary of agriculture, had just resigned. Butz immediately drew the ire of many farmers with his brash statements of what farmers should do if they wanted to keep farming. There were two that stood out. The first was "get big or get out," and the second was "plant fence-row to fence-row."[3]

Unfortunately, these statements proved to be accurate. Government farm policies began to reflect his viewpoints. Many farmers were listening to Butz. Not only was another farming boom germinating, but also bigger structural changes were occurring as well. They had already been for quite some time. There is ample evidence to suggest that some changes may have been designed to get rid of farmers. For agriculture, the 1960s were associated with crop surpluses, and wondering what to do with them. That seemed to be the feeling about the number of farmers, too. There was an oversupply. As early as 1947, *Life Magazine* published an editorial saying that the solution to farm problems was the elimination of over 3.8 million family farmers.[4] A committee report of the Agricultural Chamber of Commerce stated small farm units were economic and social liabilities.[5] Some of the committee members represented the largest corporations of that era. Most were in the business of food. The corporations included General Electric, Pillsbury, Carnation, Ralston Purina, and the meat-packing company Armour. They recommended at least one-half of all farms be eliminated even though "these farms are of the family type and apparently constitute a substantial portion of what is supposed to be the backbone of the nation."[6] According to the committee, a major problem was small farmers who preferred farm life were not large consumers. This behavior was because the value of self-sufficiency and sacrifice had been seared into many farmers by the events of the late nineteenth and early twentieth centuries, when economic survival depended greatly on doing without and getting by with what they had. The farmers counted their blessings, because they would never go hungry on the farm.

Was there an organized effort to get rid of farmers? Maybe. Most would say it was inevitable, given improvements in production technologies, such as the hybridization of corn that more than doubled yields in just a few short years. It was due in part to the adoption of pesticides and synthetic fertilizers, which were ushered into crop production with the help of the bomb-making technologies of both world wars. It was due to machinery improvements allowing farmers to harvest higher volumes in less time and encourage farmers to increase the size of their farming operations.

However, from the very beginning of farm programs in the 1930s, government farm policies also influenced and encouraged farmers to decrease and increase production simultaneously. That sounds like an oxymoron, but regardless, the net result was bigger farms and fewer farmers, justified by the results of increased efficiencies and what some critics started calling the nation's new "cheap food" policy. Poorer land may have been taken out of production by farm program subsidies, but yields increased on the land still in production.

With the advent of higher yielding crops of corn in the Midwest, wheat in the West and North, and cotton in the South, these crops came to be referred to as "commodity crops." The diversity of crops—such as oats that fed the many "workhorses" before tractors and legumes such as alfalfa and clovers that produced much of the nitrogen and slowed soil erosion— before the advent of synthetic fertilizers, came to be discouraged by the emerging corporate agribusinesses and by the government, too.

What to do about the farm problem of overproduction and low prices has vexed many politicians and farm leaders for a long time. Farmers never have been able to quite get over the hurdles of self-organization in their collective best interests to control supply and, consequentially, prices.

Government policies to affect prices and acreage controls and all kinds of mechanisms to try to guarantee farmers a livable income have been a mixed blessing. If you look at where we are at present, it becomes apparent that the policies have fallen short of protecting small producers while being more beneficial to large producers. In my opinion,

> Farm programs have a history of rewarding farmers financially for poor conservation and ecological farming practices. This in turn has allowed those farmers to have an unfair advantage in competition for land and other agricultural resources.

As stated in Chapter 3, the problem of low prices that farmers faced in the late nineteenth century, in the first boom and bust period of American agriculture, had seen limited federal government involvement and efficacy in improving the lot of farmers. By 1924, when the next bust was growing in severity, the first real attempt to address the problem of low prices and depressed farm income was made by the US Congress. It was based on the concept of parity for what farmers produced, the idea that farmers should be paid a guaranteed fair base price for their products based on costs of production and what would be considered a decent livelihood. The concept was first introduced in the McNary-Haugen Farm Relief Bill, which passed Congress only to be vetoed by President Calvin Coolidge.[7] Coolidge's legacy as president was one of small government and laissez-faire economics. The national debt fell by one-fourth under his presidency. Unfortunately for farmers, though, that attitude plunged them further into debt and hard times. They had not shared in the improved economy after the First World War. Gilbert Haugen was a Republican congressman from northeast Iowa. He served from 1899 to 1933. The McNary-Haugen bill was passed on a number of occasions, from 1924 to 1931, only to be vetoed every time by Coolidge. It had support from then vice president Charles Dawes and from Henry Cantwell Wallace, US secretary of agriculture under Coolidge, and also from Iowa. The plan was to control prices via the government purchasing crop surpluses and paying farmers a target price computed as a real price equivalent of those in the 1905 to 1914 period, which was considered a "golden age of agriculture."

By the early 1930s, the Great Depression was deepening for farmers. Gross income for farming had fallen by 52 percent between 1929 and 1932. Something had to be done. Farmers began to look to the new Franklin Roosevelt administration for relief. The *Agricultural Adjustment Act of 1933 (AAA)* is considered the first comprehensive attempt by federal farm policy to control production by trying to balance supply and demand.[8] The parity principle was used for allotment for specific crops. I was somewhat surprised to learn corn was not covered by the original concept. The 1933 bill was similar to the McNary-Haugen bill. The bill asked farmers to voluntarily cut production on wheat and on cotton specifically by a designated percentage so the effect would be to raise prices. There was a guaranteed base price based on what was considered a percentage of parity. Parity was set at 95 percent of that value computed from 1905 to 1914.

Relief could not come at that time in the form of a direct payment, which was considered a form of socialism. It had to come indirectly in the form of production controls that would allow prices to rise on their own.

In 1936, the US Supreme Court threw out the law providing price supports to farmers. It stipulated that regulation of agricultural production was outside the bounds of Congress's authority as provided in the US Constitution, and the court declared it unconstitutional. Farm leaders and Congress looked for another solution. They found it by combining production management controls with soil conservation, arguing soil conservation was well within the authority of Congress to provide for the general welfare of the nation. It probably helped that the effects of the Dust Bowl were being felt in Washington, DC, in the form of a dust cloud originating in the Great Plains and settling on the capital. The new *AAA Act* provided payments to farmers who reduced their plantings of soil-depleting crops such as corn and cotton and increased soil-building crops like grasses and legumes and even cover crops. The Soil Conservation Service began and Soil Conservation Districts were formed around the country. For the first time in the history of agriculture in our country, it looked like conservation of natural resources would be given its due. It proved to be short-lived.

Surpluses never really came down because, overall, farmers would put their worst land into crop acreage reduction programs and put more emphasis on their best land with large amounts of fertilizer investments. Fertilizers, lime, and tile drainage actually became a part of the conservation funding awarded to farmers. By 1962, 38 percent of the Agricultural Conservation Program (ACP) was used for those three practices. That seems counterproductive to the intent to lower crop surpluses.

Conservation provisions were put on the back burner during the Second World War era. Prices and crop production both went up because of the need for food during the war and then after the war because of the resulting worldwide devastation.

The rapid adoption and improvements of corn hybrids, along with the increased production and use of synthetic fertilizers raised yields rapidly and brought more oversupply of the commodity crops in the 1950s. This continued through the 1960s.

One aspect of farm programs has bothered me the most: how a producer's base acres are calculated. It goes way back to when base acres were established. Government subsidies and payments in the form of price supports, deficiency payments, and government commodity loans were, and still are, determined by how many acres of a covered commodity crop you have as your base acres in that crop. For instance, most farmers in my area had large base acres of corn to start with because they had already stopped using the more diverse crop rotations that included hay, pasture, and small grains. Farmers could increase base acres by putting more land into corn, and then a little later on, when soybean acres developed their own base and price supports, they could do the same. Dad did not give up using the diverse crop rotations of hay, small grains, and rotational pastures. So he had a smaller corn base. That meant he received smaller payments in general because, again, the payments were based on acreage of the commodity and the yield. His proven yields were somewhat smaller because less synthetic nitrogen was applied. All of this meant that total subsidies were less. What it meant in actuality was one was penalized by the government for using good conservation farming practices that contributed to crop diversity, water and soil quality, and other environmental benefits. I made the same decision as Dad did. I decided it was more important to think about long-term soil health than to work with the government subsidy program. The thinking went something like this: if you were not farming the government, then you were not a very smart or pragmatic farmer. I guess I was not very smart.

Suddenly, just about the time I came back to start farming, talk switched from surpluses to never having enough grain again to meet the world's needs and demands. According to some of the followers of Secretary Earl Butz, we could export our way into eternal prosperity because the rest of the world needed and would buy our grain. That became the mantra and rallying cry for the input sector of the rapidly developing industrial corporate kinds of agriculture, which became a permanent fixture in the 1970s.

The Russian grain-buying incident, referred to as the "Great Grain Robbery," came to symbolize for me what was happening in agriculture when I first came back to the farm. You could call it the commodification of agriculture. You could say that it resulted in a new era of agriculture that began to emphasize speculation and the futures trading market. For me personally, it could be called my welcoming party to the new world of farming when I started in 1973. Russia had poor wheat harvests in both 1971 and 1972. The US Department of Agriculture (USDA) and a couple of large grain trading companies decided to start selling wheat to Russia. The deal was supposed to be over a three-year period with a price tag of $750 million. Instead, the Soviet Union purchased that much in just one month! The US government subsidized it to the tune of $300 million, thinking it was for livestock feed to increase their herds. USDA did not have adequate crop monitoring arrangements with NASA to get good satellite data on what was really happening on the ground. Russia went on to buy one-fourth of the US wheat crop in 1972. Because of that, prices started to skyrocket for both farmers and consumers over the next few years. Wheat went up from $1.68 on the nearby futures market in the summer of 1972 to over $3 per bushel in May of 1973. Corn rose from $1.20 per bushel in July of 1972 to over $4 per bushel in October of 1974.

These kinds of rapid price hikes are what start "boom" periods in farming. I remember that Dad had only about sixty-five bushels of beans left from the previous year's harvest by the spring of 1973. He would not have had those except they were poor-quality soybeans that he decided to keep over the winter. Much of the reason for the poor quality stemmed from the single largest rainfall event in the history of our county. It rained thirteen inches in one day and night in September of 1972. The rest of the soybean crop had been contracted to a local seed company in our community of Westphalia. They had been contracted in the $3 to $4 per bushel range for the 1972 crop. We sold the rest in May of 1973 and filled our old truck, hauled them to an elevator in Harlan, and received over $10 per bushel.

It seems that when farmers get a taste of high prices, even if it is for only a short period of time, they get hooked on the idea that it is sure to happen again or maybe it is just around the corner, and maybe they will make a ton of money. When prices get higher, they are bound to go higher yet, so they are going to wait for the high. Usually, that "high" never comes back. At least not for some time; then the farmer gets hooked all over again. I am sure this is an oversimplification, but on the surface, and maybe below it too, is the notion speculators in the commodity futures markets and hedge fund buyers in the markets of today are the ones who control at least some of what happens in the market. I have always thought if speculators had to legally own the actual grain instead of owning it only on paper, maybe that would lessen their roles as speculators and gamblers to the extent they are now.

Other things were happening that made me realize farming was going through some major changes, whether farmers thought them okay or not. Some were welcomed by the farmers. One was the advent and adoption of pre-emergent herbicides that would be used to prevent weeds from sprouting in the first place. These herbicides were applied before the crops were planted. We began to apply them on our farm, too. I started to do all the mixing and spraying of the chemicals. We did not have a cab tractor at that time, which protected us from the spray drifting on us while pulling the sprayer through the field. If a spray nozzle was plugged with dirt or something else, you had to take your gloves off to get the nozzle unplugged. I did not necessarily like doing this, but did it anyway.

We still cultivated our corn and soybeans at least two times each, even while using herbicides. Dad figured we could really get good weed control that way, and I agreed. Most other farmers did the same thing, although some began to cultivate only once for each crop to save time. Even with the use of herbicides and mechanical cultivation, most farmers also "walked" their soybeans in late July or early August to get the rest of the weeds out. Many farmers hired crews of high school boys and girls to do that. Since soybeans were not as tall as corn, the neighbors and everyone else who drove past a field could easily see the weeds that were in it. Again, a good farmer would not have any weeds for drivers to see. Peer pressure helped to sell herbicides.

As soon as I started farming in 1973, Dad and I increased our farming acreage to 480 acres. Yes, we were getting somewhat larger, too. He had purchased 160 acres right next to our original 320 acres in 1970. Three neighboring families, all with the same last name lost their land in the depression years of the 1920s and 1930s. Some of it had been purchased by a person who lived in Council Bluffs. He was a beekeeper and ran a honey company. The 160-acre farm right across the road from our farmstead was part of that. An additional 320 acres next to that parcel was purchased by one of the farmers who my dad rented 160 acres to so he could get a start. It all came up for sale in 1970. I want to make it clear that, if my dad was only concerned about money, he would have purchased the entire 480 acres when it came up for sale. After all, it sold for only about $450 per acre. In 2021, Iowa land prices rose by 29 percent from the previous year. In 2022, they grew by 17 percent. Land values in 2023 rose slightly. In our area of Shelby County, a few parcels of land have brought as much as $16,000 per acre.

In 1970, corn was only worth an average of $1.17 per bushel, and hogs were bringing twenty-one cents a pound. Farmers were not exactly getting rich quickly. It is ironic I have spent most of my farming career paying for a little over 143 acres of that original 160 Dad bought for $450 per acre in 1970. I had to pay many times more than that when I started buying out my two brothers in 2010 after Mom passed. The price agreed upon was even somewhat below that of the current market price in our area.

Many farmers are wealthy on paper. They are worth a good deal of money in terms of land assets. The problem for many farmers is that they can never spend any of it until the day they die, and then it is a little late. What happened in *Animal Farm* tells us something about the nature of pigs . . .

It could be argued that the pigs did a decent enough job of running the farm policies until there got to be too many of them, and they started doing what pigs are known for doing; they became piggish.

Mortgage Lifters

The decade of the 1970s proved to be good for hog farmers; most farmers raised at least some hogs. Hogs historically have been considered the mortgage lifters. Hog enterprises had the reputation of generating positive cash flows by not only paying the monthly bills but also paying off land debt. The turnaround for pork was much quicker than for cattle. Cattle could take fourteen to eighteen months to be ready for slaughter, while hogs went to market at about six and a half months. Today, hogs are ready to market at about five and a half months or less.

My dad never raised a large number of hogs, only 350 to 450 or so per year. He had sixteen wooden farrowing pens in a 20 ft. × 40 ft. farrowing building. He farrowed about twenty to twenty-five sows twice a year in that building. In my growing-up years, we also had pigs that were born out in our hog pastures in two wooden portable buildings that we pulled into the pastures in June when the pigs were born. Those hogs remained on pasture until the fall when they were moved into the cement floor yard and barn for the final finishing. I ran after many slippery and muddy pigs in the summertime, when I was young, trying to catch them when they escaped our fenced-off pastures. I discovered what poison ivy was like on one such occasion after falling into a patch while chasing pigs.

We no longer raised pigs that way when I came back. In fact, our production was down to about 250 to 300 hogs per year. It stayed that way until I had to make a decision in 1980 whether to expand the herd.

Significant research was being done by land-grant universities, such as Iowa State University, that helped independent pork producers, especially from the 1950s into the 1990s. Big improvements were being made in many different areas in producing leaner pork, improving genetics, improving production, decreasing costs of feed per pound of pork, and so on. The ISU Cooperative Extension service helped to lead the way. More often than not, most independent family farmers made money raising pigs in farrow-to-finish operations. Large numbers of pork farmers during the 1960s and 1970s paid for their farms by adding value to fairly cheap corn and soy. However, that was not the case every year, of course. Significant attention and help were given to producers who did not raise hogs in confinement. In fact, confinements were just beginning to come into the picture. Hog confinement buildings with liquid manure pits under them started in the late 1960s and grew quickly in the early 1970s.

Many communities were thriving economically because of the high turnover of the dollars invested in independent family pork production. Pork producers bought soybean meal high in protein, premixes of vitamins and minerals, and other feed and supplies from many small companies in just about every community. Many larger communities in Iowa began to build pork slaughtering plants in counties with large numbers of hogs. Our home county of Shelby was one of them. The Western Iowa Pork Company was built in Harlan in 1964. It created many jobs for my generation still living in rural communities. They would be the last. None of us really knew in the 1970s what was going to happen to the pork industry. There were warnings, however, and other telltale signs of what could happen. Could pork production someday go the way poultry production had gone in the South? Certainly not, was the basic answer. One new organization, however, had the courage to put out that warning, that is, the Center for Rural Affairs in Walthill, Nebraska.

The Center for Rural Affairs was the first farm organization that I became involved with when I came back to the farm.[9] It became the place I would turn to for learning about the real effects of farm policy on farmers and on communities. The Center started in 1973, just as I did in farming. Their founders, Don Ralston and Marty Strange, arrived in Nebraska in 1970 from the eastern United States as Volunteers in Service to America (VISTA). They were working at the Goldenrod Hills Community Action Center in tiny Walthill, Nebraska. Their mission was to work in a federally funded effort to lessen poverty in five counties in northeast Nebraska. When the funding stopped in 1973, instead of calling it quits, they got to work and formed a nonprofit designed to address controversial questions of economic policy affecting agriculture in northeast Nebraska. They realized that, though their funding had dried up, the problems did not go away. Since then, they have never backed away from a fight nor found an issue too big for them to get involved with in the defense of family farmers. Their work quickly took on a national focus and captured national attention.

They began to look at tax policy and how it was influencing corporate investment and encouraging the construction of something new called CAFOs, which was short for confinement animal feeding operations. These were the new giant hog confinements starting to spring up in Nebraska and Iowa. Tax policy was one of the big drivers of this new phenomenon. For making investments in economic growth industries, 10 percent was taken directly off owed income taxes. It was called the Investment Tax Credit and was first enacted in 1962. New center pivot irrigation systems to grow corn in the Sandhills of Nebraska were another favorite target for the tax credit. Corporate farming interests were looking to cattle feeding and hog confinements in the geologically and ecologically fragile sandhills. Another primary intent of the investment tax credit was to encourage replacing labor in agriculture with the substitution of capital. Although the 10 percent tax credit ended in 1986, tax policy still rewards capital over labor. This should not be surprising. Farming has become capital-intensive. It is one of the biggest obstacles for beginning farmers, unless they come from a family that is already farming.

In 1982, the Center for Rural Affairs was able to convince the people of Nebraska through a ballot initiative to enact Initiative 300, which outlawed corporate farming in Nebraska. The measure lasted from 1982 until 2006, when a federal appeals court ruled it was unconstitutional on the grounds that it interfered with interstate commerce.[10]

The years following the adoption of Initiative 300 showed there were positive outcomes for families farming in Nebraska. While a 1991 study showed a decline of 11 percent for Nebraska's farmers under the age of thirty-five during the worst of the farm crisis years, the states around it fared much worse. Iowa's numbers of those under thirty-five years of age dropped by 31 percent, Illinois's by 29 percent, Kansas's by 23 percent, and South Dakota's by 20 percent. The 2002 Census of Agriculture showed Nebraska had 33 percent more farmers and ranchers under the age of thirty-five than the country as a whole. There was ample environmental evidence to indicate corporate farming had more damaging consequences in states such as North Carolina and Iowa. In 1996, the nation's largest pork producer stated, "The reason people use corporate forms of doing business is to avoid liability."[11]

None of this made any difference to the outcome in 2006. It is hard to stop the corporate ag train. The Center for Rural Affairs has done a great deal of good for farmers and rural communities whether in Nebraska or not. Chuck Hassebrook spent over thirty years at the Center and was its executive director from 2001 to 2013. He called on me many times to help in the fight against the investment tax credit and corporate takeover of the pork industry, in working on sustainable agriculture and research issues, and in providing testimony before congressional agricultural committees on family farming issues. In 2014, he ran unsuccessfully for governor. He served two terms as president of the Nebraska Board of Regents.

Making Sense of What Happened to the Hog Business

It took only one generation for independent, family-owned pork production to decline and transform the pork industry into the corporate monolith it is today. This consolidation has occurred in so many different ways, and the numbers are staggering.

In 1920, 4.9 million farmers raised at least some pigs. In the early 1970s, there were 736,000 hog farms in the country. In 2021, that number dropped to about 67,000. Iowa raises over one-third of all the hogs in the country. In Iowa, only about 5,600 pig farms are left, but they market over 46 million head annually. Our family is still one of them, barely.

Only 5 percent of hogs were contracted in 1980. Today, that number is around 70 percent. Hogs raised under contract by operations of over 5,000 head accounted for 52 percent of the US hog inventory in September of 2023. Just two companies, Smithfield and Tyson, market over 50 million of the roughly 120 million hogs marketed each year in the United States. Iowa's hogs, dairy, beef, turkeys, and egg-laying chickens generate waste comparable to 134 million humans. There are a little over 3.2 million people living in Iowa in 2024. The questions to ask yourself are:

> Half of the hogs in the United States are owned by a company with its own packing plant.
>   Who owns Iowa? Is it Smithfield and Tyson, who own and slaughter the majority of the hogs? China owns Smithfield. Who really owns Iowa?

A Producer's Interpretation of Historical Prices of Corn and Hogs—Four Periods of Corn and Hog Price Trends from 1925 to 2020

From a historical perspective, I wondered if there was a relationship between the price of corn and the price for hogs received by the producer. I put together two charts on the price of corn and prices received for market hogs going all the way back to 1925. I wanted to figure out if and how much significance there was in how those prices fluctuated over time and to see if I could make any sense of it. These historical prices for corn and hogs were reported by the Iowa Office of National Agricultural Statistics Service.[12] I also wondered if the relationship between corn and pork prices changed at all when the industry started to dramatically consolidate and become vertically integrated after 1998.

The first trend that caught my attention was that, while corn prices doubled or nearly doubled for each time frame, the corresponding prices paid to the farmer also more than doubled until the last time frame of 2006 to 2020. During this period, average pork prices paid to the producer climbed only about 20 percent. What might this tell us? To me, it shows the effects of vertical

Table 8.1 Corn and Hog Price Trends from 1925 through 2020

Year Range	Corn: price per bushel	Hogs: price per pound
1925–44	$0.66 cents/bushel	8.06 cents/lb.
1945–72	$1.28/bushel	18.1 cents/lb.
1973–2005	$2.30/bushel	44.1 cents/lb.
2006–20	$4.13/bushel	54.6 cents/lb.
1925–44	$0.66 cents/bushel	8.06 cents/lb.
1945–72	$1.28/bushel	18.1 cents/lb.
1973–2005	$2.30/bushel	44.1 cents/lb.
2006–20	$4.13/bushel	54.6 cents/lb.

Source: Iowa State University, "Iowa Cash Corn and Soybean Prices (USDA NASS)," Ag Decision Maker, File A2-11, https://www.extension.iastate.edu/agdm/crops/pdf/a2-11.pdf (accessed February 7, 2025); and Iowa State University, "Historical Hog and Lamb Prices," Ag Decision Maker, File B2-10, https://www.extension.iastate.edu/agdm/livestock/html/b2-10.html (accessed February 7, 2025).

integration in the pork industry. Why? Because the supply of hogs has not decreased even though the feed costs have risen. Why? Because vertical integrators do not care if the prices received for hogs go down, as they can raise the price of pork to the consumer to make up for the shortfall they would have otherwise experienced.

I have been raising hogs since 1973. Today we do it organically. With the present market structures, if we could not sell our pork as certified organic, we would be challenged to keep raising pork. We would have to work much harder to sell our pork locally and regionally. Instead of selling 400 head per year as we do now, we would probably be selling far fewer than one hundred head unless we were able to aggressively get more markets. I am worried that this day will come unless organic pork is able to capture more of the overall organic meat market. It does not help that organic pork from Canada comes into the United States at a 25-percent reduction right off the bat because of the high value of the US dollar. Still, though, that is not the critical reason for poor organic sales. In my opinion, it is because conventional pork continues to dominate the traditional market because of low prices. Since the pork industry has become an oligopoly controlled by only a few vertical integrators, it doesn't matter what price they get for the hogs they raise. They can control the industry from start to finish. Independent producers have a very difficult time trying to compete with that oligopoly.

How did it come to this? In the list of prices received for pork since 1925, prices paid to farmers never went over an average of 10.8 cents per pound from 1925 to 1942. That meant that, for a 230-pound market-weight hog, the gross paid to the farmer never reached $25 per hog. In 1942, pork prices were even worse than they were at the height of the Great Depression years of 1932 through 1934. In 1942, a hog was worth only $7.13 to the farmer due to a huge surplus of hogs. The number of hogs reached an all-time high that year and has not been repeated since until December of 2019. That number is over 75 million hogs at any one time.

The total number of hogs marketed annually from about 67,000 pork producers today is over 120 million head. Even though prices to the farmer were low from 1925 to 1942, in some years money could be made because of the low price of corn and other input costs. The price of corn averaged only sixty-six cents per bushel from 1925 to 1944, essentially a twenty-year period. It takes about ten bushels of corn today to raise a market hog from 40 to 260 pounds and is the highest expense in feeding a hog. The next highest expense is a high-protein source, which is nearly always soybean meal. About 25 percent of the feed is soybean meal. The old standard for a market hog in the 1970s and 1980s was a 240-pound finished weight. At that time, feeding hogs over that weight meant that more fat than lean muscle was being put on. Today, with the more extreme lean genetics in hogs, this average weight has increased substantially, going over 275 pounds.

The next time period of interest is from 1925 until 1972. Prices paid to farmers never went above twenty-six cents per pound of live hog until after 1972, the year of the great Russian grain shortage. That meant that a farmer never even received a gross price of at least $60 per hog from 1925 through 1972. That is a long time. The saving grace, again, was the relatively low price of corn from 1945 to 1972. It averaged only $1.28 per bushel and never went over $1.85 per bushel at any time during that time frame. The gross range of hogs was from a low of $30 per hog to nearly $60 per hog.

For the sake of consistency throughout the years, I used an average market weight of 230 pounds for a finished market hog for the entire price chart. From 1925 to 1972, 230 pounds was likely the average weight. After 1972 and through about 1998, the weight first went up to 240 pounds and then to 250 pounds. Since the early 2000s, it has climbed even further, where today the average weight is about 275 pounds.

From 1973 to 1998, things started to change dramatically for pork farmers when there was more money in hogs and a large number of farmers were still raising them. Expansion was on the increase. The adage during those years was that if you could get a gross of $100 per hog, then you should be able to make some money if you were doing a good job in production and management. That generally was the case for many pork producers. The price of corn from 1972 to 2005 averaged $2.30 per bushel. That average low cost of corn and the average higher pork prices paid to the farmer propelled the pork industry forward. Still, enough years occurred in which hog prices dipped under an average price of forty cents per pound, coupled with higher-than-average prices for corn and soy, which significantly reduced profits for less efficient pork farmers.

In 1980, we decided to expand our pork production. We built an eighteen-crate solar-heating farrowing/nursery building. There were many reasons for this. One was that my older brother, Mike, decided to come back to begin farming in addition to being employed as a clinical psychologist. This cut our own farm acreage by one-third, and we decided that we could make up for it by raising more hogs. Another was that the 1980s Farm Crisis, with interest rates of nearly 20 percent, was forcing our hand to do something. It was either find a way to make it farming or get a job in town, too. I vowed not to get a job in town. Compared with other hog farmers, though, we did not expand that much. The most hogs we sold in any one year was around 1,400 head.

For the most part, building the new farrowing/nursery building was a good decision. Especially for the 1980s, when the farm crisis years were in full swing. The lower corn prices during most of those years meant feeding it to hogs made good economic sense, and if we farrowed the pigs ourselves, that was even better because we did not have to buy fairly expensive feeder pigs. It was not the hog market that forced so many farmers to lose their farms over that decade as it was the high interest rates, lower land values, and high debt loads.

From 1980 through 1997, many farmers survived by raising hogs and working in town, too. That is just the way it was, especially for many young farmers who started in the 1980s. If you started in the 1970s, you were a little better off. If you started in the 1960s, you were born lucky. Your land was paid for a long time ago when land values were cheap compared to the 1970s and 1980s. From 1980 through 1997, pork prices paid to the farmer averaged close to forty-seven cents per pound for a 250-pound market hog. That weight was becoming the norm. The average price of corn was a little over $2.40 per bushel. You were getting about $120 gross per hog.

Looking for a Different Path

By the fall of 1998, the bottom was falling out of hog prices. There seemed to be a perfect storm in the making that caused at least some of that dramatic decline. There was an oversupply of pork and decreased packing plant capacity in which to slaughter them. The four plants that closed accounted for a loss of 9 percent of the slaughter capacity. Canadian imports rose by 1 percent or 25,000 hogs per day.

Prices paid to the producer reached as low as seven to eight cents per pound. Prices that low had not been seen since the 1930s, except for the year of 1942. It was ironic that prices in the supermarket did not come down at all during that time. Many pork farmers wondered if packing plants were making a concerted effort to drive smaller producers out of production. It was certainly becoming more evident that the large packers no longer needed to bid on the open market because they had a sufficient supply of hogs through contractual arrangements with large producers. We sold hogs for thirteen cents per pound that fall in Harlan when it still had the pork plant. Two of our boys were in high school and one was in middle school. They were helping greatly with the pork operation. We sold right at 700 head of hogs that year. I decided we needed to have a family meeting. Would we continue to raise hogs? The answer from Maria and the boys was a resounding, "Yes." Hogs did so much for the farm. We would find another way to make it work. Our crop acres were already USDA Certified Organic. The only organic crop we were selling, though, was organic soybeans to the Japanese. Organic corn was just beginning to happen when we began our own beef label called Rosmann Family Farms in 1999. We could not label it as USDA organic as the term "certified organic" did not yet exist for meat products. We also began to start selling organic beef to the organic dairy cooperative based in La Farge, Wisconsin. At that time, it was still better known as CROPP, which stood for Coulee Region Organic Producer Pool. It came to be known as Organic Valley. The coop had decided to get into the organic meat business.

Raising hogs in confinement versus either on pasture or on cement lots with outdoor access is like night and day when it comes to differences in behavior and genetics. Today's confinement hogs are a lot different from what they were forty years ago. Today's hogs are bred to be extremely lean with all the emphasis put on growth and muscling. That results in pork that is watery and pale-looking. Our outdoor hogs have more marbling and flavor in the meat. There is a little more fat, but not too much. It is just enough to help enhance the flavor. Some people say they can taste the "confinement" in hogs that are raised in crowded confinements and use liquid manure pits either directly below the pigs or nearby in a liquid manure storage facility of some kind. These extremely lean genetics can cause a great deal of stress in pigs. Some become very aggressive and nervous, just like we do as humans when we are forced to live in close quarters with many people. I have visited with some of the employees of large confinements, and they usually do not paint a very pretty picture of what it is like. The stress imposed on the hogs can result in hogs often biting workers in the back of their legs, coming from behind to do so. If a sow or a market hog happens to jump over a high wall into a pen of different hogs, that hog is killed almost immediately by that group.

For the most part, our certified organic hogs really are "happy pigs." They are highly social animals. They are intelligent, inquisitive, and friendly, and come up to you to see if you have something good for them to eat or play with. They have plenty of straw bedding, and we have various kinds of toys in their pens to keep them occupied, like hard rubber balls, bowling balls, and big, round bales of hay or straw. Usually the hogs are so tame and gentle that if you suffered a heart attack and fell down in the pen, they would just come up to you and nuzzle you and lay down next to you. Still, I would not like to try it in reality.

Some states, such as Iowa, have passed gag order laws that do not allow anyone other than the employees into a confinement facility. No filming is allowed. Six states have some kind of animal gag legislation. Prior to 2011, similar laws known as ecoterrorism laws were passed to protect industrialized farming operations from trespassers looking to damage property. Today's trespassers are more likely to be concerned about how the animals are being treated. That is what happened recently in Iowa when a person allegedly got a job in a hog confinement operation so that he could film and attempt to document animal welfare abuses. A federal judge in Iowa ruled that individuals may not be singled out for special punishment based on their disfavored viewpoint of agriculture.

Hog profitability greatly declined in 2023. Some industry analysts say it was as bad as it was in 1998 in terms of real dollars lost. This was due to high feed costs, record productivity in sow litter sizes, and other high-production expenses. The question is, will it continue to fuel further consolidation with the possible continued loss of smaller, less efficient operations? I wonder . . .

> There are more than ten times as many pigs in Iowa as there are people. That is fact. It may be fiction but if I were one of those pigs of "Orwellian" fame, I too would be looking for a chance to revolt and get out of my confinement pen. In fact, some already have and are now operating in guerilla warfare fashion as feral pigs. They are plotting as we speak about how to release their comrades before they reach their ultimate fate of becoming "bacon." However, the pigs in Orwell's *Animal Farm* turned out to be worse than the master that the animals originally despised.[13] The escape of 23 million pigs in Iowa should give us pause. Those cunning, ruthless pigs!

In Texas, many pigs apparently have escaped! The number of feral pigs in the state is over 2.6 million!

When I decided to come back to the farm in 1973, the future fate of the pork industry was certainly not the first thing on my mind. There was a more immediate concern, as you will see in the next chapter.

Figure 1 Farmstead, c. early 1950s, photograph courtesy of Ron and Maria Rosmann.

Figure 2 Ray and Ellen Rosmann family with 4 sons, 1962, photograph courtesy of Ron and Maria Rosmann.

Figure 3 Ron and Maria Rosmann with three sons, David, Daniel and Mark, 1988, photograph courtesy of Ron and Maria Rosmann.

Figure 4 Holding feed ration consisting of oats, wheat, barley, and field peas, 2009, photograph courtesy of Ron and Maria Rosmann.

Figure 5 Rosmann cattle and view of farmstead, November, 2024, photograph courtesy of Ron and Maria Rosmann.

Figure 6 Ron and Maria Rosmann family, January, 2025, photo courtesy of Katie Reynolds.

Figure 7 Litters of pigs and sows in hoop building, 2020, photograph courtesy of Ron and Maria Rosmann.

Figure 8 Town of Westphalia, 2014, photograph courtesy of Ron and Maria Rosmann.

Figure 9 Cultivating soybeans, June 2023, photograph courtesy of Ron and Maria Rosmann.

Figure 10 Composting straw and livestock manure, photo courtesy of Ron and Maria Rosmann.

Chapter 9
Good Times, Hard Times, Getting Married, and Beginning a Family

I think the extent to which I have any balance at all, any mental balance, is because of being a farm kid and being raised in those isolated areas.
—JAMES EARL JONES[1]

Having a social life and balancing it with farming was just about the most challenging aspect of coming back to the farm. By 1973, many of my high school friends and classmates were long gone. Some had graduated from college and taken jobs in Omaha or Des Moines. Some had been drafted into the military and were just getting out. Some were in Vietnam. Thankfully, none of them came home in a box. Some were married already, and some were soon getting there. Out of the eleven guys from my class in Westphalia, Iowa, five were in the military. One was in Vietnam. Four of us eventually started farming. One of those had been in the US Navy, one was working at the pork plant in Harlan and farming, and then there was me.

Early on, some of my college friends visited me on the farm. That began to slowly fizzle out as time went on, except for a friend and former college roommate who was in Ames getting his PhD in microbiology. We remain good friends to this day.

When it came to meeting young women, that was more of a challenge. Most of the young women who chose to stick around were married or already divorced. I figured I would have to try to meet someone who, like me, came back single or moved to the area, most likely a teacher or a nurse or someone employed in a profession of some kind. Some college education was a prerequisite. Then there was the possibility I could look for a suitable partner outside of the area and convince her that she would just love to join me in farming and being part of a small rural community. That seemed to be the more daunting possibility. There were many lonely times. Saturday nights were worse, if I was stuck at home with no place to go but one of the local bars in Harlan, where mostly nothing good ever happened.

Luckily for me, my interest in agricultural politics would come to my rescue.

Tom Harkin

In the late summer of 1973, Dad and I attended a Democratic social event in Westphalia at the ballpark. The guest of honor was a young man in his thirties by the name of Tom Harkin. He was an attorney in Ames who had run for the US Congress in our district in 1972 against the Republican incumbent, Bill Scherle. Harkin lost that race, 55 percent to 45 percent. Harkin grew up poor in the small town of Cumming, south of Des Moines. His father had been a coal miner and laborer. His mother, who was an immigrant from Slovenia, died when he was just ten years old. He went on to graduate from Iowa State University (ISU) with a degree in political science and entered the military. He became a Navy pilot, and during his five years of service, flew crippled jet aircraft out of Vietnam.[2] He went on to work for US Representative Neil Smith of Iowa in 1967. In 1970, he went back to South Vietnam on a congressional tour and took some photographs of what were termed "Tiger Cages," where the government held captured North Vietnamese Viet Cong soldiers as political prisoners. The photos created a stir in Congress. He sold those pictures to *Life Magazine*, which helped to pay for law school at Catholic University of America in Washington, DC.[3]

In 1974, Harkin took another shot at upsetting Scherle. He was not the unknown he had been in 1972, plus a bit of a Republican backlash was occurring in southwest Iowa because of Richard Nixon's resignation as president in early August of 1974. Harkin worked extremely hard to upset the incumbent by being a dogged fighter. He also had a brilliant campaign strategy coined "workdays," during which he worked side by side with people in different businesses and occupations to learn about their struggles and needs. In April, the farm got a call asking us to host Tom Harkin for one of the workday events. We said yes. Mom and Dad were still on the farm. They did not move to Harlan until 1975. I wanted to bend his ear on family farming and tax issues.

He got to the farm at 6:30 a.m. on a sunny and warm morning in early May, which impressed me because it was over a two-hour drive from Ames to our farm. He worked side by side with Dad and me for the whole day, listening and sharing what he thought should be done to help family farming and rural communities. I thought, "Wow, here is a guy who really cares."

We hosted a neighborhood get-together that evening for Tom. The neighbors, both Republicans and Democrats, came. Harkin ended up beating Bill Scherle by the slim margin of 51.1 percent to 48.9 percent in November. Representative Harkin became Senator Harkin in 1985 and served until 2015. By the time he ran for the Senate, he had put in over a hundred workdays at different jobs. We were proud to be one of them. Harkin served with distinction as Chair of the Senate Agriculture Committee from 2001 to 2003 and from 2007 to 2009.

A Lucky Man

In early October of 1974, Hubert Humphrey, a former US vice president, came to Council Bluffs for a fundraiser for Harkin. I decided to attend. It turned out a young lady was covering the event for WOW radio in Omaha through a Journalism internship. Her name was Maria Vakulskas, and she was from Sioux City, Iowa. She was a fourth-year Journalism student at Creighton University. She walked up to the bar to get a couple of drinks, and when she walked by us, the drinks were knocked out of her hands by a politician from Omaha. I reached down to help her and said, "he's been drinking, do you want to dance," all in one breath; she said yes, and that is how I met my future wife. It did not happen though the first go-round. After a few dates, we broke it off. She later said she could not see herself living on a farm. It was for the best. As it turned out, neither of us was ready for a lifelong commitment.

Three years later, things would be different. Maria moved back to Omaha from Sioux City to become the assistant public relations director at Creighton University. She had been working at the ABC television affiliate in Sioux City. I was told that someone I knew would be at a party a cousin of mine was throwing, who had, by coincidence, become friends with Maria. However, I had no idea who that person could possibly be. Maria's name never crossed my mind. I thought she was long gone. My teeth dropped to the floor when I walked into the room, and there she was. About eight months later, we became engaged to be married. Previously, I had thought I never would meet the right person to spend a lifetime with. The most beautiful part of our budding relationship was that neither of us ever had a doubt we had met the right partner. I was, and am, a very lucky guy. My prayers had been answered.

Maria has always said getting married was the easy part. Getting used to living on a farm in a small rural community was the hard part. Sometimes you have to just let things roll off your shoulders in small-town living. Some people just will not change long-held behaviors. At one of the numerous housewarmings that neighbors and friends threw for us, Maria sat down with the men who were visiting together in the living room. A middle-aged woman walked into the room, took her by the hand, and ushered her out to the kitchen. "Oh no," she said, "this is where the women are talking." The work of building a life together in farming was starting. We were learning how to be loving partners, which meant give and take and respect for one another. Sometimes the local community needed to get up to speed with the changing roles of women elsewhere. Small-town change is slower than most Saturday nights at the local bar.

Interest in saving energy and finding energy alternatives began to take off during the late 1970s and in 1980. Maria and I were especially interested. We were wondering what we could do on our own farm to become less dependent on oil. Since we were looking into building a new farrowing nursery for our sows and nursery pigs, we wanted to make it as energy-efficient as possible. The Center for Rural Affairs in Nebraska had initiated their Small Farm Energy Project a couple of years previous, and we were studying what they had to offer. They were doing a great deal of work with solar installations for grain drying and passive and active solar systems for homes and hog barns.

The ISU Extension Engineering program came up with a low-cost approach to new swine buildings that would save a lot of energy. It consisted of an active solar system in which proper ventilation for the pigs living there could be pulled through a large solar collector on the southern exposure of the building. We decided to incorporate it into our design. In our case, it consisted of an eight-foot-high wall of concrete block filled with small stones and gravel. The block was painted black and covered with fiberglass. It was ninety feet in length, so there was a lot of square footage to gather the sunlight. There was a gap of four inches between the wall and the fiberglass where the air was pulled through, stored in the block and small gravel to warm up, and then pulled through and out of the building. During the warmer months, when supplemental heat was not needed for little pigs, the air simply bypassed the solar collector. It has proven to be an efficient and cost-effective design. We still use it to this day and estimate we have saved over 50 percent of our liquid petroleum gas needs to keep the building heated for baby pigs in the winter. We were also looking into installing a wind-powered generator to take care of our electrical needs. We decided to put that on the back burner because of the rapidly changing and developing technology of wind generators. Today we are hoping to install solar panel arrays to meet our farm and home electrical needs.

We made the conscious decision not to install liquid manure storage in our new hog barn. I never liked the smell or the potential danger of the poisonous gases that liquid manure creates. Liquid manure storage tanks are typically located directly under the hogs where the solids and urine run through slatted floors where there is no straw or solid bedding for the animal is the typical storage type for hog confinement buildings. It is either that or storing the liquid manure elsewhere in a lagoon or an above-ground storage container. Various gases are given off by the microbes that break down the manure particles. This can be either in the presence of oxygen (aerobic) or without oxygen (anerobic). In anaerobic situations, hydrogen sulfide, a potentially dangerous gas, is released. At very low levels of 15 parts per million (ppm), it smells like rotten eggs. At very high levels (greater than 150 ppm), it has no smell, which can give workers a false sense of security and causes olfactory paralysis. Death occurs by suppression of the respiratory center in the brain, leading to asphyxiation. It is rare, but on occasion, you hear news reports about a power outage in a hog confinement building where the ventilation fans that get rid of the hydrogen sulfide stop running, and a person goes in to see what the problem is, and that can be the end for both the person and the hogs.

All of which means farmers absolutely need a backup generator or a safer way of storing and handling manure. Liquid manure systems are economical and easy ways to store large volumes of manure, but they carry a host of other possible damaging properties, including the creation of other harmful gases such as methane, ammonia, carbon dioxide, and carbon monoxide. Using straw in our farrowing crates required us to hand-clean the pens every other day while farrowing. This was usually a four-week period of time that included birthing and nursing the mother for four weeks. I did that for eighteen years (1980–98) before we changed the way we raised hogs. It was a routine task that gave me time to think. It became a part of chores. It took about an hour every other day.

May 10, 1980

The spoken word was not needed. We worked in the field together in silence as the sun worked its way across the sky. The farmer and the *sun* worked together as one. The farmer and the *son* worked together as one. Why did he have to die?

I took the spade out of Dad's hands the morning before. We were starting to do the soil grading work for the new hog barn in May of 1980. I had most of the corn planted, except for one field that badly needed rain because it was dry and cloddy. We were setting a fence post that had broken off in one of the small hog yards next to the original farrowing house my dad had built in the early 1940s. I was always taking a "spade out of his hands" or something like that, thinking maybe it would mean I could have another day with Dad. Maybe it was silly, but I was afraid.

Dad died on Saturday, May 10, 1980, in the early evening. I had continually been fearful of when that day might happen. Dad found out he had severe coronary artery blockage in the winter of 1974. That was the first time a heart catheterization was called for, which was prompted by a heart EKG indicating something was wrong. The heart doctors at Creighton Medical Center in Omaha, Nebraska, decided it was too risky for him to be a candidate for a coronary bypass operation. He had already had significant heart damage during what were called previous "silent" heart attacks. I had never seen Dad looking like he was depressed until that winter. The depression lasted until about spring, as I recall. I think his statement to me that you should go outside and do some work if you are depressed kicked in for him. I think it was also his faith and his growing acceptance of when it was his time, he was ready. I think he was always ready. His quiet, steadfast strength sustained him.

When Maria and I went to the hospital the morning of May 10, Dad was still conscious. I told him I loved him, and he told me he loved me too. I did not remember us ever saying those words to one another before. They did not have to be said. It was understood. He looked at Maria and said, "I love you girls too." Then I remember him saying he did not think he was going to get out of there this time. I could hardly hold back the tears. Then after a bit of silence, he said his last words to me, "It's time for you to go back home and get to work." I squeezed his hand. Maria and I left. He would pass that evening. If nothing else, I can say one thing about my life. I listened to my dad. I went home and got to work. About two weeks later, I was walking down to the site where we were starting to build the new farrowing house. It was close to the same place where I took the spade out of Dad's hands the day before he died. A feeling of deep peace came over me. I had just received a message from Dad. The message was everything was okay and he was at peace. I had never received what I thought was a message from heaven before.

Dad and I had seven years during which we got to know each other a little better as adults. What I think I valued the most about our relationship was that, without even saying it to one another, when it came to farming, we each knew the next task that needed to be done. We understood intuitively the language of farming without ever uttering the words. That had to have come from so many years of working side by side. I know I am guilty of putting him on a pedestal. But I don't care; he earned it. He was a gentle, patient, and kind man.

How do you measure a man like that? My brother Joe wrote a beautiful meditation for the funeral of Ray Rosmann on May 13, 1980.

> Kbang, sha . . . Kbang,sha . . . giddyup Pat, giddyup Maude, Whoa
> Kbang, sha . . . Kbang,sha. . . giddyup Dan, gee Bob, haw Bert, Whoa

> If you listen carefully, it is December
> A little more than three score and twelve years ago.
> The wind rustles through the parched cornstalks.
> Kbang, sha . . . kbang, sha . . . the ears of corn make
> As they pop against the bang board, so rightly named
> And then fall into the wagon with a thud.
> Kbang, sha . . . kbang, sha . . . kbang, sha . . .

> A good man, they say, picking steadily, Could fill a wagon,
> More than a hundred bushels in a day—Of the fruits of his year-long labors—
> Listen, softly, quietly

> Kbang, sha . . . kbang, sha . . . Giddyup Bill, Giddyup Sally, Whoa
> Every man had his favorite team to pull his wagon.
> Never too fast, never too slow—
> Knew when to stop and when to go.
> Gee, Haw, Which way to turn, how hard to pull.

> Listen, kbang, sha . . .kbang, sha . . .
> A little boy was born that December,

To begin his lifelong task of filling his wagon. He had his favorite team too—Jim and King—
To pull his wagon.
Giddyup Jim, Whoa King, Gee Jim, Haw King—
Giddyup, giddyup.

How do you measure a man's life?
How full his wagon?
How hard his work?
How good his harvest?—

Kbang, sha . . . Kbang, sha . . .
This little boy set out early to fill his wagon
We're getting a start—
With love for his family, and love for hard work.
Kbang, sha . . . Kbang, sha . . .

And as a young man his wagon got fuller,
With singing for the Lord, With the choir in his church. Giddyup Jim, giddyup King,
We're getting there boys! Kbang, sha . . . Kbang, sha . . .

And filling it more, with loving care for mom and his sons,
And with help for his neighbors, and care for his farm.
Gee Jim, Haw King, Kbang, sha . . . Kbang, sha

And fuller still the wagon was getting ,
With help for the community,
And those who needed it, humbly given.
Gee Jim, Haw King, We're just about there boys—
Kbang, sha . . . kbang, sha . . .

And in the fall of his life,
When his labors were done and his family grown,
There was more time to praise God,
And to pray as he wanted. And the harvest continued---kbang, sha . . . Kbang, sha . . .
kbang, sha . . .

Whoa Jim, Whoa King, I'm tired, I've got to rest
Listen, Listen,
To the stalks as they rustle and the wind as it wheels.

Giddyup Jim, Giddyup King.
Let's go home boys, my harvest is done.[4]

The Looming Farm Crisis

I had thought that Maria and I were on our own before Dad passed, but that feeling did not really sink in until after his death. I could not turn to him any longer for his advice and knowledge. There were challenges on the horizon that had never been experienced before. The first was the beginning of the so-called Farm Crisis years of the 1980s and the ensuing debt that it caused. I never had to worry much about going into debt or how to pay the bills until those years. We had to borrow nearly all of the money to have a contractor build the new hog farrowing and nursery building. As reality set in and as I gained experience in doing my own building projects over the following years, I realized I should have built the structure myself. At the time, I did not have that experience or confidence. Afterward, I had to learn the hard way by just jumping in and doing it.

Mom

Dad's passing had the biggest effect on Mom. It was very tough for her. She was twelve years younger than him and just sixty years old. She leaned on him for so much. Each of them made the other a more secure, happy, and contented person. They were a good match. She lost some of that security but still had over a third of her life to live. She struggled with it for the rest of her life. She was lonely and missed having companionship. She found some of that companionship in the volunteer organizational work she did and in some of her travels with other ladies to historical sites around Iowa. Still, much was missing. Losing the love of her life created a void that would never be filled again. She lived to be ninety-one years old and passed in 2010. Just like with Dad, I was not there when she died. I went to visit her in the hospital the evening before she died. We talked and prayed together. She was at peace. We all thought she might get better and get out of the hospital. She died early in the morning of the next day. She was a very intelligent, alert, and civic-minded person. She was a country schoolteacher at heart.

Bishop Dingman

Bishop Maurice J. Dingman served as Bishop of the Catholic Diocese of Des Moines from 1968 to 1986. Our parish is in that diocese. He and a farmer named Joe Hays were successful in their invitation to Pope John Paul II to visit Iowa in 1979. My wife, Maria, was chosen in a speaking competition to be the lector for the papal mass. Bishop Dingman worked tirelessly against the removal of farmers from their land during the farm crisis years of the 1980s, a decade associated with a devastating bust in agriculture. In the winter of 1986, he gave a speech at an annual conference series called "Religious Ethics and Technological Change" at ISU. In it, he talked about a government study from 1962, noting that even though the exodus from agriculture in the previous decade had been large, it had not been large enough. The title of the study was *An Adaptive Program for Agriculture*. It was conducted by the Research and Policy Committee of the Committee for Economic Development.[5] Again, large corporations were on the membership list. They included Ford Motors, Standard Oil, AT&T, H. J. Heinz, General Motors, and IBM. Bishop Dingman called it an embarrassment for agribusiness. This approach made it very clear to him the values, lives, and communities of family farmers were of less importance than the technological innovation and opportunities for capital investment on the part of corporations. The family farmer was forced to adapt to the needs of the corporate-dominated economy that cannot survive without constant growth, sales, and profits. This meant some family farmers had to leave their farms. He called it the beginning of the cheap food and labor movement, where mass production of cheap food would be produced by and for people earning cheap wages so that production and consumption could both be high. I was at that conference. In less than two months, Bishop Dingman suffered a debilitating stroke. He passed in 1992 and remains a prophet and a hero to me.

The borrower is slave to the lender.

—PROVERBS 22:7, NEW AMERICAN BIBLE

The extremely high interest rates during those early years of the 1980s are what dealt the death blow for many farmers. You could not expect a rate of return to match a borrowing interest rate of 18 percent. So, if farmers had taken on too much debt, it could mushroom quickly and soon get out of control. That is, if they had borrowed money to make interest and principal payments, they were going down the wrong road.

It took me more years of experience in farming to realize the word debt is not such a terrible thing in farming if the debt load is manageable. Agricultural tax laws reward some debt or at least make it seem so. By that I mean, every farmer's goal is to maximize both income and expenses so that unreasonable amounts of income tax are not paid in the current year. Interest payments on borrowed money can be deducted. By deferring income from the current year to the following year and by making purchases and paying some of next year's expenses in December, farmers can decrease their taxes and continually defer taxes to the following year. That all sounds good. However, there are a couple of rubs. The wealthiest of farmers may buy new combines, tractors, and pickups at the end of the year for that reason. Smaller farmers like my family might pay for some of next year's seed or soybean protein needs, for instance, and that is about it. And we have to borrow the money in order to do so. Since we are self-employed, we have to pay Social Security taxes on ourselves to ensure we can have at least a somewhat more secure retirement income. Even smaller family farmers must rely on someone paying rent or making land payments if they are to have a livable retirement, and that is only if they are lucky enough to own some land. Only a very few have other retirement incomes, such as, 401Ks or stock portfolios.

Something is not quite right with an economic system in which farmers make a little additional money in farming and end up having to borrow the money to pay their taxes. The wealthiest farmers can make new purchases every year, limit their tax implications, and have plenty of money left over to pay their taxes without getting a loan. Then there is the fact that the largest farms are the recipients of the largest government subsidies. The very largest farms can use those subsidies to leverage paying rent to farm more ground or use it to buy prepaid expenses like chemicals, seed, and fertilizer for the next year. It keeps the chemical, seed, and equipment dealers happy for sure. It is a game many farmers have to play.

Our Three Sons

We started our family in 1981. Our first son, David Raymond, was born on August 20 of that year. Maria and I had attended Lamaze classes all that spring and summer, learning about how to reduce the stress of delivery through rhythmic breathing exercises. Unfortunately, after eighteen hours of labor, nothing was happening, so Maria had to have an emergency C-section. Everything went fine. Mom and the new son came home. Three days later, Maria became very sick with pancreatitis and had to have her gall bladder removed. Then a day or two later came the much scarier news she had a pulmonary embolism. Her doctor told her to make her peace with her family because she might not make it. I do not think either Maria and I quite understood the gravity of the situation. We were young. The fact she was young and strong and never smoked got her through it. She did not get to be with our baby for a month. In the meantime, Maria's mom moved in with us, and we carried on as best we could.

Maria's parents became like second parents to me. Mary Vakulskas was born Mary Smith in Sioux City in 1922. Her father and mother were both immigrants from Poland. Her last name should have been "Smet," but the immigration authority at Ellis Island looked at the name and said it would be Smith. Mary was the oldest of three surviving children. She grew up having to work in the Sioux City packing house business cutting meat to help support her family and could not finish high school. She was a wonderful, smart, and hard-working woman. She loved coming to the farm. The joke became that if she and I were married, we would be farming 10,000 acres. Her husband, John, was in the Pacific during the Second World War. They were living in Portsmouth, Virginia, where John was in the US Navy working at the naval shipyard when he was drafted by the US Army for duty in the war. He worked in the boiler room of a ship that made repairs on damaged destroyers. These ships were not far from the front and faced most of the same risks as the destroyers did. Like so many veterans, he rarely talked about the war. His nightmares about it stayed with him for his entire life. After the war, they moved home to Sioux City where he worked the rest of his career in the US Postal Service. After that, he and Mary came to the farm as much as they could to help Maria and me and to be with our three sons. John died in 1999, and Mary died in 2014.

A little more than two years later, on September 20, 1983, our second son, Daniel John, was born. Thankfully, there were no complications for mom or baby this time. One of the things I remember the most was he was just three months old when the brutal winter of 1983 occurred, creating numerous wind chills ranging from fifty to seventy degrees below zero. There was a great deal of snow too. We spent Christmas that year snowed in. Even the Christmas Eve Mass had to be cancelled because of the snow and cold. We were fortunate we did not lose power during that time, although we did have an emergency back-up if needed, a tractor-driven generator.

Our last child was born on February 10, 1986, and we were blessed with our third son. We named him Mark Andrew. It has been historically difficult for the Rosmanns to have a girl. It took our family seventy-three years. We are now blessed with six grandchildren, five of whom are living. We have three granddaughters and two grandsons.

Up until our first son, David, was born, Maria had been working at our local hospital as the public relations director. She gave that up to become a full-time mother and continue her role as farm helper, bookkeeper, chef, spouse, church pianist, and part-time caregiver for my brother Larry. She still thought she was not doing enough.

The farm crisis of the 1980s meant many hard times for so many farmers, including us. Still, maybe it was a blessing for us. We were forced to watch where and how we spent money, and it made us appreciate all the blessings we had with a good marriage, healthy children, and a growing care and concern for the stewardship of the land. Little did we know that what would happen next would change our lives forever, and in a good way, too.

PART III

Restoring the Promise

Chapter 10
The Real Journey Begins

The soil is the great connector of lives, the source and destination of all. It is the healer and restorer and resurrector by which disease passes into health, age into youth, death into life. Without proper care for it, we can have no community, because without proper care for it we can have no life.
—**WENDELL BERRY,** *THE UNSETTLING OF AMERICA*.[1]

In this chapter, my journey to a new way of farming begins, in April 1982. Discussed is the formation of our organization, Practical Farmers of Iowa (PFI) and its contributions. It examines the philosophy of on-farm research, as well as the legal battles surrounding the use of the pesticides, glyphosate and dicamba. Further examined are the use of cultivation and tillage techniques and crop rotation instead of chemicals for weed control. Another technique discussed on our farm is the use of compost and its benefits, as well as a discussion on fertilizers. This chapter also highlights benefits of various methods during periods of drought conditions, which continue to increase as the planet warms. Also provided is our journey to becoming a certified organic farm for both crops and livestock.

The Thompson School of On-Farm Research

It was April of 1982. My farming career was about to take a turn that would affect us far into a future whose outcome we did not yet know. I read a magazine article about a couple who farmed near Boone, Iowa. Their names were Dick and Sharon Thompson. They had a 300-acre grain and livestock farm and had quit the use of pesticides in 1968. What caught my attention, though, was the feature described a different approach to getting answers to the questions involving raising crops. The approach was called doing on-farm research trials. The story also contained a description of something called ridge tillage, which is a type of no-till planting where the soil is not stirred up before the crop is planted. I knew virtually nothing about ridge tillage.

The magazine article stated the Thompsons had purchased a Buffalo till planter and cultivator as early as 1965. They were on a continuous corn program then and thought they could reduce both the tillage and the soil erosion occurring on their farm with continuous corn, chemical fertilizers, herbicides, and insecticides.

Dick and Sharon Thompson decided to start using organic methods of farming in 1968, especially by eliminating pesticides. Why? Dick said they were on a treadmill and had hit a brick wall. Their cattle were sick. Their hogs were sick. Antibiotic usage to treat their sick animals was on the rise. Weed control using herbicides was not as effective as they thought it should or would be. Dick was not feeling well. Something had to give. So they had an awakening of sorts where Dick decided there had to be a different and better way. But to find that way, he needed to first ask the right questions and second to become informed in order to seek accurate answers. Dick had a Master of Animal Science degree from Iowa State University (ISU) in Ames, which was located just ten miles east of their farm. He understood the scientific method necessary to begin his quest.

By 1979, Dick was wrestling with this new idea of how to set up plots on his farm so he could become informed and get accurate answers to his many farming questions. The small ninety-feet-by-forty-feet trials that ISU Experimental Stations were using did not seem practical to him. So in 1979, he made his first stab at a suitable experimental design. Finally, he found the right one for doing practical on-farm research. It had a plot width of eight rows. He had a four-row planter, so if he made a trip across the entire length of any particular field and came back to where he started, he had completed what is called a round. He replicated the trial six times, and it was randomized.

Narrowness of the experimental design lends itself to less variability. Going across the entire length of a particular field captures variability if there are differences across the different plots. The randomization and replications satisfy the statistical requirements and the necessity for making logical inferences about what is observed. It is a convincing and practical method and enables farmers to get reliable answers with just about any crop. This design has also proven over time to have low coefficients of variability, which are indicators of statistical reliability. Thus, it is a useful tool for doing practical on-farm research.

In July of 1982, three other local farmers jumped into my car, and we traveled to visit Dick and Sharon Thompson's farm. They had fewer weeds in their corn and bean fields without the use of herbicides than we did with the use of them! I was astounded. I decided right then and there if he could do it, then we could, too. We also visited an organic farm near Carroll, Iowa, about fifty miles from us. It was a beautiful farm. They were not doing on-farm research trials, but what they were doing looked like it was working. Salespeople at the tour were promoting a fertilizer product for farmers to use if they wanted to become organic. It was a mined product that was black in color like coal, which is kind of what it was. It was high in organic matter, and I am sure it made a good fertilizer. I was very skeptical, though, because they seemed to be promoting it as a "miracle cure." That part did not appeal to me.

I went home and started reading what I could find about farming without pesticides and the usual chemical fertilizers. Sadly, there was not much out there to read, and it was easy to become confused.

That did not deter me. I decided to go cold turkey on the whole 320 acres when it came to the use of herbicides and insecticides for the 1983 crop year. Everything we did revolved around that central idea. We jumped right in.

Shortly after our visit to the Thompson's farm in 1982, I started phoning Dick from time to time to get his advice on the many questions that came up as I tried to figure out how to make it all work. Thompson always graciously gave of his time through the years. He became my mentor in so many ways. Thompson's ideas about how to do on-farm research caught the attention of the Rodale Institute in Kutztown, Pennsylvania, through their magazine, *New Farm*. It was specifically for traditional-sized family farms around the country. These were the roughly 160- to 500-acre farms that still made up so many of the farms over the country. J. I. Rodale founded the Rodale Institute in 1947 as a nonprofit organization to help promote regenerative organic gardening and farming methods. The word "regenerative" is important because it refers to the use of natural farming systems, when managed properly with reliance on internal regenerative resources, can improve soil fertility and productivity over time. The concept emphasizes decreasing reliance on potentially environmentally dangerous external chemical inputs.[2]

Rodale's son, Robert, took over the Rodale Institute in 1971 and purchased a larger 333-acre farm on which he started doing research. He initiated a relationship with Dick and Sharon, and by 1984, Dick and Sharon developed a cooperative relationship to host what Rodale called "The Thompson Farm-Nature's Ag School." I attended the first tour in 1984. Six hundred people from all over the country were there. The on-farm research trials were part of the biggest draw for farmers.

The *New Farm* magazine polled farmers about which kind of agricultural research they would have the most confidence in. Was it small university trials, private company research, or on-farm research? The overwhelming majority stated it was on-farm research.

I got to know Robert Rodale after beginning this kind of farming. A highlight for me was hearing him speak in Omaha, Nebraska. He talked about how his great-grandfather knew Abraham Lincoln. He stated he thought it an honor to have shaken the hand of the man (his father) who shook the hand of the man (his grandfather) who shook the hand of the man (his great-grandfather) who knew Abraham Lincoln. I thought that was very cool. I shook his hand that day so I could get in on that connection. I came home that night and shook my wife's hand and our three sons as well. Sadly, he was killed in a traffic accident on the way to an airport in Moscow, Russia, in 1990. He was there to launch a Russian version of the *New Farm* magazine.

After what was happening to agriculture in the decade of the 1970s with consolidation, increased use of pesticides and fertilizers, and increased pollution of water in our nation's streams and rivers, US secretary of agriculture Bob Bergland requested that a study on the state of organic agriculture be conducted. The results of the study were published in July of 1980 during the administration of President Jimmy Carter.[3] Bergland was a farmer in Minnesota before entering politics, and one of his farming neighbors was a successful organic farmer. So when the scientist who was head of the US Department of Agriculture (USDA) Science and Education agency wanted to learn more about the state of organic farming, Bergland was all for it. It was a comprehensive report. One part of the study indicated farms producing organic corn, soy, and oats did better than conventional farms, especially in periods of drought. The study recommended that a USDA coordinator for organic farming be named. That did not happen.[4]

In 1971, US secretary of agriculture Earl Butz, who was well known for making controversial statements, had noted if the country went back to organic agriculture, the country would have to "decide which 50 million we are going to let starve or go hungry."[5] In 1971, there was probably a good deal of truth in that comment. At that time, most farmers nationwide were using pesticides and chemical fertilizers. It may have been a political statement as much as anything, possibly referring to the "hippie farmers" on the West Coast who had a few acres and were going "back to the land." However, by the early 1980s, many established and thoughtful farmers were starting to question the downsides of dependence on so many pesticides and overapplications of chemical fertilizers, such as anhydrous ammonia used especially as a nitrogen source for growing corn. Excess nitrogen was showing up in our drinking water.

Practical Farmers of Iowa

In the summer of 1985, the Fisheries and Wildlife Department at ISU held a meeting for farmers who were interested in the environmental effects of chemicals and fertilizers and runoff into Iowa's fishing streams, and also for those concerned about the overall lessening of water quality. It was ironic to me that this department was the first one to begin raising red flags about the possible damage coming from overdependence on pesticides and synthetic fertilizers—and not one of the agricultural departments. ISU Extension personnel were in attendance. They were grilled by farmers who were concerned about some of the things happening on their farms and asked whether ISU would do research on alternative farming methods. Dick Thompson was at the meeting. At one point, Thompson asked farmers if they would be interested in starting an organization in Iowa dedicated to finding out what was really going on with their farms. Arms shot up all across the room. There must have been one hundred farmers present.

In early February of 1986, I was one of a handful of farmers invited to attend a meeting at ISU in Ames. It was hosted by Dick Thompson and Larry Kallem who was then the executive director of the Iowa Institute of Cooperatives (IIC), which works with farm cooperatives of all kinds throughout the state of Iowa. It resulted in being the founding board meeting for a new farm organization in Iowa.

It would be known as Practical Farmers of Iowa (PFI). I liked that name from the start. It described very well what we all thought farmers should be. Our mission was to be profitable while doing less harm to the environment. How would that be accomplished? It would be through conducting on-farm research trials that were scientifically legitimate where farmers could learn from one another.

We partnered with ISU researchers wherever and whenever possible to dig deeper into farming research questions. Early on, we decided we wanted to "grow people," too, not just crops and livestock. That is, new leaders for Iowa agriculture could also be developed within this group. We would conduct field trials, hold field days, and publish the results of the trials. We would try to be better neighbors and celebrate being members of our rural communities. We started with about one hundred members. Dick Thompson was selected as president, and I had the honor of being selected as vice president. PFI's 2024 membership totals over 9,000. We have become an organization where there is both realism and hope for the future. As far as we know, we have more young farmers in our group than any other such group, anywhere. David, our oldest son, is the first second-generation board member and was named board president for 2024.

One of the most enduring characteristics of PFI is its insistence on listing both spouses' names when referring to family farms, either when holding field days or in general reference. This means a great deal because these farms are not just a "man's farm"; they are farms where both spouses are partners. This concept ran counter to what most other farm organizations did back in the 1980s and continues to be the case today.

The legacy of Dick Thompson lives on, even though Dick passed in 2013. He and Sharon hosted over 41,000 visitors to their farm over a span of thirty years. That is an incredible number of people. He never grew tired of asking the right questions and then looking for the answers through on-farm research trials. He was a father figure and teacher to me and to so many farmers in PFI when we were all much younger.

PFI slowly began to identify some of the core principles and components it would take to be successful in our mission of being profitable and environmentally conscious, too. An important governing principle we have held ourselves to is we have nothing to hide. My family has held at least a hundred or more farm tours on our farm since 1987. I tell everyone at the beginning of the tour what you see is what you get. You get to see the good and the maybe not so good. There is nothing to hide. There is no product to sell. I suppose we are attempting to sell something, though, that is very important: respect for nature and for one another.

Walking the Walk, Not Just Talking the Talk

You can talk a problem to death and never solve it. You must be willing to jump right into the fray and learn from doing and from making mistakes and to keep an open mind while you are at it. That is what we did. I was never really afraid of quitting the use of pesticides, but I was cautious. In 1983, when we stopped the use of pesticides, we were farming 320 acres. We had about fifty stock cows and were selling the calves as weaned feeder cattle at a livestock auction to someone else who would feed them out. Much of our corn was going to feeding hogs after we built the swine farrowing house in 1980.

We were prudent not to give up the longer rotations, which included hay, pasture, and oats, and plant just corn and soybeans in the early 1970s. We already had a leg up in terms of some of the biological and natural or organic ways of controlling weeds, insects, and diseases and of creating at least some of our fertilizer through the use of our own legumes and manure. We bought a tractor-driven compost turner that first year. My mentor, Dick Thompson, was skeptical. He had purchased one in his earlier days, only to sell it later because he thought he lost too much nitrogen through volatilization at the pile and loss of potassium from leaching into the soil at the composting site. He was correct on both counts, but there are some good reasons to compost, and we are glad that we stuck with it. We are creating a stable source of nutrients for both crops and the plethora of microorganisms in the soil. We are killing the weed seeds and pathogenic microorganisms with the high temperatures gained through turning the manure. I offer these two guiding principles,

> I believe strongly that there are two guiding principles to making a sustainable conventional or organic system work. The first is having some kind of livestock on your farm. The second is to have a diversified crop rotation that includes more than the predominant two crops of corn and soybeans.

Corn and soybeans are grown on well over 50 percent of all US crop acres. Nearly 40 percent of the corn crop goes for ethanol production. In Iowa, the percentage of corn used for ethanol production increased to 62 percent in 2023.[6] I have always thought farmers should be growing food, not fuel for their vehicles, or at least not in the amounts we are today. There really is no limit to the number of rotational crops if you can make them all work financially and improve the quality of your soil, too. Having ruminant livestock that can break down cellulose and lignin and graze on forage consisting of various grasses and legumes and other perennials may be the best kind of livestock because of their ability to recycle nutrients on the farm. That is not to say pigs and chickens, ducks, geese, etcetera, cannot do their part, too. Having livestock on your farm improves your self-sufficiency and resiliency in so many ways economically, for soil-quality and resource-wise, and in terms of providing more ecosystem services.

> Never does Nature separate the animal and vegetable worlds. This is a mistake she cannot endure, and of all the errors which modern agriculture has committed, this abandonment of mixed husbandry has been the most fatal.
>
> —SIR ALBERT HOWARD[7]

Sir Albert Howard was a British botanist who spent much of the first four decades of the twentieth century working and teaching in India. He consistently observed the value of compost applications in increasing the general health of plants and of the whole system of plants and animals. He was the first Westerner to document and to publish the ancient Indian techniques of organic agriculture.

One of the biggest hurdles in today's US agriculture is the ability of individual small and medium-sized farmers to sell their livestock and the products they produce. The traditional market structure of meat packers buying small numbers of livestock from many producers has all but disappeared. Throughout this book, I have documented the steady increase of pork, poultry, and beef being raised in ever-larger numbers by ever-fewer producers. This situation has resulted in one of the largest detrimental effects for diversified crop and livestock farms, which, in my estimation, were the backbone of this country until fifty years ago in the latter 1970s.

Livestock and longer diversified crop rotations have taken on even greater significance on our farm because of climate change. In terms of where we have been, there was little if no emphasis on such rotations in the early days of our alternative and then later certified organic processes. There has been much "where we have been and where we are going" thinking and talking throughout my fifty years of farming.

Until about 2010, thinking or talking about climate change was not one of them. Now, I tend to think about climate change and what we can do to slow it on a daily basis. New insights and information on how to lower our carbon footprint are popping up every day. How a particular practice or technology affects climate change is not easy to figure out, what is accurate, sound, and unbiased versus what is misleading and more inconsequential. What crops are raised and what and where livestock are raised are having a huge impact on our climate, a topic examined more extensively in later chapters.

Where We Have Been and Where We Are Going—Weeds and Pesticides

I learned how to control most weeds through cultivation when I was younger, at least that is what I thought. After forty-one years of no pesticides and now with more weather extremes and some tougher weeds than ever before, I have to be honest and say good weed control continues to be one of the biggest challenges. Even though we were using most of the herbicides that were on the market at the time of our conventional farming years up until the early 1980s, we still cultivated our corn and soybeans at least once and usually twice. We were used to using four-row cultivators that were attached to the front sides of a tractor so you could easily see where you were going. I wanted to try the rear-mounted Buffalo cultivator that Dick Thompson had, but I was afraid at first about how much corn and soybeans I would be taking out by having to look back to see what I was doing. We bought one in 1983. We still use it today. We have four more rear-mounted Buffalo cultivators. Installing large mirrors on each side of the tractor helps to keep you from looking back. Having an electronic or hydraulic guidance system helps to keep you straight on the contours of our more hilly slopes. Cameras that keep you with more precision on the row may be on our to-do list fairly soon. GPS tracking systems may also be incorporated. The cheapest and easiest-to-use method of staying on the row is the mounted mirror.

There were multiple reasons for us to justify stopping the use of all pesticides. They were costly and the precarious finances of the early 1980s justified it in my mind. They did not work all the time, and I did not like handling them. Pesticides represented to me the kind of control over independent family farmers like ours that I detested. I wanted to prove to myself, to other farmers, and to the local chemical dealers that I could do just fine without them. To me, it seemed plausible that a chemical designed to kill unwanted plants could not be entirely benign for other forms of life like bacteria, fungi, earthworms, not to mention people.

Weeds are probably the number one concern for farmers transitioning to become organic farmers and also for long-established organic farmers, especially over time. Below I share some knowledge I have learned about weeds along the way.

What is your strategy for weed control? This can be a step-by-step process to first significantly reduce herbicide use by only banding them over a row and then by using mechanical cultivation so that, as experience is gained, herbicides can be completely eliminated. What I mean by banding herbicides is that the spray pattern is only over the row, not over the entire width of the sprayer. Because of this, less herbicide is used and the weeds between the rows can be removed by cultivation.

This is for farmers in the early stages of transitioning. Unfortunately, most of the equipment used for the banding of chemicals in the 1980s and 1990s are sitting behind farmers' sheds or has been sent to the scrap metal yard. Once you make the commitment to become certified organic, you must not use pesticides for two complete crop-growing seasons before you can seek organic certification for a crop.

Whether organic or not, consider managing weeds as opposed to entirely eliminating them. I am continually amazed at how many weeds conventional farmers have who have been applying herbicides every year on the same fields for well over a half-century. You would think that, after seeing a field virtually free of weeds, no weed seeds would be left. Many weeds may still be there, however. This can easily be seen in fields where a sprayer shut off too soon or where a gap exists in the spray boom coverage resulting in some part of a field not getting sprayed. Herbicide-resistant weeds show up as the season progresses and may produce new seeds, which also must be considered.

Some farmers insist on having perfectly clean fields so that coffee-shop talk does not become a problem for them if they do not. I would ask: why worry about perfectly clean fields? Why not worry about effective control instead of total elimination? Yield reductions and net return reductions due to weeds are what farmers should be concerned about. In the early years of PFI, many farmers compared the practice of banding pesticides just over the row and cultivating in between. As farmers gained more confidence, they began to have plots where they compared managed use of herbicides with completely eliminating them. The results gave numerous farmers the courage to eliminate herbicides entirely.

One safe and nontoxic technology being pursued for controlling weeds revolves around the idea of keeping weeds from reproducing via seeds. What this means is decreasing the weed seed bank over time. A company has come out with a weed seed harvesting compartment that can be attached to a combine; it collects and pulverizes weed seed as corn or soy or other crops are harvested. This approach has shown to be highly effective on weed seeds that are not dispersed before a crop is harvested, which could be a breakthrough for both conventional and organic farmers. However, it is an expensive machine at the present time.

If I appear to be presenting a good deal of minutia about controlling weeds and other details on how to make farming without pesticides work. . . well, I am, and for important reasons: (1) Attention to details is what makes organic agriculture work. (2) You do not have pesticides to cover up your mistakes. (3) You need to put together a list or system of practices that generally have to follow one another. (4) This will not happen during one crop-growing season. (5) You will discover that there are many different "tools in the toolbox" that can be used to aid you in your journey. An important reminder:

> Remember, weeds will become resistant to herbicides over time. I am a strong proponent of this notion: the more you spray, the more you have to spray.

In 1996, Senator Richard Lugar (R-IN), who was then head of the US Senate Agriculture Committee, invited me to testify about what should be the role of agricultural research in the next farm bill. One of the other invited panel members worked for Monsanto in research. Usually panelists do not get to ask other panelists questions during a hearing; that is strictly for members of the committee. I asked for and got permission from Senator Lugar to ask the gentleman from Monsanto a question: how long he thought it would take for weeds to start to show resistance to the herbicide *glyphosate*, better known as Roundup. The gentleman's response was "not for at least another twenty years." That would have been 2016. The first glyphosate-resistant weed was documented in 1996 in Australia. It was an annual rye grass. By 2012, the Weed Science Society of America (WSSA) reported that 14 million acres in the United States were infested with glyphosate-resistant weeds.[8]

Farmers have lived off the success of Roundup for quite some time. When soybeans were genetically modified so Roundup would not kill them, farmers thought they had found the silver bullet for weed control. It was cheap and 100 percent effective with usually only one pass. That was in 1996. The gene was inserted into corn in 1998. That method changed some of the dynamics of farming. Roundup was both cheap and effective; its use opened the way to farms increasing in size, which continues to this day. No-till practices increased, meaning you do not till or disturb the soil at all for either seedbed preparation or cultivation. Many farmers also quit applying their own spray because now it was both safer and more economical to have the coop or the crop services company do it for you.

Some weed scientists think the era of chemical herbicides is coming to an end. Weed resistance is becoming a bigger issue. Rotation of different types of weed killers slows down the resistance problem considerably, but for how long is a legitimate question.

Glyphosate and Cancer

Recent research results on the cancer-causing dangers of glyphosate are raising new fears about the safety of glyphosate. Glyphosate is a long-suspected carcinogen, and in 2015, the World Health Organization's (WHO) International Agency for Research on Cancer gave it that status. Monsanto did a big push back, citing that the US Environmental Protection Agency (EPA) determined that it was safe to use and did not cause cancers. The *Journal of the National Cancer Institute* in January of 2023 published a study conducted by the National Institutes of Health that is probably the most damning report yet for glyphosate. The study noted that high levels of glyphosate found in biomarkers in the urine of both farmers and non-farmers alike were linked to hematologic cancers that include lymphomas, myeloma, and leukemia.[9]

In 2022, the US Centers for Disease Control and Prevention (CDC) detected glyphosate in the urine of 80 percent of a large sample size of over 2,300 children and adults.[10] In a 2023 study, oxidative stress was found to occur in the DNA of the participants. Oxidative stress has long been associated with causing cancer.[11] The EPA is still making the assertion that Roundup is safe, although an appellate court ruling forced the agency to halt that claim. Since Bayer purchased Monsanto in 2018, it has lost 40 percent of its market value due to having to pay very large claims in 125,000 suits brought against the company. Investors are not happy with what is going on, and the future of glyphosate is uncertain. Apparently, Bayer is planning on phasing out the lawn and garden usage of Roundup over the next few years. That is not their intent, however, for crop production.[12]

During the winter of 2024, numerous state legislatures introduced bills that would eliminate liability for chemical companies in cases involving defendant claims of severe cancer-causing illnesses from the chemical Roundup as well as any other crop pesticide. Bayer, the chemical company that produces Roundup, proposed the bill in Iowa, Idaho, Missouri, and Florida. It claims pesticide makers cannot be held liable for failing to alert people of possible health risks as long as their products have a federally approved label. The bill passed the Iowa Senate on April 4, 2024, but the Iowa House failed to take it up. I suspect it will appear again during the next session. Iowa has the second highest rate of cancer in the country. Only Kentucky is higher.

Dicamba

There are plenty of other herbicides to worry about besides glyphosate. Different formulations of older and sometimes even more dangerous herbicides seem to be one strategy occurring in the chemical world. One of the most troubling occurrences to me is the reintroduction of *dicamba* to the market to control resistant weeds in soybeans.[13] Monsanto genetically modified the soybean plant so dicamba could be used in those fields. Bayer claims if the product is applied strictly according to the label directions, it will not drift. This has not been the case in reality. ISU has determined that the number of days in a crop-growing season in which it is realistically safe to apply the chemical is very limited. Bayer has not been able to completely control the volatility or vaporization of dicamba to prevent it from unintentionally traveling to other soybean fields that were not genetically modified to withstand it.

Off-target dicamba drift can occur in one of three ways. The first is tank contamination where the spray tank is not emptied completely and cleaned adequately to prevent contamination to the next-sprayed crop. The second is through particle drift, which occurs primarily when windy conditions allow spray droplets to blow downwind of the sprayer to contaminate another crop. The third way is the most problematic. It occurs when there is a temperature inversion. A temperature inversion occurs when light, warm air rises into the atmosphere and cooler and heavier and more damp air settles in below it. This generally occurs on warm, quiet, and humid days late in the afternoon. There is no mixing of the air when this occurs. The mixing does not occur until the following morning when the wind begins to pick up and the vapor begins to move, perhaps drifting onto sensitive crops that are not immune to plant damage from dicamba. Label instructions do not allow for the spraying of dicamba during temperature inversions or on days when the wind blows above a certain speed. Still, it happens anyway.

Since purchasing Monsanto, Bayer has been able to force many farmers to buy their genetically modified soybeans so the farmers' traditionally planted varieties would not be damaged or killed by the drift of dicamba. So far, they have escaped blame for drift damage. ISU Extension reported in the summer of 2021 that at least 40 percent of Iowa's soybean acres were adversely affected by the drift of dicamba.[14]

Dicamba is considered carcinogenic. It was first registered as a broad-spectrum herbicide in 1967. I remember using it in the 1970s on our farm, under the brand name Banvel, to control weeds in corn. Increased cancer rate ratios and positive exposure response patterns have been reported for dicamba in a review of data gathered in the National Institutes of Health's Agricultural Health Study.[15]

In February of 2024, a federal court ruling in Arizona found that the EPA unlawfully approved the use of dicamba.[16] An earlier case resulted in a Court of Appeals overturning the agency's prior approval of the pesticide. The EPA reapproved the same use of the pesticide in 2020, leading to the most current lawsuit. The ruling in February of 2024 outlaws dicamba products sprayed over emerged soybean and cotton crops that are genetically engineered to withstand the spray.

The chemical was approved for use in 2017. Since then, 15 million acres of drift damage to non-genetically engineered soybeans has occurred, according to the USDA. It has also damaged orchard, garden, and tree crops on a scale unprecedented in US agricultural history. Numerous pollinators, such as monarch butterflies and some bumblebee species have also been negatively affected. The EPA has determined existing stocks of dicamba can be used until June of 2024. Apparently, that is supposed to be the end of dicamba in the United States, but I would not bet on it.

During about the first ten years of PFI, some of us thought we could actually reduce the overall amount of pesticides used by all of agriculture, because during the farm crisis decade of the 1980s and even into the 1990s, farmers were looking for ways to save money so they could keep farming. The banding of pesticides over the row only with the combined use of mechanical cultivation became popular.

We were naïve to think that we had the capacity to change all of agriculture. The introduction of genetic modification in 1996, so glyphosate (Roundup) could be applied to soybeans and not kill the crop, changed that and the opposite occurred. Chemical use began to skyrocket. Roundup is still the number one herbicide used around the world. As of 2020, 9.4 million tons of Roundup had been sprayed around the world. This is enough spray for one-half pound of Roundup on every cultivated acre in the world. Nearly 300 million pounds of glyphosate are applied annually in the United States. Mexico's president has ordered glyphosate to be totally phased out in that country by 2024. Bayer is trying to reverse that decision by convincing USDA to impose crippling trade sanctions on the Mexican government if that occurs.

Where We Have Been and Where We Are Going— Ridge Tillage

There was a significant learning curve in understanding how to control weeds organically through cultivation. The Buffalo brand has been my choice because of its flexibility and the ease in setting it in so many different ways to take out as many weeds as possible. We have two cultivators set up for first cultivation, where the disc hillers are set as close to the row as possible with a tent-like shield completely covering the small crop or raised up some, depending on the height of the crop of soybeans or corn. A twenty-two- to twenty-four-inch-wide steel sweep eradicates weeds in the center part of the row. We are still on thirty-eight-inch-wide rows for corn and soybeans at this time, which is really considered old style. It still works, though that may change when and if we purchase a different combine and different corn and soybean heads. In fact, we purchased a newer and larger combine during the 2022 cropping season. We decided to keep the row spacing at thirty-eight inches because of the ease of having more room for cover crops compared with a thirty-inch or less row configuration. This change also allows the combine's large front tires to run between the ridges and not disturb them. The tires straddle three thirty-eight-inch rows. We changed our four-row equipment to six-row, thirty-eight-inch-wide row planters and cultivators so that we can do the work more quickly and save on energy costs. We do not feel at this time that it would work to go wider yet with our planters and cultivators. Some organic farmers use eight- and twelve-row equipment where the land is much flatter than ours. For the second and final cultivation process, we build a ridge of soil with the sweeps and with what are called ridging wings. They push the soil up around the stalks of the corn and soybean plants.

I describe ridge tillage to people who cannot picture it as being a form of raised-bed gardening, which is a little easier to visualize. In our case, the raised beds have thirty-eight-inch spacings between them and are done on large acreages, not like a typical small garden, but the principles are the same. We use mirrors and hydraulic or electronic sensors to help us cultivate as close as we can to the crop to remove the weeds. It has proven to be an extremely effective way for acceptable weed control. We can say that because of doing on-farm research trials comparing ridge tillage without herbicides to conventional tillage and no herbicides. We did so on four different occasions with PFI over a twenty-year period. Other PFI farmers have done the same research trials on their farms. We all used soybeans as the crop for our research. No herbicides were used on the research plots for weed control. The trial consisted of comparing ridge tillage (raised beds) to conventional tillage (meaning discing or pre-planting tillage of the soil in some form). The broadleaf weeds were hand counted in the field trials. Weeds consist primarily of two types, broadleaves and grasses. Broadleaves are part of a group of flowering plants known as dicotyledons, which means they have two seed leaves at emergence. The veins on the leaves are branched or netlike. Grasses are known as monocotyledons, which means they have one seed leaf (cotyledon). The veins on grass leaves are narrow and parallel to each other. There are other differences, too, such as the flowers of grasses are generally less conspicuous than their counterparts. Five to seven times fewer broadleaf weeds have consistently been found where a ridge tillage system was used, compared with using a more conventional tillage system of seedbed preparation.

Grass plants were not counted in the trial because, if done correctly and if the weather cooperates, grasses are nearly completely eliminated in the ridge tillage planting system due to the design of the planter. The crop seed is firmly planted into a narrow groove that the planter makes in the soil. The soil around that groove is left loose and not compacted. Grasses do not typically germinate in those kinds of conditions. They prefer compacted soils where there is not as much oxygen. If, however, a heavy rain occurs before the crop comes up, soil that is moved by the planter into the center of the row may wash back into the row, where weed seed has the opportunity to germinate close to the crop seed, which makes it difficult to get rid of those weeds throughout the crop-growing season.

Ridge tillage counterbalances conventional no-till as much as possible without using chemicals. It is what my mentor Dick Thompson spent so many years developing and perfecting as far as possible. In this system, soil is not disturbed after the final cultivation until you plant the crop the next growing season. The ridge tillage planter has a steel sweep that takes part of the ridge or raised bed of soil off and puts the soil and previous crop residue and/or cover crop into the center part of the row, which is about twelve to fourteen inches in width. The field is once again almost level after this two to four inches of soil is taken off and flattened where the crop seed is planted. Research at ISU has shown that the majority of weed seed is in the top two inches of soil. The final cultivation is the last time the soil is disturbed until the next growing season, with the exception of the planting of a cover crop, which is now recommended for many reasons.

When you buy a piece of equipment to control weeds, it is a one-time purchase. It can be used year after year if it is well built. Herbicides, conversely, must be applied every year, unless you decide to do otherwise, such as doing weed counts or some sort of surveillance. Integrated pest management (IPM) is supposed to rely on scouting and taking into account crop damage, insect egg numbers, and other measurements to determine whether a pesticide is warranted. I have rarely, if ever, heard of it being applied to weeds. With herbicides, the same amount or more are applied year after year.

Buffalo cultivators have become harder to find. The company that built them, Fleischer Manufacturing in Nebraska, sold the company to Apache Manufacturing in Norfolk, Nebraska. You can still buy a new Henke-Buffalo cultivator. Because many of them are still sitting behind conventional farmers' machine sheds gathering rust, many people try to find used ones. Today, due to the growing resistance of some weeds to pesticides, more conventional chemical farmers are pulling those cultivators out of their iron piles behind the shed to perform at least some cultivation.

Cultivation is certainly not the only way to control weeds. Some farmers have worked on and perfected what is called flame-weeding where propane gas is used to burn the weeds at a very specific growth stage so that the crop is not significantly affected. In some cases, high-voltage weed zappers placed on the front of tractors are being used to kill larger weeds later in the growing season. These machines are quite expensive. This technology requires a great deal of energy and a large tractor to make it work.

Where We Have Been and Where We Are Going—Crop Rotations and Cover Crops

Diversity of crops and crop rotations and the time of year when crops are planted all play important roles in pest control. Weed and insect pressure can be diminished by using a systems approach to deciding what crops to plant. This is why it is not a good policy to only think about single components in farming. Nature does not work that way, and growing crops and raising livestock do not either.

The books I was reading in the winters of 1982 and 1983 did provide a framework for where to start. We were already using certain practices that made organic farming work. They were crop rotations that involved diversity, the use of legumes, and having livestock on the farm. We grew oats that had an underseeding of legumes such as alfalfa and red clover that would become hay for at least two to three years, followed by corn and soybeans. The length of the rotation was flexible, but it always had to be more than just corn and soybeans on our farm.

Oats and other small grains, such as hard red spring wheat and barley, are all early spring annuals. The sooner you can get them in the ground in early spring, the better. We like to plant oats in March if weather conditions allow. Early planting means the crop is headed out and already maturing before hot and dry weather that sometimes occurs in late June and early July. Lower test weights can be caused by hot and dry winds that limit ideal growing conditions, resulting in less weight and density of the grain. Oats do not like "wet feet," either, meaning growing conditions with too much standing water. We have also had good success with a mix of small grains that includes hard red spring wheat, barley, oats, and field peas. The blend is simply called succotash.

This has resulted in a high-protein feed for all the species of livestock we raise, that is, hogs, chicken, and beef. That is the species order of preference for feeding the mix as well. For hogs, with the field peas, a 14 percent protein finishing ration can be easily attained. This is because field peas are high in protein, at around 22 percent. Our succotash comes in at around 17 percent protein. We have reduced our purchase of organic soybean meal by over 50 percent by feeding succotash. We do not just feed succotash to pigs as they find it more palatable to have some corn in their ration. Organic dehulled soybean meal is 47.5 percent protein and is costly, too, at generally over $800 per ton ($1,400 to $1,500 per ton in 2022). The four ingredients in succotash are all planted together and are harvested at the same time, making this a practical cropping endeavor.

You may be wondering why we purchase soybean meal since we are growing soybeans as one of our crops. Raw soybeans need to be heat-treated during the production of soybean meal in order to denature the trypsin inhibitor that otherwise would interfere with protein digestion. Soy meal is a dried product that has had the oil removed. Older animals, such as breeding sows, can be fed raw soybeans without detrimental effects on protein digestion. It is not recommended to do so with younger livestock, such as pigs under 300 pounds. Most soybeans that are processed into soy meal protein have the oil removed, which is used primarily to make soy oil for human consumption in the cooking of foods. When there were many more pork producers in the 1980s and 1990s, some purchased their own on-farm protein processors called extruders that cooked the beans and removed the oil to either be put back into the feed ration to increase the desirable fat content or possibly to be used for on-farm soy-biodiesel fuel production. The time requirements, cost, and efficiency challenges all contributed to the lack of wide adoption of on-farm soy extrusion.

Corn and soybeans are considered early summer annuals. After oats and succotash and hybrid rye harvest (hybrid rye is planted in September and October of the previous year, making it a winter annual), cover crops are planted, usually after mid-July. Either the oats or succotash would already have legumes like alfalfa and red clover and grasses such as orchard grass planted with it in early spring. If not, a cover crop could be planted in mid-July or so. There are an almost unlimited number of mixes of different cover crops to fit your needs. The particular species to be included in each mix is being experimented with and determined by some cover-crop companies that have come into existence over the past twenty years. There are cover crops for grazing, such as turnips, cover crops for breaking up hard layers of soil like oilseed radish, legume cover crops like hairy vetch for adding nitrogen to the soil through the work of the rhizobium bacteria on the roots of the legume plant, and many more. Late summer annuals like millet and sorghum Sudan grass perform well under hot and dry weather conditions. Sometimes cover crop mixes, which are referred to as cover crop cocktails, are specifically designed to meet wanted specifications and desires. A dozen or more species may be in them.

It is not uncommon for us to grow three different crops in one field in one year's time. I find this very satisfying. Here is how it might work for us. Let's start with crop number one, which is hybrid rye. Hybrid rye was the first small grain to be hybridized in the world, by plant breeders in Germany who began to work with the cereal rye genome to create hybrid rye. Compared with common rye, hybrid rye has many characteristics that make it a better choice, both as a feed and a cover crop.

Rye is an ancient crop as are barley and wheat. Hybrid rye is not as susceptible to the fungus ergot that has been a serious problem throughout much of history in Europe and areas of the Mediterranean. Epidemics caused by high amounts of ergot in rye used to be common and are referred to as far back as the 800s and as late as 2001 in barley in Ethiopia. Ergot can occur in various small grains such as barley, but has been most associated with ingesting the fungus through rye bread flour, which results in a type of dry gangrene where blood flow is restricted to the outer extremities such as the toes. In epidemics of the disease, limbs slowly deteriorate leading to eventual death. One of its historical symptoms has been associated with what are known today as psychotic reactions.

Hybrid rye yields about twice as much as common rye. It has superior stalk strength and does not grow as tall as common rye varieties. It requires no pesticides because it chokes out weeds and because of chemicals produced by the plant that have what is called an allelopathic effect on weeds. Most conventional farmers who plant rye kill it with Roundup early in the spring. This approach does not give the rye sufficient time to produce enough biomass to make much of a difference in carbon dioxide (CO_2) being used by photosynthesis and sequestering carbon in the soil. Common rye is probably the most popular cover crop planted in the fall and survives as a winter annual and begins to grow again early in the spring. It is used to keep the ground covered through the late fall, winter, and early spring. Hybrid rye holds much more promise if it can be planted early enough in the fall after chopping corn silage or combining soybeans. Earlier maturing soybeans with good yielding capability would help in its adoption. It is being touted as a third crop for much of the corn and soybean growing regions of the country. It can be used as a valuable feed grain. It is considered to have 90 percent of the feeding value of corn and actually has a more favorable amino acid profile than does corn.

Rye straw is of high quality and can be used as bedding for livestock. There is growing interest among some to promote it as a practical way to move from liquid manure systems in pork production to solid ones where the rye straw and pig manure could be composted to improve soil quality. A solid manure system in which manure is composted would have a number of stacked benefits for combating climate change. It would eliminate most of the methane, which is a far more potent greenhouse gas than CO_2. It would eliminate the nauseating odor of liquid manure. It would dramatically reduce the runoff of liquid manure into our streams and rivers and thus improve water quality. The compost would improve soil quality by its ability to enhance microbial life in the soil.

After the first crop of hybrid rye is harvested around the middle of July, a second cover crop or mix of cover crops can be planted, as has already been discussed. If you plant a legume as a cover crop, you can enhance the production of nitrogen for the next year's crop. In our case, we usually try to graze the second cover crop that is grown. Often it is turnips. Turnips grow quickly and by the end of August are ready to be grazed by cattle or by any forager. Cattle will eat the leaves, but they particularly go after the turnip itself, which has a high sugar content. As the fall season proceeds and freezing temperatures increase, more sugar is concentrated in the turnip. Cattle will pull up the plant to get to the turnip. If the weather cooperates, we plant a third cover crop for late fall grazing, such as rye or oats or legumes like hairy vetch or forage peas. The cattle are often fed supplementary hay if needed on this ground throughout the winter. They spread the manure on this field themselves. We don't have to. The following crop is usually corn.

Because of the ongoing drought conditions in the Midwest the last few years, we have also had excellent success in planting sorghum-Sudan grass, which is a cross between these two grasses.

It is a warm-season crop that grows rapidly and does relatively well in hot and dry conditions. In 2024, we were able to get two grazings where we planted it in late July after oats and succotash.

Where We Have Been and Where We Are Going—Manure and Composting

We purchased a compost turner in 1983, the Wildcat brand from South Dakota. I was a little concerned at first because it was expensive. It cost $7,000 purchased new. It was tractor-powered and very heavily built. We still use it today. The money needed to purchase it in 1983 has proven to be very economical compared with what it would cost new today. It turned out to be a good decision. The amount of composted manure annually has increased, especially over the past fifteen years. We now produce around one thousand tons of finished compost annually. In reflection,

> One of the most striking characteristics of our farm is that we have purchased very little additional sources of nitrogen for our crops since 1982 (the one exception was limited amounts of liquid 28 percent urea from 1990 through 1993, if the late spring soil nitrate test called for it).

Some farmers find that to be unbelievable. Our yields have been very good, and our net return per acre is even better. In 2021, our organic corn yields in the plot of six different varieties ranged from 173 to 205 bushels per acre in a dry year, a testimony to having good soils to start with and to the potential of recycling nutrients on a farm. Some would say it is because of having livestock on the farm. Dick Thompson had a keen observation about livestock on a farm. He said if the livestock leave a farm, the farmer eventually goes with them, and so then do rural communities.

Making good compost is a skill learned over time—how to do it with the right amount of moisture, the correct temperatures, and the correct carbon-to-nitrogen ratio. To officially be able to call composted manure compost by the National Organic Program (NOP) standards, you have to meet a number of requirements. We do not try to meet those requirements, so we cannot officially call our manure composted. That is okay for us because we are not trying to sell compost, and we are able to meet the basic requirements for length of time to apply manure before human consumption of the product produced.

The initial carbon-to-nitrogen ratio (C:N) of the manure with straw must be between 25:1 and 40:1 for official compost. Official compost requires a temperature between 131 and 170 degrees Fahrenheit for fifteen days total in a windrow situation. It must be turned a minimum of five times. Official certified compost can be applied at any time to the crop. Raw manure must be applied a minimum of 120 days before human consumption if the crop is growing in the soil, such as potatoes or carrots or lettuce. Ninety days is sufficient for crops not grown in contact with the soil, such as corn and soy, and small grains like oats, wheat, and barley.

The high temperatures of the pile greatly decrease or eliminate the potential for weed seeds to germinate. Pathogens such as certain harmful bacteria are also greatly reduced. These include *E. coli*, *Salmonella*, and *Campylobacter*. Viruses and protozoa are also reduced through composting. Composting itself has some criticism if the process is anaerobic where the production and release of methane becomes an issue. If the manure is being turned as often as it should, then that should not be an issue. However, aerobic composting can release much ammonia.

An extensive controversy surrounding the use of manure in organic agriculture has to do with currently allowing raw manure from conventional concentrated animal feeding operations (CAFOs) to be used on certified organic farms. This manure may contain a good deal of harmful pathogens and other undesirable compounds. These could include antibiotics, hormones, parasiticides such as Ivermectin, and pesticide residues. Our farm has never purchased raw manure from any other farm.

Some organic farmers who do not have livestock purchase both raw manure and stockpiled or semi-composted manure to provide fertility for their farms. I think this is problematic. Stockpiled chicken litter or cattle and hog manure with straw for bedding is certainly better than straight manure coming directly from a cattle feedlot, for instance. The pile has a chance to heat up to some degree. If all livestock operations were somehow convinced or encouraged to use straw bedding and to compost their animal manures, then progress could be made on various climate and soil quality fronts. One of the most important challenges is reducing methane production. Methane could be reduced significantly in liquid manure systems, which are the standard in the hog confinement industry, if solid manure systems were used instead. Confined livestock operations that produce liquid manure counter with the argument that they are now beginning to capture methane through the installation of methane digesters that capture methane from liquid manure and turn it into a usable fuel. The digesters are very expensive, costing often as much as $250,000 per confinement operation.

At the time we were contemplating making these changes, questions about fertilizer were the most confusing ones and where indecision and doubts resided. If I am being perfectly honest, that is still the case, along with a great deal of ongoing controversy and questions about what kinds and types of fertilizers should be used to grow a crop. A national list exists today on what you can and cannot use to be certified organic. What makes for a healthy soil and how to achieve it over time is a question not easily or quickly answered. It is on a continuum, to say the least, which begs two questions that need to be answered: (1) Are you soil testing to know what you need and do not need for your soil to grow the optimal crop? However, probably no two soil testing labs will come up with the same results for the same soils. (2) What are the methods used by the lab to test the soil? Since the pretense is always looking for a prescription from the lab to get the highest yields possible for the next season's crops, it becomes just a numbers game for many farmers.

Some organic farmers would say that they do not need purchased fertilizers to make up for what their soil lacks to grow a healthy and abundant crop. Today, we are close to being able to say that ourselves, which is much different than when we first started in 1983, partly because of the larger amounts of animal manure created and composted on our farm. Our good soil fertility also comes from the use of cover crops, green manure crops, legumes, and multiple kinds of rotations of different crops. It certainly helps we were blessed with fertile soils to start with.

We have a beef stock cow herd of around ninety red angus cows and feed out all of their progeny as certified organic beef. We have had our own label, "Rosmann Family Farms," since 2000. Meat could not be labeled as certified organic until 2000, when the first USDA certified organic meat label appeared.

We raise 450 head of hogs annually as certified organic pork that is also Global Animal Partnership (GAP) certified. GAP focuses strictly on standards for animal welfare, not organic practices or materials. The organic food conglomerate Whole Foods started GAP certification and requires us to send them only GAP certified meats. We do not sell directly to them but through the Organic Valley Cooperative (an independent farmer-owned cooperative) for which we are growers of both beef and pork.

In looking back at those early days of organic farming, the word organic was problematic in different ways. First, I would say that we adopted some of what were commonly agreed upon as organic methods. That did not necessarily make us organic farmers. We did not become certified by an independent third party until 1994, eleven years after we quit using pesticides and acidic synthetic chemical fertilizers.

You do not have to be certified to be considered an organic farmer. First of all, the NOP does not require certification if you sell less than $5,000 worth of products. It also depends on what your customer is comfortable with, which usually means farmers who sell directly to consumers who know and trust the farmers and know the kind of operation they are running. Organic certification through the USDA and the NOP can be very expensive and detailed, depending on the size and kind of business. Most organic farmers today do choose to be certified.

Back in the early 1980s, it did not make sense for us to be certified by a third party as organic. We did not have anything we thought would sell as organic, and very few markets were in the Midwest. The markets for organic corn and small grains such as wheat and oats were very limited, and none that we knew of for soybeans. Growing vegetables for farmers' markets was just starting in our area. Whatever markets there were for organically grown crops back in the early days were filled by the early pioneers in the organic business. We were told that if the few organic companies out there of any size needed more grain they would contact us. That never happened.

My opinion was that stopping the use of all pesticides, including herbicides, insecticides, fungicides, nematicides, or anything with "cide" at the end of its name, was the most important consideration on the crop-growing end to be considered organic.

Where We Have Been and Where Are We Going— Fertilizer

Nitrogen, phosphorus, and potassium (NPK) are the three most important elements plants need for optimal growth. Conventional sources of synthetic fertilizers for nitrogen, phosphorus, and potassium are prohibited in certified organic agriculture.[17] I was of the mindset during my first ten years of farming (1973–1983) that I had to apply these three fertilizers and to do it every year. I did what the fertilizer people recommended. The same tendency was there at first in 1983. That was especially true for phosphorus and potassium. On the nitrogen side, I already knew about the value of legumes, manure, and crop rotations. What I did not know was how to increase and improve it. When we started, I did not fully appreciate the importance of two of the basic tenets of organic agriculture, that soil fertility should rely mostly on the recycling of organic matter and I should be most concerned about feeding the soil, not the plant.

Synthetic nitrogen (anhydrous ammonia) was the form of nitrogen we used until we changed our basic farming practices. The last year we used it was 1982. The history of producing ammonia is a particularly fascinating one. Few people understand the significance of what is called the Haber-Bosch process.[18] It has been called the detonator of the population explosion. Detonator is an appropriate term because two German scientists, Haber and Bosch, developed the process for extracting nitrogen out of the atmosphere by the chemical reaction of nitrogen (N_2) combining with hydrogen (H_2) to form ammonia (NH_3). The H came from the heating of oil, coal, or natural gas. Almost all of today's ammonia comes from natural gas. Ammonia is used to make explosives. Until the First World War, black powder for gunpowder and explosives came from the natural source called saltpeter. Saltpeter's chemical formula is potassium nitrate (KNO_3). At the time of the First World War, Germany could not get the amount of saltpeter it needed to make bombs because of the Allied Powers' blockade of the saltpeter mines in Chile. The Haber-Bosch process uses high temperatures (500 degrees Celsius) and high pressure (100 atmospheric pressure) along with a catalyst to produce ammonia. Ammonia as a nitrogen source for growing crops "exploded" after the Second World War.

In 1918, Prime Minister Winston Churchill made some very astute remarks about the discovery of how to make sodium nitrate and nitrogen fertilizer:

> It is a very strange thing that for the invention of Professor Haber, the Germans could not have continued the War after their original stack of nitrates was exhausted. The invention of this single man has enabled them, utilizing the interval in which their accumulations were used up, not only to maintain an almost unlimited supply of explosive for all purposes, but to provide amply for the needs of agriculture in chemical manures. It is a remarkable feat, and shows on what obscure and accidental incidents the fortunes of possibly the whole world turn in these days of scientific discovery. [During World War I, Fritz Haber and Karl Bosch invented a large-scale process to cause the direct replacement of sodium nitrate in the manufacture of explosives and fertilizers.][19]

In 1900, 1.6 billion people inhabited planet Earth. By the end of 2018, that figure rose to 7.7 billion and is predicted by many to reach 9 billion by 2050. Historians have written that this growth in population occurred due to the ability to produce synthetic nitrogen for crops. While that may be somewhat or even mostly true, I think that assertion could be contested to some extent. It is estimated today about 50 percent of the world's use of nitrogen to grow food comes from mass-produced ammonia fertilizer. The other 50 percent comes from natural processes, including bio-fixation (*Rhizobium* bacteria and cyanobacteria), atmospheric deposition, and the recycling of crop residues and animal manures.

Having said that, I would argue that the 50 percent coming from natural processes could have been increased to some extent over the past seventy-five years. Farmers could have placed greater emphasis on producing their own nitrogen through legumes and cover crops; the production of perennial food crops such as fruits, nuts, and berries could have been increased; better grazing methods around the world could have occurred, resulting in more meat and dairy products; there could have been improved recycling of nutrients from plants and animals and reductions in food waste and storage losses and, in general, conservation of soil and improved soil quality and health through diverse crop rotations. Instead, the emphasis began to be placed on the government-supported commodity crops of corn, soybeans, rice, wheat, and cotton. Corn and soybean production and yields per acre increased dramatically because of the emphasis on plant breeding of the major commodity crops. Convenience and eating out replaced cooking meals at home. Highly processed foods replaced high-nutrient foods. High fructose corn syrup made its way into just about every processed food and drink imaginable. The results were increases in obesity and type 2 diabetes. The government wanted to keep consumer food prices low. They mostly got their wish.

Over 90 million acres of corn are grown in this country every year. Forty million acres, or 40 percent, go to ethanol to feed our vehicles. Most of the rest is fed to livestock. There are just under 350 million acres of land where crops are grown in the United States. Take away 40 million acres for ethanol, and you have a 11.4 percent reduction in land that could have gone for food production.[20]

Where We Have Been and Where We Are Going—Phosphorus and Potassium

After nitrogen, phosphorus is the second most important limited and essential soil nutrient. Phosphorus is needed for the growth, maintenance, and replacement of all tissues and cells in the human body and for the production of genetic building blocks, deoxyribonucleic acid (DNA) and ribonucleic acid (RNA).

Phosphorus occurs naturally in sedimentary rock close to the Earth's surface. Limestone and mudstone contain a good deal of phosphorus. To increase the amount of available phosphorus in conventional agriculture, phosphorus rock is heated, pulverized, and treated with either sulfuric acid or phosphoric acid. An alternative source of phosphorus was a big headache at first in terms of meeting organic certification requirements. Super phosphate, or what is called diammonium phosphate (DAP), is not allowed in organic operations. This is because of the acid treatment. Soft rock phosphate in its natural untreated form is available for organic use. Animal manures also contain a significant amount of phosphorus.

Availability of phosphorus could become a greater problem for agriculture as phosphate rock is a finite resource. About 70 percent of the world's known phosphorus is contained in the Western Sahara in Africa, followed by China, followed by Algeria. Phosphorus and potassium are critical elements needed for life that are not present in the atmosphere like nitrogen, oxygen, and water vapor are. Unfortunately, only 20 percent of the phosphorus contained in mined phosphate rock ever reaches the end consumer. It is lost during mining and processing, and only one-half of the world's manure is recycled back into the soil.

The movement of phosphorus from farm fields through leaching and runoff, and in manure runoff, along with detergents and soda going down drains, is resulting in a very serious problem down steam wherever the rivers take it. In the United States, the runoff goes into the Missouri River and other major tributaries of the Mississippi River. where it ends up being deposited into the Gulf of Mexico. This problem is called *eutrophication*. The fertilizer causes phosphorus algae blooms where a great deal of available oxygen is used up, so when the algae dies, aquatic life suffers, too. It is called the Gulf of Mexico Dead Zone (an area about the size of New Jersey) because there is little life there.[21]

In Iowa and other farm states, no mandatory actions have been put in place to lower these fertilizer losses. It is all voluntary. Some reductions have occurred, but the decreases have been too small to make enough difference. With more extreme weather and rainfall events, the problem is only getting worse. Another problem is that overapplication of phosphorus and nitrogen keeps on occurring. Fertilizer companies want to sell fertilizer. Farmers want to get the highest yields possible—if 150 pounds of nitrogen are called for, then 250 pounds will be even better.

The third most basic elemental fertilizer is potassium. Potassium is sourced from mined potassium-containing minerals and from salt lakes and brines. The most commonly used potassium source is potassium chloride, which contains more salt than other forms. It accounts for about 90 percent of all potassium used in the fertilizer industry. Another potassium source is potassium sulfate, which can be more desirable because of its lower salt content, which is useful for soil already high in chlorides. Sources of potassium that are allowed by the NOP are controversial.[22] Some people think that potassium sulfate should be allowed. It is allowed in its purest form from the Salt Lake City, Utah, region but not where it needs to undergo further refinement. Muriate of potash is acceptable for organics. Luckily for us, our soils tend to be very high in naturally occurring potassium. I have not applied a potassium amendment since 1982, other than what is recycled through our manure and compost.

When it comes to micronutrients—such as the important ones of zinc, sulfur, copper, boron, manganese, selenium, molybdenum, iron, and cobalt—synthetic forms are allowed by the NOP. The most limiting ones for us are zinc and boron.

Where We Have Been and Where We Are Going: What Happened to Our Yields the First Years and What We Did about It—Adding Tools to the Toolbox

After the first couple of years, I saw our yields on corn decline some. If you look at the average for the first thirty years, from 1983 to 2013, average whole farm yields were respectable at 140 bushels per acre. The first two years of 1983 and 1984 and the severe drought year of 1988 did not help. The 1983 average was 115, and the 1984 average was one hundred bushels per acre. We still had a 100-bushel yield in 1988. In retrospect, this was in part due to the result of transitioning away from applying higher amounts of nitrogen. Nitrogen historically has been perhaps the biggest limiting factor for organic yields in corn and small grains such as wheat, oats, and rye. Nitrogen is generally not applied to soybeans or other legumes such as alfalfa and red clover. Legumes produce their own nitrogen through the presence of rhizobium bacteria on the roots of the plant. They are able to extract nitrogen from the air for their own growth and the growth of the host plant they are attached to. When using conventional farming methods, we never applied more than 150 pounds of anhydrous ammonia in any given year. But when we made the switch, no additional nitrogen was applied the first few years. We relied totally on manure and compost, which we did not produce enough of at that time. We did not know how to fine-tune our crop rotation sequences to increase the availability of nitrogen for the growing crop. We were still learning. We still are. We will always be on a learning continuum.

Remember, we did not have to follow any organic rules because we were not officially certified organic until 1994. Other farmers in our PFI group began to experiment with more precise applications of nitrogen in cornfield trials. We began to do that as well. The most economical and efficient way to do so was by split application of nitrogen, with some applied at planting time and the rest at cultivation time, which could be done by putting what are called saddle tanks on each side of the tractor to hold 28-percent liquid urea nitrogen. We chose that kind of nitrogen because it was much easier to work with than anhydrous ammonia, which became a gas as soon as it came into contact with air. We were able to put on precise amounts by dribbling the liquid nitrogen within a couple inches of the corn seed and not significantly harm the seed. The salt in the urea could burn the young seedlings. We did this for about four years, from 1990 until we became certified organic in 1994. Urea nitrogen, anhydrous ammonia, and ammonium sulfate are prohibited in organic agriculture because they are synthetic forms of nitrogen.

Many PFI farmers were able to become even more precise, efficient, and environmentally conscious about nitrogen application when Dr. Alfred Blackmer, a soil agronomist at ISU, developed the late spring soil nitrate test.[23] This test, which I learned to do myself, determined the amount of nitrogen needed for optimum corn yields when the corn was six to twelve inches tall. First, nitrogen was applied as a starter fertilizer when the corn was planted. Then, by taking a soil sample and either sending it to a laboratory or testing it yourself, you could determine how much additional nitrogen was needed. The N that was measured was a combination of residual N from the previous year's crop, any fall or early spring application of nitrogen, and mineralized N from soil organic matter. PFI farmers became the test subjects for determining the accuracy of the test for Dr. Blackmer. After a period of years, in the early 1990s, this came to be calibrated to an acceptable range of amounts of nitrogen applied when the crop needed it the most (six-to-twelve-inch stage). Another tool to help farmers determine more precise nitrogen needs was through the use of the late-season stalk test. This test measured how much nitrogen was left in the stalk after harvest. Large amounts indicated that too much nitrogen was applied that previous year.

At first, it appeared the late spring soil nitrate test would really take off and be used by farmers to do a better job of fine-tuning the needed amounts of nitrogen for that year's corn crop. Regrettably, a number of developments stood in the way. Its broadest appeal seemed to be for farmers who were still cultivating, which meant they could do two jobs at once. Farm size was increasing at the time the test materialized. Many large farmers wanted to make sure they could get all their nitrogen needs applied in the fall if the weather conditions were good. That way they would not have to worry about a cold, wet spring keeping nitrogen applications from occurring the following year. Another factor seemed to be that fertilizer companies did not push for adoption of the test. The other most disturbing reason was that apparently not enough farmers cared about fine-tuning the application of nitrogen to decrease the leaching into ground water.[24]

Dr. Blackmer was dedicated to helping farmers better control their nitrogen applications so that groundwater would be less contaminated. Dr. Blackmer died in 2006 from pancreatic cancer. Had he lived, I think he would have found even more novel and effective ways to manage nitrogen, especially now with its overapplication and application continuing into the late fall, when there are no growing plants to use it. This results in the addition of nitrous oxide to the Earth's atmosphere, which is nearly three hundred times more powerful as a greenhouse gas than CO_2.

Where We Have Been and Where We Are Going—Practical Farmers of Iowa Research Trials Over the Years

We have completed more than forty-five research trials on our farm in conjunction with PFI, ISU, and the USDA Sustainable Agriculture Research and Education Program (SARE).[25] Most were sponsored by PFI until the past few years when the ISU Organic Program, under the leadership of Katheen Delate, PhD, began to accelerate its on-farm testing of organic corn hybrids. Most of the trials have been fairly simple trials, comparing only two variables. In the 1980s, some trials keyed in on tillage in soybeans where conventional tillage (discing or stirring up the soil) was compared with ridge tillage (no-preplant tillage). Other trials compared composted manure with raw manure in corn. Different seeding rates in oats was done in 1990. The early 1990s saw trials focus on different nitrogen fertilizers to increase yields and the use of nitrogen with the late spring soil nitrate test. We did our first livestock trial in 1996, where we compared normal finishing swine rations with those in which alfalfa was added. After we became certified organic in 1994, the research switched to improving yields and management strategies for organic crops.

For instance, in 1994, we did a trial important for us, comparing different planting populations for organic corn hybrids. Organic seed does not contain synthetic insecticides or fungicide seed treatments. We were asking ourselves how much seed should be planted in order to get the most satisfactory populations of corn plants for the growing season. With insecticides on the seed, we could plant sooner and perhaps use less seed because more would germinate in colder soils. The opposite was generally the case for organic conditions. In 1994, seed corn was generally being planted at significantly lower planting populations than is the case today, over twenty-five years later. Seed planting populations have grown along with better and higher-yielding corn hybrids.

We compared three different planting populations in 1994: The standard at the time, which was around 22,000 seeds per acre; a medium-high rate of 24,500 per acre; and a higher rate of 28,200 seeds per acre. The highest planting rate resulted in the highest yields. We decided to begin planting heavier populations after that, usually over 28,000. As we moved into total certified organic production, we found through two years of trials we needed to go higher still, especially if we wanted to plant corn before May tenth. We bumped up the corn populations to 30,000 to 32,000 seeds per acre. Now, with gradual improvement in soil quality and organic seed genetics, we generally plant between 32,000 and 34,000 seeds per acre, depending on the field and the genetics of the corn. However, new research reported in *National Geographic* in early 2022 shows that increased corn yields may not be due to improved genetics after all.[26] It may be tied more to longer growing seasons and more photosynthesis due to climate change and increased levels of CO_2. That may not be good news for the long term because higher-yielding genetics continues to be promoted as a primary means for growing more food for a future population of 9 billion.

It has taken longer for organic corn and soybean seeds to make the same kinds of improvements that the conventional industry has seen. Conventional seed companies that emphasize biotechnology methods for creating crops that are resistant to herbicide damage have no interest in organic yields. Why would they? It would mean unwanted competition. However, they are improving every year in both the organic seed companies and those seed companies that are willing to produce both conventional and organic seed varieties. It would help if we could reinvigorate public plant breeding departments at land grant universities.

The winter of 2023 brought some new hope in that regard. Maria and I were invited by the two public corn breeders from ISU and another from the University of Illinois to travel to Puerto Rico to do some hands-on learning and pollinating of corn at the USDA corn breeding nursery there. I was excited to learn that they are working on some of the concerns that organic corn farmers have about the development of better organic corn hybrids, such as better stalk strength and standability, better nitrogen uptake with lower applications of N, higher levels of an essential amino acid called methionine, better understanding of biotic communities in organic soil and their relationships to corn roots, and, finally, quicker release of new organic hybrids. One of the hoped-for outcomes of this trip was the beginning of a formal organic farmer-plant breeder relationship where organic farmers can work with public corn breeders to test, develop, grow, and market their own hybrids adapted to their own specific locations.

We have learned a great deal through both research trials and practical experience. We would not have had the confidence or the required information to make these kinds of decisions without on-farm research trials, PFI, and, of course, Dick and Sharon Thompson.

The same could be said about the many other trials we have done since 2000. We began to diversify even further with experiments aimed at planting aphid-resistant soybeans and planting dates and seed populations in soybeans; at different cover crop plantings at second cultivation; at improving feed efficiency in swine; and at 100-percent grass-fed beef compared with beef finished with only one-half the amount of corn fed to finish grain-fed beef. We felt we were in the process of fine-tuning our operations as we became more diverse and complex with our regimen of organic practices, both for crops and livestock. What stands out is as we did this, the integrated systems approach to managing and making the system more resilient has always paid off. Our yields and profitability per acre have consistently improved as time has progressed. So has our soil quality, although it is still not where we would like it to be.

We can honestly say that we have never lost a crop or even come close to losing one due to pests, disease, or mismanagement. We have tried to learn from our mistakes so we would not make the same ones again. When it comes to the changing weather and the extremes it brings, farmers have to be ready at a moment's notice for all cropping tasks. Timing is everything. Spreading the risk over numerous crops and seasonality of planting and harvest makes timing and flexibility easier to handle as well.

Where We've Been and Where We Are Going— Organic Certification

The decades of the 1980s and into the 1990s centered on lowering costs of production by lowering input costs and machinery costs. The USDA even had an acronym for that kind of agriculture that other farmers like us were doing, called LISA (low-input sustainable agriculture).[27] The farm crisis years of the 1980s forced some farmers to lower their input costs if they wanted to continue farming. The problem with that rationale is that you could lower your input costs to zero and still not be able to make a living unless you could get more for your product. We were learning how to lower the input costs without reducing yields through the PFI trials. We could not do anything, though, about raising prices for the products we grew. I finally said enough. We had been "almost" organic for ten years, but if we wanted to make the final plunge, we had to become certified organic.

The Japanese helped us make the decision to do so in 1994. Some Japanese buyers came to ISU around 1990 looking for organic soybeans. There was a new and growing demand for certified organic soybeans for making tofu. Tofu is made from condensed soy milk mixed with salt extracted from seawater to form a solid, block form. The Japanese were becoming increasingly concerned about pesticides and the potential for genetic modification of soybeans. A farmer's cooperative was formed by my cousin Ken Rosmann and another farmer in southern Iowa, Jim Boes. They successfully started selling soybeans to Japan via shipping containers through ports on the West Coast. Certified organic soybeans were starting to reach as high as $24 per bushel or about three times that of conventional soybeans. I decided it was time to put aside my concerns and questions on which fertilizers could be used and take the final steps toward certification. Organic certification became a learning experience similar to learning how to farm organically.

The paperwork for certification can seem daunting at first and is seen by more than a few to be a barrier to entry into organics. However, the paper trail is there for good reasons. There are fraudulent claims of being certified organic around the world. Some countries have more issues than others. Many claims involve the use of prohibited products such as pesticides, fertilizers, fumigants, and antibiotics. A common area of concern involves the comingling of conventional grain, not raised organically, with certified organic grains. There can be issues of fraud up and down the organic food chain, from the producer stage, the transporter and handling chain, and the processor and marketing chains. Why is this such a big deal? There are at least some monetary premiums for growing and raising organic foods. These premiums flow down the food line to the consumer. There must be trust in the integrity of the USDA certified organic label. The absence of pesticides and other chemicals and prohibited substances is what makes certified organic stand apart from conventional foods. Trusting the organic meats and dairy and egg products you buy were raised without antibiotics and hormones and were fed certified organic feeds is critical for the integrity of the organic process.

Organic certification requires keeping detailed yearly records on what you are doing with every field, animal, and organic product. An audit is conducted on-farm by a certifying agency. Certifying agencies can be both public and private. An increasing number are being administered by state agricultural departments. Crop rotation sequences must meet the standards set by the NOP of USDA.[28] That is a basic requirement. For example, you cannot continually rotate just corn and soybeans as conventional agriculture does; such systems do not adequately maintain soil health and quality. You need to have at least one year of rotating other crops, such as small grains like oats or wheat with an additional cover crop to go along with it. The cover crop should be a soil-building one such as a legume. Every time you make an organic sale, a paper trail must accompany it along the way. This ensures that if anything goes wrong or if an end user finds something wrong with the product, it can be traced back to the producer, which is rare, but it can happen.

Where We Have Been and Where We Are Going— Organic Livestock

We knew it would be fairly easy to raise certified organic cattle for meat. That is one reason we sought organic certification for beef cattle in 1998 before there was a USDA label for organic beef. We knew we could do it. Cattle rarely get sick or require intervention by a veterinarian, that is, if you provide healthy feed, water, and adequate shelter from weather extremes. Cattle are better off outside in the field than in close confinement in yards or buildings. Cattle do not need routine antibiotics or growth hormones to stay healthy. Vaccines are allowed for all species. We use vaccinations for our pigs and cattle. In the case of cattle, we have around ninety head or more of mother beef cows at any given time. This is because we tie the number of cattle to the crop and grass production resources of the farm. We rarely bring in cattle or pigs from an outside source, which might bring disease with them. If we do so in the case of breeding stock, they are isolated from the rest of the herd for at least three weeks. In other words, our herd is almost a completely closed herd, separated from contact with any other stock in the area. In the case of cattle, on average only about three to four animals are treated with antibiotics annually. This is out of around 180 to 225 head of cattle present at any given time.

Organic livestock must be identified by a numbering system.[29] If antibiotics or other prohibited substances are administered to an animal, that animal must have a means of identification so it is not sold as organic. Organic livestock producers do not withhold needed medications for their animals if they are sick. Sometimes *certified organic* natural remedies are used to treat sick animals. Sometimes you have to use antibiotics or health interventions that do not fall under the organic method. That is the way it should be. The health of the animal comes first. The animal then has to be marketed through a noncertified organic channel, such as a conventional market or sale barn.

We did not seek certification for our pigs until 2003. We routinely raised and marketed around 1,000 head of hogs annually until 1998. That is when the markets crashed and drove so many small producers out, as has been previously discussed. It seemed to be that the more hogs we raised, the more disease issues occurred that required antibiotic interventions. We tried to raise many hogs because that is how we coped with the farm crisis years and with cash flow and paying bills. I was afraid that we could not raise hogs without the use of antibiotics. After 1998, we decided we would take that plunge as well.

First, we produced for Niman Ranch, which was not certified organic, but instead was considered as natural pork. The word natural can be problematic. USDA says it means the product must be minimally processed and contain no artificial colors, flavors, preservatives, and other artificial ingredients. It also says the product must be minimally processed. This definition does not apply very well to natural pork, which means there are no antibiotics or hormones and the animals are raised with strict animal-welfare guidelines. This means no confinement crates for mother sows and their babies. It does not, however, say anything about what the animals are fed. This means that nonorganic feed stuffs can be fed, which does allow GMOs and pesticides, for instance, in the feed crops fed to the animals.

For three years, until 2002, we raised the Berkshire breed of pork for export to Japan. The hogs had to contain at least 50 percent Berkshire in their genetics. Berkshire pork is more heavily marbled and darker in color than its counterpart "the other white meat," which is the pork that comes out of confinements that emphasize white-colored pork breeds. It is also considered more flavorful, as it contains slightly more fat than the extremely lean white breed genetics. However, that program did not last, so we sought organic certification for pigs as well. Once again, we took the plunge into a new learning curve.

In 1998, Maria and I decided to start our own label for meats and began to sell the meat ourselves. That was a big step for us, as we had never tried to market our own products. It seemed like the natural thing to do. We could not label the meat as organic, but we could label it as being fed organically grown feeds. With the initial help of my cousin, Ken Rosmann, Maria and I began to deliver pre-ordered meat bundles to individual customers in Des Moines. The cattle were butchered at a small federally inspected beef plant in Des Moines. We remained with them until they closed in 2018. We kept the meat in a cold storage plant in Des Moines, which meant over a 200-mile round trip every two weeks or so. We did this for a couple of years, and then new opportunities arose in grocery stores in Ames and Des Moines.

The largest incentive became the ability to label the meat as certified organic by the USDA in 2000. It was a long-overdue regulatory advancement in the right direction. Small steps forward led to other bigger steps forward when we put five freezers into a state-inspected and licensed room in our farm shop, which meant we could take our meats out of cold storage in Des Moines and begin to sell locally as well. This went on for about ten years until 2012. That is when we decided to take the next step. Maria and I put up a store on our farm Maria proudly named "Farm Sweet Farm."[30] We decided to put our money where our mouth was and try to move toward the end target of creating more of a local food system for our products. The store is still going today. We still sell to stores in Ames and Des Moines. Our largest customer base is within a geographical area of about sixty miles or so. The store includes many items that are produced locally and also specializes in coffees, teas, spices, and other food ingredients and preparation items. We also sell our own labeled popcorn and eggs there.

Maria chose not to sell our frozen products long distances through online sales. This would have required a substantial amount of dry ice and created another layer of time, energy, and increased shipping costs for the frozen products and perhaps also hiring another person to help with that venture.

In today's world, most venders seem to think fresh beef is the way to go, as opposed to selling a frozen product. Selling fresh requires selling volume so the whole carcass can be sold. The problems are various ones. Many restaurants, for instance, think they must have fresh cuts of meat and are generally willing to buy only certain cuts like steaks and other primal cuts. That leaves hamburger and other cuts looking for another market. We believe we have a superior product, frozen or not, because the carcass is hung and dry-aged at a controlled temperature for two to three weeks before it is cut and packaged. It is in the dry-aging process that tenderness and flavor have an opportunity to occur. The typical large, mass-production packing plant does not have time for something like that. They have to push through as many cattle as they can every day. Their idea of dry-aging is same-day, Cryovac packaging and shipping, where the meat might sit in the meat counter for only a few days at most.

We sell meat in various ways. You can buy a half or a whole beef and have it cut up the way you want, or you can buy it by the individual cut if you like at our store or in some of the grocery stores. Stores that require fresh meat may have considerable waste if the meat does not sell in twelve to fourteen days. Some throw it out, while others donate it or possibly refreeze it. That provides another unnecessary path for the detrimental effects of climate change to occur.

Let's do some math. The average American in 2020 consumed nearly 264 pounds of meat comprising mostly beef, chicken, and pork. That is much more per person than is recommended by dietary recommendations. Mind you, we raise beef, pork, and egg layers. I will be accused of biting the hand that feeds me if I say I think we should lower our meat consumption per person. I said lower, not eliminate. Eating just eight ounces of meat per day, seven days a week, lowers the amount per person to 183 pounds annually. Let's say you reduce it to six ounces per day, which is close to the recommended dietary amount. Then you come up with 138 pounds per person. If you eat just six ounces of meat once a day for six days a week, then you reduce the amount to 118 pounds per year, or over 55 percent of what it is today. That is without considering replacing meat once a week with fish or by serving more meatless meals.

I presented at a conference in Washington, DC, in 2010 sponsored by the National Research Council of the National Academies called: "Toward Sustainable Agricultural Systems in the 21st Century." A fellow Iowa farmer made the public comment that not everyone could do what we were doing with a diversified organic crop and livestock operation, and we were only filling a niche market not every farmer can get into or make work. Arguments could have been made on both sides of those important points, and the comments have become more valid as the organic industry has grown and changed over time. It is not an easy task for certified organic farms to market their cattle, pork, chicken, or eggs in large enough volumes to make a decent living. Only large organic chicken and egg companies and a few large organic dairies (which all operate at least somewhat the same as their conventional counterparts do) have been able to create the economies of scale needed to sell large volumes of organic products. There continues to be ongoing controversy and challenges to some of these entities as to how much they are following the organic standards that smaller operations must abide by. Organic beef and pork numbers have lagged behind, especially pork. Much of organic hamburger being sold comes from countries like Australia, New Zealand, and Uruguay. It is grass-fed and can out-compete most American organic hamburger in price.

Organic Valley, where we also sell beef and pork, is the only cooperative entity making meat sales work at any significant level. Its strength came and continues to come from family-owned, pasture-based organic dairies where it has performed many great functions for the success and health of thousands of family farms across the country. The standards are strict but fair and exemplify the best of what organic should and could be.

In Iowa and in a few other states, an effort, with some bipartisan support, is being made to create more smaller-type lockers for beef, pork, and chicken where local farms, both organic and conventional, could begin to harvest and sell more of their products locally and regionally. It is a small but important step in the right direction. The effects of climate change, food production, and how far food will travel will all be critical determining factors in the future. This seems like a very appropriate time to say I do not have all the answers. I hope I am asking a few of the right questions.

Chapter 11
Growing People, Not Just Crops

Grace is the beauty of form under the influence of freedom

—FRIEDRICH SCHILLER[1]

This chapter outlines some of the legislative evolution of organic farming and founding organizations. It also includes a cameo on our family and the involvement of the next generations in the pursuit, both near and far. It ends with questioning the correlation of the use of pesticides and the occurrence of cancer.

Opportunities as a Spokesperson for Practical Farmers of Iowa

I consider myself a lucky man in so many different ways. I have had many interesting and exciting opportunities that I likely would not have had without the Practical Farmers of Iowa (PFI) becoming a part of my life. I do not think I would have been totally satisfied just farming. I also liked and needed to be around people. At least some of the time anyway. I also felt the need to be around people with whom I could share common goals and feel a sense of community. Maria and I have always had one employee for all the years that we farmed until two of our sons started taking over the operation. Working alone is a good way to get hurt when trying to complete tasks that require more than one person. We had the same employee for the first seventeen years that we farmed. I do not like to use the word employee because I feel that anyone who chooses to work on our farm is really more of a trusted partner than a hired laborer. I would not have been able to travel to the many meetings that I participated in without a competent and trusted person being there in my absence. I certainly could not have done it without the help and support and hard work of Maria who handled all of the family matters and also some of the farm business when I was gone. However, I wanted to make a difference in the areas I thought were important. Maria has always supported me in that. For me, that naturally meant being able to try to affect agricultural policies.

PFI's volunteer board members had to figure out immediately how it could be funded. We all donated our time, the mileage traveled, and hearts to board meetings. But we obviously had other expenses, and the small membership fees of $50 per year did not go very far. We wanted to hire an Iowa State University (ISU) agronomist named Rick Exner, PhD, who had been working with

Dick and Sharon Thompson with their field trials. We hoped he could assist PFI cooperators in conducting the on-farm research trials, hosting field days, conducting meetings and conferences, and publishing the results of our research. PFI owes Rick Exner a great deal for his tireless efforts in helping us to establish our presence. The same holds true with the dedication and assistance of Jerry DeWitt, PhD, from ISU Extension. He was a passionate believer in what we were trying to do, and he provided the organizational and initial financial framework that allowed us to turn baby steps into running a new and exciting organization. Rick Exner ended up working halftime for PFI and the other half for ISU extension. ISU and PFI had a good working relationship with this formal partnership for the first fifteen years of the organization. In 2001, the first executive director was hired, and the formal partnership with ISU ended. That did not mean our partnership with ISU research ended. On the contrary, PFI remained and grew in its relationship with various departments within the College of Agriculture and with professors wanting to do research projects with PFI farmers.

We immediately began to seek funding from foundations that were beginning to promote more sustainable agriculture. These foundations experienced a steep learning curve as they began to consider what the words "sustainable agriculture" really meant. I had a simple definition, which was it sustained the soil, the water, the air, and the people. The well-being of rural communities and farm families had to be just as important in the equation as the environmental components.

I took on the responsibility of representing PFI at various meetings, mostly around the Midwest, and in helping to secure funding and exposure through grant writing and program presentations. To be honest, I loved that part. I loved to travel to universities and other state organization meetings that were noticing what our organization was able to do in partnership with a land-grant university. Various states wanted to start organizations that would also partner with their own land grants. Unfortunately, very few succeeded in that goal because some universities did not have the buy-in or the people to make it happen like we had with ISU. The responsibilities and opportunities that PFI brought helped me to grow personally. PFI grew people, too, not just crops and livestock.

I was often the only farmer at some of the meetings in Washington, DC, involving research and commodity groups, and often it became apparent some people did not appreciate having "real" farmers attend their meetings. It was far easier for them not to have farmers present than to risk them not adopting whatever action was proposed. One such instance occurred when I was invited to attend a grain user's advisory event sponsored by the National Research Initiative of the United States Department of Agriculture (USDA). At the reception before the meeting started, one individual gave me a warning—he hoped I was not some radical from a group of hippie farmers, or something like that. I had a suit and tie on so did not look any different than him. The next day centered around the discussion of the new National Research Initiative competitive grant program that was to center on food safety, environmental quality, and the relatively new area of genomics. Some of the discussions centered around making progress in mapping the corn genome. I talked about the role of on-farm research trials that could provide farmers answers to some of the very practical but also very important financial and environmental quality questions. I thought it was generally well received. The chief scientist of USDA was there, and he talked about how sad it was that his brother lost the family farm in North Carolina due to his inability to grow anything but tobacco. The farm had been in the family for over 300 years. At that meeting, I asked the National Research Initiative Advisory Committee to consider appointing a farmer on their committee. That request was not well received. Sometimes, unless you have a PhD behind your name, you are not considered qualified. I was an outlier, as was more often than not the case.

Participating in these opportunities was sometimes "heady" events. I would come home from Washington, DC, all fired up about progress that could be made on this policy or that policy only to find that the realities of running a farm business quickly took over again, which was probably a good thing. I might have grown tired of hitting my head against a brick wall. Sometimes I felt like I was in the minority or the one who had to fight the hardest to even be heard. Those focused on such things as small farms and rural communities, organic agriculture, local foods, on-farm research, and fair markets always had to fight for every little thing they got to make even small inroads. It was the first time some of these terms had been used at the USDA and federal government levels. This was especially the case for government farm programs administered through USDA. Trying to get programs that do not discriminate against small, resource-conserving farmers has always been an uphill battle. I have great respect for those organizations that work every day on these issues and do not get burned out in the process. You have to really believe in the mission you are trying to accomplish. Organizations working in the sustainable agriculture sector have been doing that for over forty years. Many started in the 1980s, both at the state and national level. They deserve our utmost gratitude, support, and cooperation. It is not easy work.

In light of the sad events that took place on January 6, 2021, at the US Capitol, one prior experience deserves to be shared. I was at a meeting in Washington, DC, to discuss the new Sustainable Agriculture Research and Education program (SARE).[2] It was held in the latter part of July, as I recall, the year possibly 1992. It was one of those perfect summer evenings with a full moon. At about 11:00 p.m., I decided to take a walk from my hotel to the Capitol. I was the only one around at the top of the steps on the west side. No guard rails or guards were there back then to prevent people from walking around what I believe to be the second-floor balcony on the west side, at the level where presidential inaugurations take place. I walked all over looking at every nook and cranny, admiring not only the building but also the freedoms that allowed me to be there and do this in the first place. I thought, where else in the world would such a thing be possible? I felt a great sense of humility that an ordinary citizen could have such a great privilege. We all know that those unrestricted days are over. They began ending after September 11, 2001, when planes crashed into the Twin Towers in New York City, followed by the insurrection and destruction that occurred on January 6, 2021, at the Capitol. I fervently hope and pray it is not an omen of even worse things to come for our fragile democracy. Yet, I feel blessed to have had the chance to fully experience it when I did.

National Sustainable Agriculture Coalition

Some of the important national organizations that I have the privilege of being involved with are best known by the people who founded them, particularly the National Sustainable Agriculture Coalition (NSAC)[3] and Ferd Hefner. NSAC represents 116 different groups from around the country as their voice for sustainable agriculture in Washington, DC, and at the grassroots level. Ferd was its policy director from 1988 to 2016. NSAC is located in the United Methodist Building in Washington, DC, and is very close to the US Supreme Court building and the Hart Senate Office Building. The Methodist building has the distinction of being the organizing site of the 1963 March on Washington led by Dr. Martin Luther King Jr. Ferd knows more about federal farm policy and farm bill legislation than just about anybody who has ever been involved with it. It was his leadership efforts that paved the way for the most important legislation and USDA programs to benefit sustainable agriculture, its practitioners, and its rural communities. Without Ferd and NSAC, we would be far from where we are today in terms of agricultural policy that promotes and rewards sustainable farmers for the ecologically sound practices they employ on their farms. He is still involved in the now critical work of identifying and advocating for climate solutions for environmentally conscious farmers and rural communities.

Organic Farming Research Foundation

The second such group is the Organic Farming Research Foundation (OFRF) based in Santa Cruz, California.[4] Its first executive director was Bob Scowcroft, who founded the group in 1990. Bob came to California from the East Coast to work for the San Francisco environmental group Friends of the Earth.[5] In 1987, Bob was hired to be the first executive director of California Certified Organic Farmers (CCOF).[6] CCOF began with fifty-four farmers in California in 1973 and was the first organization in the country to form a group devoted to organic farming practices. Just before, the Rodale Institute in Pennsylvania had written some of the very first organic standards that were totally voluntary in nature in 1972.[7] CCOF helped to create the first organic standards governing the growing and sale of organic foods in California. It also served as the certification agency for organic growers in California. In the 1970s, when the first CCOF was formed, a two-page bill was written governing the first recognition that organic food even existed. The word "organic" had to be struck and replaced with different wording due to the political uproar created over the use of the word. Nonetheless, it was a start. In 1979, the *California Organic Food Act* governing the definition of organic practices was signed into law.[8] There was, however, no means of enforcing it other than through an organization such as CCOF that could take disputes, if any arose, to the courts.

Remember Alar? In 1989, a pesticide scare occurred about the chemical Alar being used on apples. The actress Meryl Streep and the Natural Resources Defense Council appeared on a "60 Minutes" program (CBS) on February 26, 1989, questioning the safety and presence of the chemical in children's baby food. Other reports surfaced about growing concerns over dangerous chemicals in our food. Those concerns created an almost immediate increase in demand for organic foods. CCOF grew from 300 to 800 certified organic farmers in a very short time. The momentum built off this led to the first federal legislation that created the framework for the first National Organic Program (NOP) and for the National Organic Standards Board to determine what the rules for certified organic would look like. It took another twelve years to get those first rules finalized and approved. In 1997, the first draft of the proposed rules came out. Organic growers and advocates were outraged because of three different processes or substances that the NOP was offering as acceptable for organic food production: irradiation, genetically modified organisms (GMO), and the use of sewage sludge as a fertilizer source.[9] Comments from the public on this proposed rule turned out to be the largest number of comments ever sent to USDA at that time (more than 325,000). In the end, the "big three," as they were known, were thrown out in the final draft of the rule in 2002.

In 1992, Bob Scowcroft decided there had to be a way to generate organic farming research so that farmers could get more information on how to make organic farming work. He wanted organic research to be information-based, not product-based. Product-based information seemed to be the dominant approach being taken by agricultural research generated by universities and the USDA. It centered around chemical and fertilizer solutions to pest and fertility problems. In response, he proposed to start a foundation that would secure funding for farmers and institutions dedicated to organic farming research. The results of this research would be disseminated to whoever wanted it, and farmers would have educational and learning opportunities available on the issues of organic farming. The foundation was launched in 1992.

In the summer of 1997, Bob Scowcroft called and asked if I would be willing to serve as a board member for OFRF. I would be the first board member east of the Missouri River. Wow, I thought. I had never been to California. I had been in most states in the western part of the country, but never to California. California had always been sort of an enigma to me. It was a mysterious place where people were different from those from the rest of the country. That was my initial and misinformed perception of what it would be like. Of course I said yes, although I felt very ill prepared to begin talking about anything like California agriculture. I did not have to. The organization was about more than California agriculture. It was a serious attempt to get more dedicated organic research in other states like Iowa, the Midwest, and all the other regions of the country that would be new territory for this sort of funding.

I immediately found the work of OFRF to be very worthwhile and interesting. We held two board meetings somewhere in California every year. Those meetings provided me the opportunity to learn a great deal about how organic agriculture worked in the state and how organic fruits and vegetables were grown, packed, and sold. They were so much further advanced in all aspects of organic farming than we were in the Midwest. They had the infrastructure and the markets. They had the processors and the packing sheds and the wholesale and retail distribution for the great variety of organic fruits and vegetables being grown there. We still lack so much in Iowa in the processing, marketing, and infrastructure processes it takes to increase a market for organic foods. We are not alone, of course. Many states are far behind us. We also held board meetings in Colorado, Texas, and Georgia, which provided opportunities to learn about organics in those states. During his career, Bob Scowcroft became a very talented and gifted organizer, fundraiser, and tactician for organic agriculture in all its aspects. He was especially adept at securing funding that could be used to fund organic research through a competitive grants program that really took off and ran during the seven years I was involved. We had our own in-house research and education review committee that determined who would be funded.

OFRF is a farmer-led organization just like PFI, which was what drew me to it and was what I enjoyed and valued the most. OFRF is not a top-down, run-by-the-executive-director group. I do not think those kinds of boards are as effective, nor do they have the needed buy-in and the direct work and participation of board members needed to accomplish what they could do if they had that kind of ownership. As I recall, it funds about fifteen to twenty proposals every year and has awarded more than 350 grants over its nearly thirty years. The foundation accomplished a great deal during the years I was a part of it. Perhaps the biggest one centered around the effort to get more dedicated organic research into the federal agricultural research budget. Organic products were growing at the phenomenal pace of over 15 percent growth every year. Yet organic's share of the federal research budget was miniscule. We wanted to get our fair share, and so we set out to do so. Interest in organic research began to explode through the grants that we began awarding.

Through the efforts of OFRF's long-time policy director Mark Lipson, we conducted the first scientific congress in the United States on the state of organic farming research. We continued year after year to work on convincing the US Congress that more funds were needed for the many worthwhile research projects that we could not fund. We finally did win an important part of that battle, due to the advocacy of both the foundation and NSAC. In 1990, Congress created the Organic Research and Education Initiative (OREI), which is the competitive grants program within

USDA that is supposed to fund organic research, but it had never been funded.[10] Finally in 2004, it was funded for the first time, with a total funding of $15 million over a five-year period. By 2020, the funding had risen to $25 million annually.

Things were rolling along mostly okay in October of 2000. The board expected me to assume the role of vice president and then become the president in 2001. I did not know what to think about that. Being the president of a board headquartered in California seemed like it would require a level of capacity that I was not so sure I had. There was enough stress in my life already. The stress usually caused pain and numbness in my left shoulder that would then go down my arm, which made me wonder if it could be my heart. Or was it just the usual aches and pains of lifting too much or working too many long hours intrinsic to my occupation? I did not know for sure.

I was at a local football game in Harlan on a Friday night in early October. Our oldest son, David, was a freshman at ISU in Ames. Daniel was a junior in high school, and Mark was a freshman. They were both performing in the high school marching band at the game. That evening, all of a sudden, a bolt that felt like lightning went through my body. I wanted to run. I was scared. More scared than I had been since that late January night in the dorm at Iowa State way back in 1970. I told Maria that I was going to drive home because I was not feeling the greatest. I did not tell her that I did not know what was going on. I had driven to the football game alone anyway, getting there a little late as usual because it was harvesting time and much work had to be done.

I almost drove myself to the emergency room of the local hospital, but thought, no, I would not do that because I felt quite stupid about it all. I finally just went home and waited for Maria to get home. I told her what happened but did not think too much about it the next day, except being afraid that the feeling would come back. Sure enough, it did. And it started coming back a little more often, too. It happened at night when, after falling asleep for maybe an hour or two, a jolt would wake me up. The fear that it invoked was unlike anything I had ever experienced. I just wanted to run, but there was no place to run to. I would begin to shake like a leaf. I couldn't calm myself down. I agreed to seek help and learned that I was having panic attacks. I was good at hiding them in public, but they were obviously no fun. I learned how to control them mostly through learning how to use biofeedback and coping mechanisms that involved repetitive phrasing and through assuring myself that I was not going to die.

This went on for a long time. The panic attacks became less frequent and more sporadic. I coped for the most part. There was also this nagging feeling that simply would not go away. What if there was something wrong with me? Was I afraid of having a heart attack? Over the next three years of 2001, 2002, and 2003, life went on. I became the board president of OFRF. Maybe I would get a slight panic attack every now and then but nothing serious. I got checked out a few times with EKGs and treadmill tests to try to convince myself that nothing was wrong with my heart. Those tests indicated that there was nothing wrong. Still, I felt vulnerable and self-conscious.

It was the first part of February 2004. I woke up feeling ho-hum. It had snowed about four inches the night before. It was winter, and time was dragging. I went out to scoop some snow so I could open the shop door to get a tractor out to start the morning feeding chores. A very slight pain went up my left arm. Then a little panic followed. I went to the house and told Maria and right away called our family doctor's nurse. I described to her what happened. She knew that I had the panic attack issues. She advised that I get my heart checked out. Only this time she said that I should call the Creighton Cardiac Center in Omaha, Nebraska, and tell them I was coming. She did not want me to stop at the clinic at the hospital in Harlan.

So we went to Omaha. I told an emergency room attendant about my symptoms, which were atypical for someone with a heart issue. I told the cardiologist about heart disease being in our family and my dad dying from it. Finally, he said this: "I am 99 percent sure there is nothing wrong with you, but you will never know unless we perform a heart catheterization." That statement hit me like a ton of bricks. I said a prayer to God and Dad, and there was an immediate thought of, yes, let's do this. I was somewhat awake during the procedure. I remember the surgeon saying I was lucky and that it was a good thing I had the dye test. My left anterior descending coronary artery was over 95 percent blocked. Had I had a heart attack, I most likely would have died right then and there. They put a medicated stent in me, and I went home in two days.

I understood then the why of the anxiety and panic attacks. I understood what my body was trying to tell me. I understood that both God and Dad had intervened on my behalf. It was not my time yet. I know we never quit learning in this life, or at least we should not quit learning, and I did learn not to be so afraid of having heart issues after that. I became much more accepting of whatever was in store for me, in that category anyway. I learned to be so much more trusting in God's love and care and also in Dad's love and care. For the second time in my life, I felt like I had witnessed a small part of what heaven must be like. Since that time, I have had another stent put in a lesser coronary artery. I have also had a procedure known as an ablation that has effectively dealt with an irregular heartbeat. I feel like I have been given a great gift by God through the intercession of my dad who did not receive that particular gift himself. That is what I want to think and believe. Whether it is true or not I do not know for sure, but I want to live out my days or years on this Earth paying for that gift as best I can. I suppose just as in farming, it is a debt that I can never fully repay.

Where We Have Been and Where We Are Going— The Next Generation Is Growing Up

As a new millennium was ushered in on January 1, 2000, I remember its celebration in our home with Maria and our three boys, David, Daniel, and Mark. I think that might have been the evening we watched the 1962 epic film, *Lawrence of Arabia,* all in one setting. (Technically speaking, millennium means a thousand years and should not have started until January 1, 2001.)

Maria and I encouraged our boys to be involved in many activities. That meant music, 4-H, youth church groups, and some sports.

All three boys participated in Westphalia baseball, our township 4-H club called the Lincoln Leaders, and all the music programs in the Harlan Community School District. They went to Shelby County Catholic School in Harlan from kindergarten through sixth grade. During some of those years, Maria was employed as the public relations director and chief fundraiser for the countywide parochial school system. That was when the towns of Earling and Panama still had an elementary Catholic school in their towns. Today, only Harlan has a Catholic grade school, a very strong one at that.

David is a drummer, and Daniel and Mark both play trumpet. All three sons are accomplished musicians, and Harlan won the state jazz contest a number of years while they were playing in the band. All three played in the ISU marching band. All of us are big ISU sports fans in basketball and football. I was a 4-H leader for twelve years, and for their projects, the boys showed our home-raised cattle and pigs at the county fair. They had many other projects and enjoyed the program.

All three learned how to work on the farm at a very young age. Maybe they did not have to do as much as I did when I was growing up, but that depends on who you are talking to. Maria had the sound opinion that all three should also have experience working for someone other than their dad and off the farm. They were introduced to money management, and back then, savings accounts earned a respectable rate of interest. David and Mark chose restaurant work and seemed to enjoy it. Not so Daniel. He wasn't so keen on this kind of employment. Now, he and Ellen own and manage an amazing local foods restaurant in Harlan called "Milk and Honey."

David graduated from high school in 2000, Daniel in 2002, and Mark in 2004. All three graduated from ISU in agricultural curriculums. David received a Public Service and Administration in Agriculture degree in 2004. He met his wife while working in Minnesota. She is a clinical psychologist at Boystown in Omaha. He worked the last two years in Minnesota for the Land Stewardship Project, a group similar in scope and purpose to the Iowa-based PFI. In 2014, David and Becky were married. He decided to come back to the farm to work in the family operation. It is a blessing.

Daniel decided to major in Agronomy. Maria and I had envisioned him becoming a scientist and researcher. He did not. He told us a little later that he always thought he wanted to come back to start farming. He graduated in 2006 and came back to the farm in the spring of 2007. He met his wife Ellen through PFI. Ellen also studied Public Service and Administration in Agriculture and International Agriculture at ISU. Ellen and Daniel were married in 2010. They live in the house where my dad was born. It is another blessing.

Mark obtained two undergraduate degrees (in History and Agronomy) at ISU. Although Maria and I considered him to be the one most tethered to home, as he was the youngest, his career choices have led him the farthest from home, beginning in 2008 when he entered the Peace Corps and was stationed in Honduras for two years. His work centered around agricultural and community capacity building, and he has continued to pursue such goals. He has been with the Foreign Agricultural Service of USDA since 2012, initially working at USDA in Washington, DC. His first job was in coffee. Much of this work was in coffee cooperative development and coffee rust disease mitigation strategies in countries primarily in Central America. He met Courtney Dunham in 2013, who was a returned Peace Corps volunteer also stationed in Honduras. Courtney was in charge of the Partners of the Americas' Farmer-to-Farmer program, which had its beginnings during the John F. Kennedy administration. Courtney had the tragic misfortune of being diagnosed with non-Hodgkin's lymphoma in 2015. She always thought that she could beat it as she was young and strong. Mark and Courtney were married for only thirty-four days when she passed in late 2016. We have always wondered if she developed this cancer while working in Honduras on farms that used pesticides banned in the United States but allowed in Honduras. We will never know.

Mark served as a foreign service attaché in New Delhi, India, focusing on trade of all kinds, both traditional and organic, from 2020 to 2023. He was fortunate to meet another wonderful returning Peace Corps worker, Virginia Lehner. Mark and Virginia were married in the fall of 2021. Virginia works for the State Department as a problem solver for US citizens overseas. They were recently reassigned to Bogota, Columbia, where they are both serving in their former capacities.

All three of our sons are engaged in food, farming, and agricultural pursuits. We try to count our many blessings every day. None of us know for certain what the future holds.

One of the most basic benefits that I learned early to cherish about farming was the independence of being my own boss. I freely admit that I would have had a hard time working for someone else all of these years. I am a fiercely independent person in many ways. I am also a self-driven person. I was not always that way, but because I had to make it in farming on my own after my dad passed, I gradually became one. It had much to do with financial challenges, helping to raise a young family along with Maria, and the challenges of farming without pesticides. I learned to accept the long hours of outside work doing everything associated with farming—fixing machinery, managing water systems, welding, carpentry, etc. My confidence grew as I gained skill and familiarity with the details of solving problems. I have also been driven to be a leader and to make a difference in agricultural sectors, which has meant many hours of organizational and policy work over the years.

I also freely admit that it has been and continues to be hard for me to let go of some of our farm's management, for some valid and some not-so-valid reasons. A valid reason is that, as I learned to work with nature on a daily basis, knowing what actions to take became almost automatic. I learned to work with the weather. During good weather, I worked out in the fields much of the time. When it rained, I worked with livestock and fixed things. I learned after many years what works, what doesn't, and what needs to be the priority. That is not always the case, but most of the time it rings true.

A not-so-valid reason is that farmers, including me, can quickly become married to their farms and do not adequately allow the next generation to find their own way, do their own thing, and make their own choices and mistakes.

Maria and I will most likely live on our farm as long as our health permits, because David and his family have settled in the town of Avoca, fifteen miles from the farm, and Daniel and his family will continue living just to the north of us on the farmstead where my dad was born.

This means that I need to gradually semi-retire more over time—it will be a work in progress. Doing so will allow me to explore new interests and adventures more seriously. Writing is one of them. I hope to be able to keep driving a tractor, skid loader, and other farm implements to help our two sons. I hope to continue to be a "Go-fer" (go for this, go for that, etc.). I hope to continue planting and caring for trees and shrubs. I hope to continue fixing things. As long as there is good communication, acceptance of one another; mutual love, respect, honesty, and forgiveness; and a shared love of the land and farming, life cannot be all that bad! In fact, it is a blessing.

November of 2023 brought another health challenge to me. I was diagnosed with prostate cancer. It was not entirely a surprise, as my dad and my two older brothers have both gone through it. The prostate organ was robotically removed in April of 2024. I was informed in the late summer of 2024 that not all of the cancer was removed. As of this time, I have completed thirty-nine sessions of radiation. The radiologist and I are both confident that this will be one hundred percent successful in removing the cancerous cells. Seventy percent of all men have some degree of prostate cancer when they reach eighty years of age. The majority die with it, not from it. However, it appears that prostate cancer is increasing and also in its severity. There is growing evidence that Iowans fall into that category. It also appears that it may be due to increased exposure to substances found in the environment. Again, could it be linked to all of the chemicals used in agriculture?[11]

Chapter 12

Matters of Church, Religion, Science, and Technology

A Layperson's Perspective

Where there is hope, there is faith. Where there is faith, miracles happen.
— AUTHOR UNKNOWN[1]

Since the 2016 election and the outbreak of Covid-19, institutions of many kinds have been hard hit. The institutional Church has been one of them. In today's society, attending church services does not have the same significance, nor does it carry the sense of obligation that it once did for many people, both urban and rural. However, in a rural area, it is much harder to keep a small parish like ours at St. Boniface in Westphalia, Iowa, operating when there is a declining population and when perhaps less than a third of the people who do live here attend services on a regular basis.

In this chapter, I bring a comparison of my upbringing in the Catholic faith and what responsibilities I believe that brings me with respect to being a good steward of the Earth and of our environment. While this sentiment is echoed in many faiths and for those ascribing to a sense of spirituality, here I offer my thoughts as they relate to my belief system.

The hopes and dreams of the people who settled here in 1872 have been greatly altered. We were 150 years old in 2022. There were hardly enough people left to put on a celebration.[2] However, some in the parish worked very hard to make it a success in spite of the sparse population.

The centennial celebration in 1972 brought nearly 20,000 people to the little town of 160 people over a three-day period revolving around the Fourth of July. The quasquicentennial in 1997 brought about the same numbers. For the sesquicentennial, numbers were estimated to be about one-half that number, at 10,000. That is still a great deal of people for a tiny community.

In 1997, we had just finished straightening the church steeple, which had been leaning eleven degrees to the southeast due mostly to the deterioration of some of the soft brick in the bell tower. It was literally finished the day before the celebration was to begin. The brick had been fired on-site back in 1881 when the church was being built. There was not quite the consistency of hardness needed to stand up to the rains and moisture that came through the louvers in the tower. The bricks were turning back to clay powder, and the massive timbers were rotting. The two bells that were cast in St. Louis were becoming precarious, with one weighing 2,100 pounds and the other

1,700 pounds. We put on a copper roof, shored up the limestone foundation, put in a new stained-glass window, and much more. We raised over $250,000 to pay for it all.

I was co-manager of the project and helped to take the cross off the steeple standing in a hydraulic bucket 117 feet in the air. It took us three years to complete the building project. It was the happiest three years of my life in terms of being a member of the St. Boniface, Westphalia, Iowa, parish. We did much of the work ourselves as volunteers. It brought the parish and community together like it had not been since the centennial celebration in 1972. Much of that spirit was due to the young priest from Council Bluffs who came to Westphalia to be the pastor of both the Westphalia and Defiance parishes. He had a college degree in history and appreciated the historical achievements that our town had attained over the years. Our entire campus, which included the church, cemetery, school, rectory, convent, baseball field and stadium, clubhouse bar and grill, first fruits corn crib, and a few other smaller structures, was all selected to be listed on the National Register of Historic Places in 1991. This was due to the campus's contribution to the settlement of the western part of Iowa by German Catholic immigrants. I, and many others, felt the same kind of belonging and of working for a common purpose we once had. The Catholic high school closed in 1964. The grade school closed in 1976. Change becomes all the harder when we perceive it as being more negative than positive.

Both the town and the parish face an uncertain future. The town may always be there, but the parish, I am not so sure about. When you start losing that will, it becomes a terminal cancer just waiting for the final days and hours to occur. I have heard it stated when a town loses its school, that is the day it begins to die. The sad reality is that now in Iowa and in so many other rural areas of our country, much larger towns and schools than Westphalia have closed or consolidated with other towns in order to survive. Many of these schools are the ones that still had enrollments of forty to fifty students or more in each grade back in the 1970s. Schools that consolidated in the 1980s and 1990s have had to reconsolidate again in order to keep going. Some schools may now include as many as five or more communities that once each had their own school.

What is to blame for all of this? I hope that it is evident by what I have attempted to describe in previous chapters and by what has transpired ironically since the day my great-grandparents arrived in this country and then settled here; that is, the community and many of the farms began an imperceptible decline soon after they arrived, which happens when there are too many losers and too few winners in the competition for land, wealth, and influence. It could be called the inevitable consequences of progress as defined by agricultural economists. In the case of Westphalia, the lack of a railroad was probably the primary reason why the town did not grow appreciably over the past 150 years. There is something to be said, though, for the fact that our community is still going, even though the numbers have declined. Credit needs to be given to those who do care about our community. The parish of St. Boniface and the volunteer fire department come to mind first.

Being Catholic, we were taught that attending Mass once a week on Sunday was an obligation. Now, I do not go because I have to. I attend weekly Mass because I both need to and want to. I need and want the opportunity to pray and to participate in the celebration of the Word and the Eucharist as an individual but also as a member of a community of believers. There is an opportunity for solace in just being in our church. It makes me feel at home. There is comfort and a real sense of peace just being in its familiar, simple, yet very beautiful interior space. Only a fool would say you have to attend church in order to be a good person. However, in my community's situation, attending church should be considered simply because we are in such grave danger of closing our parish doors. We just do not have enough people participating in parish life to guarantee we will have a pastor and a viable and fully functioning parish in the future. For years, we have shared a pastor with the Panama and Portsmouth parishes, which are significantly larger than we are. Closing would be hard to take after 150-plus years. Can we find and build our way back?

Some people have left the Catholic and other mainline churches because they do not find them personally fulfilling. Some now attend evangelical churches. I personally find a great deal of meaning and significance in the 2,000-year-old celebration of the liturgy of the Word and the liturgy of the Eucharist. The historical significance and tradition of this remembrance cannot be all wrong in my estimation. Ambivalence surrounding the questions of Jesus's divinity and that he conquered death and rose from the dead probably plagues many of us. I can appreciate a doubting Thomas. But now in our time, we have to do more than remember and believe. We are being asked to not only be aware of our current responsibilities in caring for our common home but also to act before it may be too late. We are being asked to ascertain what we must do both now and in the future for the integrity of humankind and our planet. With the pace of technological changes, we must ask ourselves whether we have developed the necessary ethical and spiritual framework to judge its merits in a clear and open-minded way.

I do not have any formal education in matters of religion or theology. I do think about these things, though. They are important to me. I will start with some of the basic questions and observations that I have had as my life has progressed. I suspect many of us have had the same ones. Martin Luther King Jr. said:

> Science investigates, religion interprets. Science gives man knowledge which is power; religion gives man wisdom which is control. Science deals mainly with facts; religion deals mainly with values. The two are not rivals. They are complementary.[3]

These words of Martin Luther King Jr. are important to me as a farmer. I want science to inform me how to be a better steward of the land and of our natural resources. The values I bring to farming have to coincide or be bonded with the spiritual stewardship principles that my faith and spiritual convictions aspire to.

Why are we here? I pondered that question when I was just a kid. I still do. But haven't we all, in one way or another? My early Baltimore catechism training said that I was created in the image and likeness of God. I believe those words to be true. But just what do those words mean? What is the image and likeness of God? First of all, I believe there is a God. That is an important start. I cannot come to grips with not believing that God exists. I believe there is a higher power. But I do not have proof that God exists. St. Thomas Aquinas developed five proofs of God's existence, but many philosophers after him have pointed out the many flaws in his reasoning.[4] I do not feel like I must have proof, but I do have to have faith, and that is my choice and is also a sort of intuitive response to how I view the world around and beyond me.

I, like other Christians in the 1950s, was taught God created all life at one time and not all that long ago, maybe somewhere around 5,000 years ago. It coincided with the biblical account of creation in the Book of Genesis. I do not remember thinking too much about these things until I went to college. My studies of Charles Darwin and his theories of evolution began to change my thinking, and having grown up on a farm, being so close to both plants and animals and their processes helped set the wheels in motion.

Fossil discoveries from around the world and the dating of them through the use of radioactive carbon began to reveal the age of fossilized plants and animals. That is, if you bought into the science of it. I did because I studied it. It made sense to me. All plants and animals contain organic matter in the form of carbon, which got there in the first place through carbon and oxygen in the form of carbon dioxide, which is used in photosynthesis to form food and energy for plants and animals. The older a fossil, the less carbon it retains. By measuring what is left, the relative age of a fossil can be revealed. How much carbon decays over time is determined by measuring what is called the radioactive isotope portion of the fossil. This was first accomplished in 1941. Carbon has a half-life of 5,730 years. The precise measurement of objects that were around 50,000 years old or more could not be done until accelerator mass spectrometry came along, which could measure very small amounts of carbon accurately and quickly.[5]

As I began to understand and study more about science and Charles Darwin's theories of natural selection and survival of the fittest, I realized there was a big difference between what the Book of Genesis says and what science says. Who is right, and does it matter? In a way, I have come to believe that it is similar to what Jesus said about giving Caesar his due. Give to God what is his due and give to science what is its due. Science seeks to reveal how the Earth and all life started and evolved to what it is today and how it continues to evolve.

However, it does not explain why. Some would posit that the "how" is based on the big bang theory. In spite of the evidence moving in that direction, scientists are not yet settled on this theory as the most plausible one for how our universe started. The theory proposes that all matter in the universe, current or past, came into existence at the same time through a massive outward explosion. At that time, all matter was concentrated in a very small point with infinite density and infinite heat called a singularity. Suddenly, the singularity began expanding, and the universe as we know it today began. The scope of this event is beyond even our imagination. Maybe you can get the starting point down to a singularity. But that still does not explain how it came to be in the first place, and certainly not the why.

The Big Bang Theory posits that the universe began at Planck time (10^{-43} seconds after the big bang). Planck time was thought to be governed by what are called the quantum effects of gravity dominating all physical interactions and that no other forces were of equal strength to gravitation.[6] The concept could be thought of as being the opposite of a black hole, which then could mean the existence of white holes. A black hole is described as where all matter is absorbed by gravitational force into a point so dense that nothing can escape, including light. According to physicists, though, white holes do not exist. Black holes can be formed by dying stars. There is now consensus that black holes exist in the centers of most galaxies. The presence of black holes can be inferred through their interaction with other matter and with electromagnetic radiation such as visible light. The theory that black holes could exist was conceived as early as the late 1700s by the natural philosopher John Michell (1724–64). He was a student of gravitational force and proposed it was possible for objects to have so much gravitational force that it would be impossible for light to escape.

All of this still begs the question, for me, of where did this original singularity come from. According to the Big Bang Theory, the first elements of hydrogen and helium were created by the tremendous heat and energy released along with small amounts of lithium and beryllium. That does not tell us how the singularity came to be. Where did helium and hydrogen come from? Where did the scientific laws that govern the universe come from? Can something be created from nothing? To me, only a higher being could be responsible. To me, only a "loving" higher being could be responsible, given all the beautiful forms of life born from those clouds of hydrogen and helium.

Perhaps I am off the mark, but my thought is God has allowed all of this to happen through the freedom of God's creation to evolve to where it has come today. Chance cannot account for how the universe began or how the original elements came to be. To me, there has to be a starting point. To me, it is God's love for all creation that has now evolved into the wondrous life of not only humans but also of all species. Science instructs we were not always fully human compared with what we are right now in this time and place. I am not afraid of any of this freedom we have been given. Science and religion should not be at odds. In fact, they should be infinitely linked and bonded together. Humankind has now evolved to where we play a more powerful and influential role in how evolution will continue to develop. It is our responsibility to recognize we can affect it for better or for worse. We have already done so in dramatic ways and especially since the dawn of the Industrial Age. Deciding what is for the betterment of the world is not to be taken lightly.

I believe I have a soul, a spiritual aspect. Some of that comes from my belief in God and a life after death. I believe in heaven, while I also hope there is one. I think I believe in some sort of hell, too, but it is harder to hope there is one. I have already mentioned the thought and feeling of having some sort of glimpse of heaven on at least two occasions in my life. I have a hard time wrapping my head around the creation of the soul because it does not fit nicely into a timeline science has for the evolution of modern man. When and how did humans form a conscience?

God takes on a whole different meaning when you consider belief in the Trinity doctrine or the triune God. God then becomes the Father, the Son, and the Holy Spirit. God then becomes steeped in scripture and its interpretation, culture, tradition, and history, not just in science and creation. You have to choose to believe this. It requires a leap of faith. Again, I have to admit I can at times be a doubting Thomas. Why? It requires one to believe Jesus is the Son of God who took on complete human identity to teach us how to live and to redeem us of our sinfulness. The Old Testament prophesizes about the coming of Jesus. The New Testament fulfills those prophecies. To me, that gives it a degree of consistency and legitimacy. Jesus is the only man who has claimed he was the Son of God. Has any other religion or philosophy claimed this for their founder?

If one takes the Old Testament completely literally, then it seems to become an either/or situation. It makes everything black or white. All matters of what it says become absolute. I believe there is much gray area to try to make sense of. Do you believe only the historical parts that make sense to you, or do you believe God literally spoke to Moses, or do you believe the Old Testament is based less on a historical recording than on the Jewish people's struggle to understand what and who God is? The idea of a single or monotheistic God was not only believed by the Jewish people. There are historical indications ancient Egyptians and Persians believed in the concept of one all-powerful god.

It is hard to reconcile the question of why people who call themselves Christian are considered worthy of heaven and have been given the gift of redemption when people growing up in different cultures and religious traditions are not. Many wars have been fought and lives lost due to differences in religions and beliefs, and some continue to be fought. If one truly believes the two most important things Jesus wants us to do are to love God and to love our neighbor, then how can one reconcile trying to force or convert people to one's faith? We actively have done that for many centuries. It is part of how we justified the settlement of America and the removal and attempts at conversion of the Indigenous peoples, who were considered so much less intelligent and civilized. The word "savage" was designated for them, as they were thought to be untamed, uncivilized brutes.

How does one reconcile the fact so many people die or are killed every day at the hands of terrible acts of injustice, accidents, starvation, diseases, or weather-related catastrophes? Why am I still living, but another dies so young? There are many mysteries. I choose to accept that. Not only does it require faith but also hope for a place where there is justice and happiness for all. Does that place require accountability for how we live our lives on Mother Earth? I like to think how we live our lives on Earth, using the talents and gifts we are given, is a little like how heaven is too.

It is comforting to dream God wants us to have a little bit of heaven here on Earth so that we can aspire to it in the next life. I like to muse and say there has to be a heaven so I can do all the things I never had time for in this life, like perfecting the banjo and violin.

I value the traditions in the Catholic Church, especially the celebration of the Eucharist and the belief in the bread and wine as being the body and blood of Jesus. There is so much value in the lessons of the Gospels. Jesus taught us how to live. I value the sharing of this communion and the celebration of coming together as a community to praise and to pray for both ourselves and one another. We are called to ask for forgiveness for ourselves and one another. These beliefs and traditions support and provide meaning in my spiritual life. Prayer has to be an important part of my life. How to become better at it is a lifelong endeavor because, much of the time, I feel very inadequate at praying well.

Tolerance and acceptance of one another are needed if we are to see God as covering and protecting us like a big tent, or as Cardinal Joseph Bernardin called it, "a seamless garment."[7]

To me, it is important to think and learn more about who God is, based on the complexity of our evolving humanity. During my lifetime, but especially now in the latter part of it, I sense God's wake-up call for what He is asking of me and other believers: a call for justice for all of God's people and God's creation. It means ecological justice in how we view and use the gifts of nature that God has given us for our sustenance and well-being. It is recognizing we have abused so much of the Earth and its resources too often in the names of greed, selfishness, and questionable technologies arising from conflicting new knowledge. Is that the same as eating from the Tree of Knowledge as Adam and Eve did in the Old Testament story of creation, thinking that now we are like gods and know as much as God does?

> It is recognizing that there has to be a new, sacred union between humankind and our environment, as Pope Francis so eloquently and powerfully elucidated in his encyclical *Laudato Si'* written in 2015. I now think of it as a sacred union of human, social, and ecological justice.[8]

Pope Francis has done the Catholic Church and all mankind a huge favor if we take his words seriously on taking care of God's creation and our common home, Earth. How many people realize that the book of Genesis includes two accounts of creation, Genesis 1 and Genesis 2? In Genesis 1, God is referred to as "Elohim," Hebrew for God. In Genesis 2, God is referred to as "Yahweh," His name too sacred to be spoken. In the first story of creation, the words "dominion" and "subdue" are used to indicate man's superiority over all living things, plant and animal. This is in sharp contrast to the words "care" and "cultivate" to describe man's relationship to the Garden of Eden in the second story of Genesis.

The Old Testament is seen by some scholars as a story of a people in search of a homeland. You will find both poignant and beautiful statements about how land is to be viewed in terms of ownership of it. The idea that land was entrusted to the Jewish people by God as their common inheritance and that they were entrusted to be good stewards of it is clearly indicated in the book of Leviticus. Leviticus was written by Moses and refers to the priestly tribe of Israel or the law of priests. When the Israelites finally settled in their homeland after years of wandering in the desert, the land was apportioned by casting lots, meaning no special privileges of claiming the best land were granted. Leviticus described what God wanted their attitudes to be regarding the land as property: "Land must not be sold in perpetuity, for the land belongs to me and to me you are only strangers and guests."[9] In fact, it was stated in prescriptions of the Jubilee Year that every fifty years all the land was to revert back to the original families that first entered and settled in the promised land. Slaves were also set free and could return to their communities. That way, the poor, the indebted, and the dispossessed could have the opportunity for a fresh start. The Jubilee Year occurs after every seventh Sabbath year, thus every fifty years.

In the Gospel of Luke, Jesus reads from the book of the prophet Isaiah: "The spirit of the Lord is upon me; therefore he has anointed me. He has sent me to bring glad tidings to the poor, to proclaim liberty to the captives, recovery of sight to the blind and release to prisoners. To announce a year of favor from the Lord."[10] This refers to the Jubilee Year.

If you are looking in the New Testament for a list of precepts or statements on the environment and ecology and how to take care of planet Earth, you will be hard-pressed to find clear examples. Jesus does address social justice issues and the foolishness of piling up wealth as a useless endeavor but does not speak directly about such things as land stewardship. Jesus lived as a human being in a historical time and place that was defined by the knowledge and science that existed in that time and space. Societal structures and what was considered to be fair and just and right had already come great lengths, but it too was limited by the times and constraints of where we were in our human history. What else should we expect? Jesus turned what was considered right and wrong on its head by teaching us what is really important and gave us a New Testament based on the precepts of the Beatitudes and emphasis on the commandments of love of God and love of neighbor. How we treat our environment and the natural world around us should be no different in its interpretation than that of loving God and neighbor. In summation, I offer these thoughts:

> To love God is to love and respect creation and the natural world. To love thy neighbor is to share creation justly and equally. To love God is to love and respect science and the processes that govern creation. Science also has to respect creation and what that means for the good of mankind.

Perhaps it is time to embrace a different creation theology. If we are to come to terms with a reconciliation or conversion of respect and care for God's creation, then we must recognize that a scientific knowledge of how creation works has to be integral to the healing of the Earth from the negative effects we have caused. These have occurred in less than 500 years of our industrialized economies. The wounds have been many—loss of soil and soil quality, destruction of many different species, acidification of oceans, removal of people from their land and homes, pollution of water and air, wasting of food, and throwing away things that could be recycled. It is exploitation of resources at the expense of poor people and poor nations. If we are abusing creation, we are abusing God.

Why can't science, God, creation, and the texts of the important books of all faiths and religions all be talked about in a new context, in the same breath, and with the same mutual love and respect? It is time to view God in a broader appreciation and perspective. Maybe it is time to give God an update! God has revealed so much more of who and what God is through scientific discovery. We owe it to God to learn and appreciate what has been revealed. That is not to say we should embrace every new scientific discovery that comes along. We have to be able to discern what is for the good of creation and mankind. However, we cannot do that if we do not understand it.

Jesus may not have spoken directly about it; there are, however, remarkable principles of land stewardship to be gleaned from both biblical and Christian traditions that describe what people's relationships should be with the land and with each other. One place these have been considered is in a regional Catholic Bishop's Statement on Land Issues called "Strangers and Guests: Toward Community in the Heartland," May 1, 1980, as shown here:

(1) The land is God's;
(2) People are God's stewards on the land;
(3) The land's benefits are for everyone.
(4) The land should be distributed equitably.
(5) The land should be conserved and restored.
(6) Land use planning must consider social and environmental impacts.
(7) Land use should be appropriate to land quality.
(8) The land should provide a moderate livelihood.
(9) The land's workers should be able to become the land's owners.
(10) The land's mineral wealth should be shared.[11]

This statement became the basis for the stand that some very brave bishops and leaders in the Catholic Church took in response to what was happening to farmers during the farm crisis years of the 1980s. It fell on deaf ears of those who controlled its outcomes. The messages given to farmers in the two prior decades implored farmers to get big or get out. Many did get bigger. Many had to get out because land prices, production costs, overproduction, and interest rates just kept going up while crop and livestock prices paid to farmers mostly went down. There were 20 million farms in 1950, 10 million in 1970, and just 2.2 million in the late 1980s. I contend both political parties have not cared that this occurred. Lately, it has been said liberals abandoned rural America and conservatives exploited it. As long as food remains "cheap," farmers are expendable and of little value. But then food has not really been "cheap." The loss of farmers and rural communities, pollution of our soil, air, and water, and the new destructive forces of our climate are causing debts we may never be able to pay off.

Historically, some in the Catholic Church saw the vocation of farming as an endeavor worthy of support and praise. The same holds true for the uplifting and support of rural communities and rural Catholic Churches. For that reason, the National Catholic Rural Life Conference was established in 1923 by Rev. Edwin O'Hara, who grew up on a Minnesota dairy farm.[12] At the time of the organization's beginning, he was a rural pastor in Oregon. He felt that the rural Church was being underserved, revealing a pattern of overall neglect by church authorities who gave little attention to the socioeconomic or religious problems of rural Catholics. Father O'Hara advocated for rural social justice concerns and for rural parishioners to learn more about their faith. He conducted rural religious vacation schools and initiated traveling medical and dental clinics in rural areas of the Northwest.

Father O'Hara's work caught on, especially in the rural Midwest. In the 1930s, its most influential and famous executive secretary became the leader of this movement for lifting up Catholic rural life. His name was Monsignor Luigi Ligutti, a native of Italy who emigrated to Des Moines, Iowa, as a young man.[13] Ligutti studied for the priesthood and for a time served as pastor of the church in Woodbine, Iowa, just thirty miles west of us. It was his appointment in 1926 as pastor to the small parish of the Assumption in Granger, Iowa, fifteen miles northwest of Des Moines, that gave the National Catholic Rural Life organization national prominence. In 1933, he started the

Granger Homestead Project. He secured a loan from one of President Franklin Roosevelt's New Deal programs called the Federal Subsistence Homesteads Division to settle fifty displaced and underemployed coal mining families from southern Iowa.[14] The program granted the families two-to-eight-acre subsistence plots. In actuality, these were more like small farms. Combined with the creation of numerous cooperatives, all fifty families were able to get off the relief rolls in just two years. It was such a success that Eleanor Roosevelt visited the project. The Catholic press extensively publicized the effort, and it helped link the Catholic Church and the Catholic Rural Life movement to the New Deal. It also came to be linked with the efforts of fighting Communism after World War II. We learned about it in grade school from our pastor, Father Duren, who was both a friend of Ligutti's and fellow leader in the National Catholic Rural Life movement. He stated that if you had five acres and a cow, then Communism would lose its appeal.

After the Granger success, Ligutti gained more influence when he became executive secretary for the National Catholic Rural Life Conference in 1940, serving in that capacity until 1960. He also became the official Vatican observer to the Food and Agriculture Organization of the United Nations in 1948. In 1962, he played a dominant role in forming the first International Catholic Rural Association, with 124 organizations participating from forty-nine countries.

At its peak in the 1950s, Catholic Rural Life had sixty programs on rural living and philosophy with an enrollment of 1,700 priests, 9,000 women religious, and 12,000 laypeople. These programs were cosponsored and housed in Catholic colleges and universities around the country. In 1941, 15,000 people attended the annual conference of Catholic Rural Life in Bismarck, North Dakota.

Msgr. Ligutti used his influence and personal friendship with three popes (Pope Pius XII, Pope John XXIII, and Pope Paul VI) to ensure that rural concerns were being paid attention to by the universal Church. I had the opportunity to personally meet and converse with Msg. Ligutti on one of his return trips to Des Moines in 1975. I drove our pastor in Westphalia, Rev. Jake Weis, to visit him. Rev. Weis grew up in the Woodbine, Iowa, area when Ligutti served there. Msg. Ligutti died in 1983 in Rome and is buried in Granger, Iowa.

Just as rural America began to unravel in the 1970s, the National Catholic Rural Life organization did to some extent as well. So did the relationship with Catholic colleges and universities in terms of rural life and agriculture. Of the 244 Catholic degree-granting institutions of higher learning in the United States, today not one offers a program of study dedicated to agriculture.

National Catholic Rural Life is still in existence, known as Catholic Rural Life. After a period of some decline, it is once again working to regain some of the influence and prominence that it once had. Headquartered at St. Thomas University in Saint Paul, Minnesota, one of its more innovative programs has been to hold summer training sessions for seminarians who may be interested in serving in rural areas. This is an urgently needed curriculum, especially for urban seminarians who know very little about rural parishes or agriculture. I hope that it can grow to be a national program for all seminarians. I served as a board member for Catholic Rural Life for six years, from 2012 to 2018.

The rural Church is again falling on hard times just as it did a hundred years ago. Today's problems, though, are centered around even more difficult issues. Some rural Catholics are choosing to no longer participate in parish life, and their loss cannot be afforded given the decline in the sheer numbers of people in rural America. So many of today's rural parishes are fractured and being pulled in many different directions. Most of these pulls are the same ones that are pulling people everywhere, whether urban or rural. Church attendance and involvement are not perceived as important as they once were. I was somewhat surprised to learn the number of priests in America has not really decreased from what it was fifty years ago. What has changed is the same number of priests are being asked to serve twice as many Catholics in the country as a whole. It is also ironic that in the early days of Westphalia, some young men from its midst went on to serve as foreign missionaries in countries such as Africa, while today three out of four pastors in our Shelby County area come from African countries. As Pope John Paul II in a 1988 letter to Fr. George Coyne wrote,

> Science can purify religion from error and superstition. Religion can purify science from idolatry and false absolutes.[15]

We must be careful that science and its applications through technologies do not become ends unto themselves. It seems like we grant technology godlike status without hardly blinking an eye. Why do we have so much blind faith and acceptance of most new technologies? We are quick to say yes to new technologies that pull us in by promising instant gratification, promising to make our lives more convenient and comfortable, filling a want or need for something we think we must have. The use of technology itself can create more problems. However, we have faith that even newer technology will solve those problems.

Just because we can do something does not make it mandatory to do so. Ask yourselves the following questions: How appropriate is the technology in terms of who are winners and who are the losers? What are the costs and benefits of the technology? How does the technology affect our natural resources? Does it put people out of work? Does it further concentrate wealth into fewer hands? Is the technology ethical? Who decides that question? The words "appropriate types and sizes of technology" need to become a part of the discussion. Technologies that promote bigger as better seem to be a concept Americans want to embrace, including farmers. Just look at today's massive combines that can harvest sixteen or more rows of corn at a time or planters that can plant forty-eight rows at a time.

One of the consequences of ever-larger and more expensive machinery is that farmers have to plant more acres to pay for the machines. Does that mean finding a way to outbid or rent land away from a farmer who may not have the stomach for that kind of cutthroat, competitive behavior? I do not know how many of today's massive combines are owned by farmers versus being leased. Agricultural machinery companies find all kinds of ways to get new combines and tractors on the land. Maybe a farmer leases them, and then they are either sold as used equipment in this country or in another country where the same trends are occurring with ever-larger farming operations—much like buying rental cars that are only a couple years old. Maybe the farmer buys a new combine and then has an arrangement where it is traded every two or three years so that it keeps its higher value and can be constantly traded in a sort of game where the equipment company is able to sell new equipment and big farmers get to keep using new equipment in an ongoing, self-perpetuating scheme. We smaller farmers end up buying used combines that may be ten to fifteen years old or older but still cost as much as $50,000 to $100,000 or more depending on their size, condition, and so on. That is still more doable than buying a new combine for $500,000 or more or even a three-year-old one at $250,000 or more.

Most of today's combines, tractors, planters, and other essential agricultural implements are computerized, meaning most farmers do not have the knowledge or ability to fix them. These technologies are expensive and are becoming the targets of farmers and hackers who want to find less expensive ways to fix systems and problems. A step in the right direction occurred recently. Courts decided in a ruling for the *American Farm Bureau Federation (AFBF) v. Deere & Company*, in early 2023, to allow farmers and other non-John Deere repair services to fix computerized

John Deere equipment. John Deere had been restricting repairs to be done only by John Deere service centers, or their warranties would not be upheld.[16] A few companies are now making new models of tractors that do not have all the "bells and whistles" (meaning all the computerization and artificial intelligence systems that are so expensive to buy and to keep working). I bought two such tractors, one in 2007 and the other in 2016. They are sound tractors and have most of the comforts, conveniences, and mechanical technologies that are needed. They were just built a little more economically in terms of certain mechanical and computerized systems such as transmission, brakes, fuel, and hydraulics.

Artificial intelligence, also commonly now referred to as AI, has penetrated our lives whether we like it or not, and it will continue to do so. Some applications are appropriate, but maybe some are not. Who does not rely on global positioning system (GPS) directions for traveling? Still, I do not want to lose my sense of knowing my surroundings and my sense of direction when driving in a new place. I want to know, within reason, where I am, its history, and what I am doing there. I want to know and have a sense of place. Total reliance on GPS can make us lose touch with the "common sense" of having good driving skills. Robotics, driverless cars, computerized surveillance, and tracking systems are all steeped in possible unintended consequences. One consequence of the many forms of AI is vulnerability to cyberattacks. For instance, if a ship or any means of transport that moves people, materials, or weapons is totally dependent on it, what happens if that capability is completely neutralized? Do the pilots or drivers or captains have the skills and does the machine have the basic capability to operate independent of the technology?

It is speculated that more and more robotics will become the norm in the agriculture of the future. It is being touted as bringing new and improved efficiencies and more comfortable and easier lives for farmers.

A 2022 National Young Farmer's Coalition survey discovered that 86% of young farmer's practice regenerative agriculture and farming more in tune with nature. Ninety-seven percent use sustainable farming practices of some kind.[17] We are talking, of course, about a different kind of agriculture than is the norm. It is farming on small acreages where young people can replace hard-to-obtain capital with their own labor, which is something most have available. This is not conventional agriculture where it appears inevitable that AI and robotics will most likely incorporate into American agriculture as the conventional agricultural workforces ages. US farms are making a rapid push into AI.[18] I do think AI could have benefits for smaller farms if they can work together cooperatively in data management to improve soil quality and use less fertilizers and pesticides, for instance.

Granted, I can see some appropriate uses for robotics. If they are agile, not too large, and affordable, then maybe they could pull weeds, and we would be able to get rid of toxic herbicides. However, will they be affordable? How many natural resources will it take to produce them? Again, the critical point for me is I want to see more farmers, not continually fewer. I will ask it again and again: If technologies get rid of the millions of small farmers left around the world, what jobs will they be able to have to support themselves or their families? What will be their fate in the continually growing urban environment? What kind of meaning will they find in their lives? My summation is this:

> I see much more hope in a future world where people can find meaning, purpose, and fulfillment in growing their own food and food for others in their local areas, and regions and countries where food sovereignty is both respected and is the dominant way of life for many more farmers than we currently have.

As Pope Francis said in his encyclical Laudato Si, "We are not advocating for a return to the Stone Age. We are advocating for a new relationship with our environment."[19]

Drones and robotics are already being promoted as a way to fine-tune the spraying of pesticides and fertilizer applications for best management practices. I suppose this could be a step in the right direction if it results in more judicious use of these materials. Global positioning systems (GPS) have taken the guesswork and tediousness out of planting and harvesting. They can make cultivation much easier for organic farmers, too.

No one should automatically be accused of being a *Luddite* when it comes to questioning new technologies. The name Luddite is thought to have been used by Ned Ludd, an apocryphal weaver from Leicester. The term Luddite refers to what were actually the skilled artisan workers of their time in the textile industry in England, according to an article written by Richard Conniff in *Smithsonian Magazine* in 2011.[20] What they were protesting were the conditions in the new machine-based textile mills of the late 1700s that were replacing them and also forcing them to transition into working in much poorer and less satisfactory working conditions. They were asking for what workers then and today have always asked for, a fair wage, better labor standards and working conditions, and the ability to negotiate with the owners for their rights as workers. The term Luddite has become a negative term implying that these kinds of people are against all new technology because they are either confused by it, or do not understand it, or are backward. Will technology alone solve the problems and sometimes unintended consequences previous technologies have created? I think not. In 1940, C. S. Lewis wrote,

> To every soul, God will look like its first love because He IS its first love. Your place in heaven will seem to be made for you and for you alone, because you were made for it, made for it stitch by stitch as a glove is made for a hand.[21]

The human mind is a wonderful thing. There have been great scientific advances to describe our ability to think in terms of the firing of neurotransmitters in our brains. It requires a sense of humility and respect badly needed if we are to continue to thrive as humans and as a planet. But what does the physical brain reveal about our inner self, our soul?

Various surveys indicate that a substantial majority of people believe in an afterlife. Where do the ideas of having a soul and a spirit fit into these questions of the existence of an afterlife? The human spirit can be thought of as including our intellect, emotions, fears, passions, creativity, and so much more. I think of it as a motivating force. Some models, such as the one put forward by Daniel A. Helminiak and Bernard J. F. Lonergan, both Catholic priests, say it is the mental functions of awareness, insight, understanding, judgment, and other powers of reason.[22] Lonergan refers to it as intentional consciousness. Spirit is, to me, somewhat different from spirituality. Spirituality involves our relationships, values, and life purposes. It looks beyond us to a power greater than us that we want to be a part of in a meaningful way. It looks to the Holy Spirit for meaning and guidance. In Christianity the Holy Spirit is the third person of the Trinity and became a key outcome of the first council called by the Church to address all its members. This was the Council of Nicaea in 325 AD. It established that Christ was divine, not created, and the Holy Spirit was the creator spirit, present before the creation of the universe. The words in the "Glory Be" prayer state this emphatically: "Glory be to the Father and to the Son and to the Holy Spirit, as it was in the beginning, is now, and ever shall be world without end."

The word soul may conjure up different images that refer to the spiritual part of the body left after we physically die. It cannot be explained merely by materialistic or scientific experiments.

Many different religions believe human beings have a soul. Early Christian philosophers adopted the concept of the immortal soul from the Greeks. Just as the vast majority of people believe in a higher power greater than us, the majority of people also believe we are made up of more than just a physical body. While the practice of attending a church is down significantly, the high number of people who pray and believe in God and the soul is encouraging.

All of this thinking about who and what God is, our universe, faith, having a soul, the practice of religion, and on and on can end up with one being very tired at the end of the day. Still, I take solace in the fact God does not expect us to do more than we can as human beings. As human beings living on planet Earth, I believe God does expect and hope that we love Him and all of creation on this Earth and beyond. I also believe He expects us to love our neighbor as ourselves. Those are big challenges for us, who are merely humans, not gods. Again, turning to C. S. Lewis,

> Aim at heaven and you will get earth thrown in. Aim at earth and you will get neither.[23]

We still have a long way to travel on our journey to heaven. Maybe we can get a glimpse of it, but it is beyond our grasp on Earth. We need to remember that, but we should still aim for it.

Chapter 13
Agriculture and a More Livable Climate

The earth is protesting for the wrong that we are doing to her, because of the irresponsible use and abuse of the goods that God has placed on her.
—POPE FRANCIS, ENCYCLICAL LETTER, *LAUDATO SI*[1]

In this book, much has transpired since Chapter 1, when our journey began on the Missouri River in 1714. The river has changed a great deal since then. The Missouri River and the Loess Hills on the eastern side were both defined by the last periods of glaciation that occurred some 30,000 years ago and then finally just 12,000 years ago. For the last one hundred years, mankind has been the defining force. Through channelization and dam building, the once slow, meandering river has become deeper, straighter, and faster, which was intended to decrease flooding over large areas, promote barge traffic, increase farming acres, and produce electricity by the thirteen dams built on the river. The river has been shortened by a total of seventy-two miles. The channel area has been reduced by 80 percent and the water surface area by 66 percent. The results have been mixed. Vast areas of timber have been cleared, wildlife and aquatic habitat have been lost, wetlands have been drained for farming, and, ironically, flooding has in some instances been made worse because of the speed and magnitude of the water flow. Decisions made upstream at the dams about how much water to release and more intense rain events due to our changing weather have taken on more significance.

Enormous amounts of soil and excess fertilizer (particularly nitrogen and phosphorus) washing off farm fields and into waterways continue unabated. Excess water from our farm flows into Keg Creek, which has its start just a half mile to the north of us through springs coming from the ground. From there, Keg Creek flows some sixty miles to the south directly into the Missouri River, then on to the Mississippi River, and finally to the Gulf of Mexico at New Orleans. Over the past five years, the average size of the low-oxygen, or hypoxic zone in the Gulf has been nearly 5,400 square miles. This is nearly three times larger than the five-year average goal of 1,900 square miles set by the Mississippi River/Gulf of Mexico Watershed Nutrient Task Force by the year 2035. Iowa alone contributes 55 percent of the nitrogen loss to the Missouri River and 45 percent to the Upper Mississippi River watershed.[2]

Maybe we would not waste so much nitrogen if we better understood the nitrogen cycle. Nitrogen comprises 78 percent of the Earth's atmosphere. Nitrogen is the basic ingredient of amino acids that make up the proteins of all living organisms. Nitrogen is the structural component of muscle, tissue, and organs. Nitrogen is the key element in the nucleic acids DNA and RNA that carry the genetic codes for reproduction. Nitrogen is also the major component in chlorophyll, which is necessary for photosynthesis to occur.[3]

Nitrogen is in an inert form in the atmosphere, which means that it is in a form that plants cannot use unless it is converted through a process known as nitrogen fixation. In this process, nitrogen gas is deposited into soil as ammonia, mainly through precipitation. Microbes in the soil are able to convert the ammonia to a form of nitrogen that plants can use. The roots of leguminous plants contain nitrogen-fixing bacteria that do this conversion. An enzyme called nitrogenase also plays a role. This is the first part of the nitrogen cycle. Legumes include alfalfa, clovers, peas, and beans.

The second stage is mineralization. In this stage, nitrogen moves from decomposing plant material or by the addition of manure in a form that plants (except for legumes) can use. Ammonia is the form of nitrogen signified as NH_3. NH_3 then reacts with water to form ammonium (NH_4).

The third stage is nitrification. This stage produces a bonus of sorts for plants as ammonia is converted by bacteria into compounds called nitrites (NO_2) and nitrates (NO_3) that plants can use.

The fourth stage is called immobilization. This occurs when there is not enough nitrogen in the soil from plant decomposition for bacteria to thrive. In this case, plants begin to take nitrogen from the soil, which can cause a nitrogen deficiency for other growing plants.

The last stage is when nitrogen returns to the atmosphere as nitrogen gas. This occurs when bacteria convert nitrates back into atmospheric nitrogen. It is called denitrification. This is an undesirable circumstance for growing crops. Fall applications of synthetic nitrogen fertilizers increase denitrification as a whole.[4]

In only the past one hundred years has humankind learned how to break the tight bonds of inert nitrogen in the atmosphere and convert it into a usable form for plants. This is the Haber–Bosch process covered in Chapter 10. More nitrogen is made today by man-made processes than by all of the nitrogen-fixing bacteria present in the soil and leguminous plants and trees.[5]

A radio ad heard by listeners of Iowa State University (ISU) football and basketball games is telling. Sponsored by the Iowa Corn Growers Association, it goes like this: "Iowa grows corn and corn grows Iowa."[6] You cannot argue with the first part. Iowa grows more corn than any other state in the country. Nearly 37 percent of the state is covered in corn. That number is over 13 million acres with an annual crop of around 2.5 billion bushels. We grow over 30,000 bushels annually on our farm. Iowa farmers apply about 1.6 billion pounds of nitrogen to grow that humongous crop. I stated elsewhere that about 57 percent of Iowa's corn crop goes to produce ethanol fuel. Much of the rest goes to feed animals. Some are used for human corn sweeteners, oil, and starches.[7]

If we are being honest, however, Iowa also grows much of something else besides corn. We are also growing the Dead Zone in the Gulf of Mexico. The size of the Dead Zone varies but, in the past, has been about the size of New Jersey.[8] We are wasting a great deal of nitrogen and applying way more than can be used by the annual corn crop. A corn plant can use only about 50 to 60 percent of the nitrogen that is applied every year. Most of the rest stays in the soil in some form, volatilizes in the air or leeches out. By leeching, it is going either into groundwater or running off fields and into streams making its journey down the Missouri on the western side of the state and eventually into the Mississippi, which borders the eastern side of the state. Not all of that applied nitrogen comes from synthetic nitrogen. Livestock manure not properly applied is also a big contributor to excess nutrients. And Iowa grows plenty of livestock, with 3.7 million cattle, 23 million hogs, and over 50 million chickens annually. The chickens are chiefly egg layers versus broilers for meat.

Exactly what does an excess of nitrates and phosphorus do to groundwater? Nitrogen introduces nutrient-rich nitrates and nitrites that cause an increase in algal blooms. While some phosphorus is necessary for optimal aquatic health, too much of it increases algae blooms and less available oxygen for other marine species. In marine waters and estuaries, the algae blooms are often red or brownish in color and are caused by phytoplankton types and dinoflagellates and diatoms. These algae blooms become so dense other living organisms are deprived of sunlight. When this algae growth begins to decompose, marine species are deprived of aquatic oxygen and in turn also die. This has greatly affected the numbers of aquatic species and also the economic livelihood coming from fishing and shrimping in the Gulf of Mexico. Algae blooms also occur in fresh water and are usually cyanobacteria types, which are blue-green algae. It is becoming more dangerous to swim in Iowa's lakes and rivers in the latter part of the summer. There are at least two reasons for this. The first is it takes that long for the nitrogen- and phosphorus-rich compounds to cause the rapid acceleration of algae blooms. The second is that, as the climate is getting warmer, algae growth is increasing as well.

Other water runoff problems begin and stay in Iowa. Perhaps the greatest is the cost of the removal of nitrates and other harmful compounds or chemicals to keep drinking water safe. A study in 2021 by the Union of Concerned Scientists noted that Iowans will have to pay as much as $333 million over the next five years to keep excess nitrates out of their drinking water.[9] Nearly all of the violations of the US Environmental Protection Agency's (EPA) nitrate limits have been occurring in small, rural water systems with fewer than 3,300 customers. These same small water treatment systems are under financial pressure to supply safe water. It can cost rural Iowans as much as $1,200 per person for removal of nitrates that may be costing large urban centers only $2 annually per person.[10]

Drinking water with high nitrates has long been associated with "Blue Baby Syndrome," which arises when there is not enough oxygen in the blood. Numerous other risks are now being discovered, including increased rates of thyroid and bladder cancers, birth defects, and other serious health issues.[11]

Creation and Demise of the Leopold Center for Sustainable Agriculture

The concerns about the health and safety of Iowa's drinking water and the role that nitrogen plays have been around for a long time. They really began to ramp up in the early 1980s with the increase in corn acreage and the amount of fertilizer being applied. Along with that concern came an awareness of what was happening along the Gulf of Mexico with the creation of the "Dead Zone." Iowa was fortunate to have some legislators who knew this had to be addressed both here in Iowa and farther downstream. Three names come to mind: Paul Johnson, David Osterberg, and Ralph Rosenberg. The *Iowa Groundwater Protection Act of 1987* was passed.[12] The *Act* created three centers, one at each of Iowa's state-supported universities. The Center for Health Effects of Environmental Contamination was established at the University of Iowa. The Iowa Waste Reduction Center was formed at the University of Northern Iowa. Its purpose was to help small businesses properly handle and dispose of solid and hazardous materials. Both these centers remain in operation today.

The last of the three centers was the Leopold Center for Sustainable Agriculture at ISU.[13] It became the most well-known of the three and also gained the most notoriety because of those who, soon after its creation, began to oppose it, mainly the fertilizer and pesticide sales businesses and dealers who were required to provide funding for the implementation of the *Iowa Groundwater Protection Act*. Fees paid on nitrogen-based fertilizer sales, license fees on pesticide dealers, and registration fees from the sales of pesticides raised about $1.7 million annually for the three centers. From just 1998 to 2017, the Leopold Center funded over 500 competitive grants coming from every county in the state. For every dollar awarded by the center, an additional $4.60 was leveraged for the projects. Six of the most important areas of research were:

- Low-input, high-diversity crop and livestock systems
- Long-term agroecological research
- Hoop barns for alternative swine production
- Regional food systems
- Riparian buffer projects
- Practical Farmers of Iowa on-farm research projects in cooperation with ISU.

In a move that left many feeling blindsided, the Leopold Center was defunded by the Iowa Legislature in 2017.[14] Some decried it as both a political move and an agribusiness and commodity agriculture move because of complaints it was doing too much organic farming research. That simply was not true, which they would have known, had they bothered to look at the list of research projects. The Leopold Center did fund many measures to conserve and protect Iowa's most precious resources, those being soil, water, and diversified family farms. It was also thought by many to be a move that ISU either wanted, or at the very least, endorsed, and thus did not put up much of a fight to keep the center alive. Dennis Keeney, PhD, who was the first director of the center, noted the biggest enemy of sustainable agriculture was the agricultural college itself. The College of Agriculture and Life Sciences at ISU gets nearly 20 percent of its annual funding from businesses, corporations, and commodity organizations. Forty-nine percent comes from the federal government and just 6 percent from the state government.

A university is a place to learn. Researchers do research. One farmer legislator went on record as stating it was not necessary to conduct additional research about sustainable agriculture. It had been completed in his mind. It becomes a laughable but very disturbing statement when you consider how little real progress has happened since 2017 in improving Iowa's soil and water quality and conservation due to fertilizer, pesticide, manure, and tillage practices, and a multitude of other careless farming practices. In retrospect, all of us, myself included, should have and could have done so much more to keep the Leopold Center from being defunded. Perhaps it could have been saved if enough people from every corner of the state had, in a quick, coordinated way, confronted their legislators and ISU. I was named to the Leopold Center Task force, which conducted meetings around the state the following year to listen to stakeholders and identify possible ways to keep the Leopold Center running. Looking back, it seems that I was just one of many who felt powerless to do anything at all to keep what happened from happening. In actuality, the Leopold Center was already dead, we were just prolonging the agony of proclaiming the outcome of the autopsy.

Iowa's Nutrient Reduction Strategy

The Mississippi River/Gulf of Mexico Watershed Project was initiated in 1997 to coordinate efforts to reduce the nutrient load and flow into the Gulf. The increasing loads of nutrients were increasing the size, severity, and duration of the annual hypoxia, or Dead Zone. The project created a plan in 2008 that asked each of the twelve states along the Mississippi to develop their own nutrient reduction strategies. This occurred in Iowa in 2013 with the Iowa Nutrient Reduction Strategy. A goal of a 45 percent reduction in both nitrogen and phosphorus loads was set by the Gulf Hypoxia Reduction Action Plan.[15] The nutrient reduction strategy is meant to address both point and nonpoint sources of nutrients. The major point source strategy is to reduce the nutrient loads after they have been removed by Iowa's largest municipal and industrial wastewater plants so the water is safe to drink. The nonpoint source nutrient reduction strategy is to reduce nutrient loads from leaving Iowa's farm fields. Iowa's nonpoint strategy is a totally voluntary optional plan, and presently only a very small percentage of farmers have voluntarily begun to get serious about nutrient overloads. Although the strategy does ask farmers to consider the timing and amounts of nutrient applications, there are no requirements for actions or enforceable standards and limits. This seems ludicrous to me because wouldn't you want to find ways to limit the excess applied nitrogen if the corn crop cannot use it anyway? Why put on too much and not at the right time and then have to deal with it after the fact? According to the Iowa Environmental Council, Iowa has already spent over $560 million trying to deal with the issue of excess nutrients and related issues. Most of this money has been supported by taxpayers.[16]

It also stands to reason that some of this excess nitrogen is percolating into rocks and into deeper water. This will only increase the likelihood of high nitrates as an ongoing phenomenon even if we find ways to effectively lower nutrient losses in the near future. Might it be a ticking time bomb beneath our feet?

ISU agronomist John Sawyer, PhD, developed what is called the "Maximum Return to Nitrogen Calculator" over twenty years ago. It calculated the optimum levels of nitrogen to apply based on the cost of nitrogen and the price of corn. Generally speaking, 125 to 150 pounds of nitrogen was the optimum range for corn following soybeans grown on the same field the previous year.[17] In 2022, an average of 171 to 189 pounds of N was recommended by the ISU Extension program.

Currently, the Iowa Nutrient Reduction Strategy appears to be emphasizing three major focus areas to reduce excess nutrients. The first is the planting of cover crops to keep the ground covered throughout the winter and to allow the cover crops to use excess nitrogen. The second is the development of wetlands to capture water drainage, and the third is the creation of bioreactors and saturated buffers. A bioreactor is typically a buried trench along the edge of a field that is filled with wood chips. The wood chips act as an environment where bacteria convert nitrate nitrogen back to nitrogen gas that is ubiquitous in the atmosphere. Currently, the average cost of a bioreactor is $15,000.

The Iowa Environmental Council issued a report in 2022 highly critical of the snails-pace progress in reducing Iowa's excess nutrient loads. The council estimated it will take eighty-five years to meet the goal of over 12 million acres of cover crops. To date, there are around one million acres. They estimated it would take 942 years to meet the goal for wetlands. The goal for bioreactor- and saturated buffer-treated acres would take 22,325 years![18] To me it is obvious what is being done now to motivate Iowa farmers to adopt more nutrient reduction practices is not working. This is demoralizing for all farmers and water municipalities and districts already doing the right thing without government assistance. First of all, we need to identify universal practices that could be adopted by most, if not all, to decrease nutrient loss. These include the following (not necessarily in order of priority):

- Terracing: Refers to pushed up dirt, usually by a bulldozer, from the downslope of a hill that slows and holds water from going down the slope and aids in prevention of soil erosion.
- No-till: Refers to not performing tillage or stirring of the ground after a crop is planted, until the next crop is planted. Continuous no-till farming is practiced by some farmers planting only corn and soybeans.
- Headlands: Refers to areas at both ends of a field where farm implements are turned around. However, most farmers plant these headlands to row crops as well. Headlands could be planted with perennial vegetation.
- Buffer strips: Refers to permanent vegetation along streams and field borders and headlands.
- Grass waterways: Refers to ditches where water naturally collects as runoff from crop fields. Keeping ditches covered in permanent vegetation slows the flow of water, prevents gullies, and slows soil erosion.
- Nitrogen stabilizers: Refers to compounds that slow down the microbial conversion of ammonia and urea to nitrate, especially in the fall.
- Rotationally grazed pastures: Refers to moving livestock from one pasture to another after the pasture is grazed to a level that will not deter regrowth of the pasture.
- Planting of perennial crops, trees, shrubs, and prairie plants: Refers to practices that are especially desirable on headlands, buffer strips, and stream banks.
- Restoration of wetlands and riparian areas: Refers to restoring wetlands and riparian areas drained by tiling to remove the water and prairie grasses and wetland trees and shrubs from these low-lying areas.
- Extended rotations: Refers to planting crops, such as small grains, alfalfa, and other hay crops, rather than continuous row-cropping of corn and soybeans.
- Applying fertilizer in the spring and using nitrogen tests already available to indicate what is needed: Side-dressing nitrogen when the corn is six to twenty-four inches tall has been a practice used by some farmers for over seventy years. That growth period is when the crop can use the greatest percentage of nitrogen.
- Cover crops: Refers to crops planted into a standing crop or a harvested crop to keep the ground covered throughout the winter, to sequester carbon, and to enhance grazing and nitrogen in the soil.

The Iowa Environment Council has called for mandatory participation in a flexible framework of practices from which a farmer could choose. Participation would not be optional.

What this mandatory participation would look like is up for debate. There needs to be a "carrot" approach for crop insurance programs and/or the Conservation Stewardship Program (CSP) to help incentivize farmers to do the right thing.[19] If they decline to participate, then the "stick" approach would have to be considered, at which point their crop insurance premiums would be significantly higher. I believe in making the polluters pay, and that it is the fair and right thing to do.

Yet time and time again, efforts here in Iowa to try to regulate the overuse and misuse of nitrogen have been thwarted by the Iowa Farm Bureau, the fertilizer industry, commodity organizations, and others opposed to government regulation no matter what the cost may be to the environment or the health and well-being of citizens. Their approach of choice has been only for voluntary action. That is simply not working. I wish all farmers and landowners would have the land ethic needed for voluntary action. In truth, that ethic is just not strong enough to make it work.

The lack of any soil loss penalties is becoming more evident with each passing growing season and the worsening storms and weather extremes accompanying each growing season. In my estimation, the spring of 2024 was the saddest example I have seen to date. In early June, vast swaths of fields revealed devastating soil erosion while driving along the Missouri River valley all the way up to the South Dakota border and through the hills on the Nebraska side of the river. There was nothing to stop it. There are fewer pastures left to stop its ravaging path. There are no grass waterways to slow it. The big sprayers make sure of that. They do not want to have to bother with working around them or to have to shut the sprayers off to maintain a single blade of grass. The terraces are being removed from steep slopes under the guise no-till practices will suffice and is all that is needed to preserve our precious soil. It is revealing itself as a big lie! But that is only apparent to those who choose to see it, know what it is they are looking at, or even care. In this case, it appears that ignorance is bliss.

The storms in April and May 2024 were very devastating in our immediate farming neighborhood. Some of our neighbors had their entire farmsteads destroyed by a series of massive tornadoes. The magnitude of these has never been seen before in our area. Our own farm missed the devastation by less than one-half mile. One of the severe lightning events did kill eight of our cattle. These were 950-pound animals huddled under some trees along the creek south of our pond. The strike was so powerful that it blew them over the electric wire fence and into the creek. This happened without breaking the wire. Nearly six inches of rain fell in a short period of time. Still, our farm fields took the hit very well because of all the conservation practices put into place.

What Hath Man Wrought!

In 1845, Samuel F. B. Morse dispatched a telegraph message from the US Capitol to a railroad station in Baltimore, making the statement, "What hath God wrought!" taken from the Book of Numbers 23:23 in the Bible. On August 17, 1945, shortly following the dropping of the nuclear bomb upon Japan, the *United States News* (later becoming the *US News and World Report*) headline was, "What Hath Man Wrought!"[20]

Today a far greater force is at play that rivals nuclear annihilation. It is a force that is already changing our lives like nothing else before it. That force is climate change. At least with nuclear holocaust it is an either-or situation. You either destroy most, if not all, of mankind or you do not. We have already made many of the choices to alter our planet. Will it lead to the ultimate destruction of our world as we know it? We do not know. However, we already recognize that climate change is affecting our food, water, air, and weather. As species and human health decline through disease and hunger, and as critical ecological systems and functions decline, the effects of climate change are more than a worry. They are a real threat.

It is not a problem where we can just employ a group of economists to come up with a solution. Our industrial economies have been living off carbon that took millions of years to form. We have burned through the easily obtained carbon in less than 500 years, and we did not begin to understand until quite recently what we were doing to the climate and our planet.

We, by our very nature, are obsessed with material things. Many of us are always striving to make more money, increase possessions, become more influential and powerful, smarter, and better looking, and avert the effects of aging—which was easier to do when we thought we were living with never-ending resources. Vast amounts of wealth have been made off our seemingly endless energy and mineral resources, our water and soil, the backs of our fellow humans, and our capacity to dream up new technologies that add to this wealth and make our lives easier and more entertaining.

Now, something else is happening that goes far beyond what we thought we could ever do. Did we ever think we could change the planet to such an extent that we may actually not recognize it in a generation or two? Of course, some humans did know through scientific inquiry that the climate was changing because of what humans and the Industrial Age caused over the past five centuries. Probably none of us foresaw how quickly the global average temperature and its effects would increase. In 1981, corporations like Exxon knew that burning of fossil fuels was beginning to warm Earth beyond the effects of normal weather or natural causes.[21] They chose to hide and ignore it for their short-term financial benefit or long-term economic interests, depending on how you look at it. They were not the only ones.

When it comes to oil and gas production, the top two are from China and are traded on the New York Stock Exchange (NYSE): Chinese Petroleum and Chemical Corporation at number one and PetroChina Company, Ltd, at number two. Number three is the Saudi Arabian Oil Company, which is not traded on the NYSE. Numbers four, five, six, and eight are the ones we Americans may know best: Royal Dutch Shell, BP, Exxon-Mobil, and Chevron, respectively.

One of the first climate scientists to warn politicians and the general public about what was happening was Dr. James Hansen, who grew up in the town of Denison, Iowa, just twenty-five miles north of our farm. As a climate scientist working for the National Aeronautics and Space Administration (NASA), he warned numerous presidential administrations and Congress in the 1980s and 1990s that planet Earth was beginning to heat up dramatically due in large part to the increased amount of CO_2 being released into the atmosphere by the burning of fossil fuels.[22] Dr. Hansen was one of the first to start talking about even more alarming possibilities if we did not address what was happening. The first NASA Goddard Institute for Space Studies global temperature analysis was published in 1981. Hansen and other Goddard scientists concluded carbon dioxide in the atmosphere would lead to warming sooner than previously predicted. That is also the time when the ideas of climate forcing mechanisms arose, such as temperature increases and carbon dioxide levels reaching tipping points where certain climate and earth-changing consequences could not be reversed. Only forty years after Dr. Hansen tried to warn us, it appears we have already reached some of those tipping points, where all we can hope to do now is slow down greenhouse gas production to stabilize Earth's warming at a goal of a certain temperature. The long-term temperature goal of the Paris Climate Agreement reached in 2015 is 1.5 degrees Celsius or 2.7 degrees Fahrenheit.[23] It has already warmed over 1 degree Celsius or about 2.5 degrees Fahrenheit since 1880.

When the term "climate change" came into being, it was an anathema to some people. Saying that the weather is changing or has changed is far less politically charged. Nevertheless, the question is, why is the weather changing?

We need to start from the beginning, with photosynthesis. Much of the basic understanding of how we as humans have influenced the weather or climate comes from the understanding of photosynthesis. The chemical equation of CO_2 (carbon dioxide) + H_2O (water) (in the presence of sunlight) > $C_6H_{12}O_6$ (sugar or food energy) + O_2 (oxygen) tells us so much about how we could be changing our climate.

Photosynthesis is what keeps most living organisms on Earth alive (a few organisms do not rely on photosynthesis for their livelihood).[24] The concept is this: a plant is able to use carbon dioxide in the atmosphere and combine it with water in the presence of sunlight to create the food energy containing the carbon and oxygen it needs to maintain itself and grow and produce offspring to help ensure its future survival. Animals, including us of course, eat the plants and use the nutrients from the plants and oxygen produced for life-sustaining purposes. Some animals eat other animals, but the energy and chemistry loops are all the same. Food is broken down through energy consumption and is eventually released through respiration back to water and carbon dioxide. It is the beautiful, seemingly simple, but intricately complex way living organisms work. When plants and animals use the food energy, they are burning up the energy (carbon) and releasing some of it as CO_2. That is what is happening when we breathe. We take in oxygen and release CO_2 with much going on in between. Carbon bonds contain energy; breaking carbon bonds releases energy, whether for industrial purposes or biological metabolism. Photosynthesis uses solar energy to form carbon-bond energy.

To fully understand how it works, we need to go a few steps further. It turns out that CO_2 is a heat-trapping gas when it is released into the atmosphere. Consequently, it is called a greenhouse gas. We need to have a certain amount of CO_2 in the atmosphere to aid in the delicate balance of infrared light absorption when the sun is not shining. Otherwise, we would turn into an icy state. CO_2, a necessary greenhouse gas, contributes about 20 percent to the greenhouse effect. Water vapor and clouds account for 75 percent. Other gases and aerosols, such as nitrous oxide, methane, ozone, and other fluorocarbons, account for about 5 percent. During the periods of ice ages, total greenhouse gases were pegged at about 180 parts per million.(ppm). During warmer periods they were around 280 ppm. Today they are at 412 (ppm) and climbing. Future worst-case scenarios put them over 600 ppm and beyond. As long as we were not burning up too much of the stored or sequestered carbon from the millions of years of plant decay, we were okay, but the dawn of the Industrial Age changed all that, when we began to burn ever-increasing amounts of carbon to fuel our growing economies, population, and cities around the world.[25]

All of that CO_2 has to go somewhere. It is either going into the atmosphere, oceans, forests, grasslands, or soils of the planet. Oceans have been the primary long-term sink for human-caused CO_2 emissions. Now, however, the increased CO_2 in the oceans is resulting in increased acidification by binding with hydrogen, causing disruptions and destruction of carbonate skeletons, coral reefs, and ocean ecological food chains. The terrestrial sequestration of carbon in plants and soil conservation has offset about 30 percent of US fossil fuel emissions. But all of that CO_2 in the oceans has caused the ocean water to warm up more, too. A warmer ocean cannot hold as much dissolved oxygen marine life relies on for survival.

There is a growing body of evidence that indicates we could run out of oxygen someday.[26] However, it is estimated it will not happen for a billion years. What we need to be concerned about is what is happening now and have some control over. Over the past 150 years, we have reversed the planet's capacity to regenerate CO_2 to O_2 through photosynthesis by the increased burning of fossil fuels and by land degradation. To me, land degradation is loss of soil quality for growing food, which can have many sources, such as deforestation, soil erosion, nutrient loss, overgrazing, and urban sprawl.

When oxygen is returned to the atmosphere through photosynthesis and vegetative biomass, it is called a profit, and CO_2 is reduced. When more CO_2 is released to the atmosphere from the burning of fossil fuels and from land degradation, it is called a loss. The first law of thermodynamics must be obeyed. It demands that, for each absolute change in CO_2 content either positive or negative, a given amount of oxygen must take part. This is the carbon cycle more fully explained because it also includes the oxygen cycle.

The best way to decrease atmospheric carbon dioxide is not to emit carbon dioxide in the first place. That is proving hard to do with a world economy based mostly on fossil fuels. We can reduce consumption and change our lifestyles. We can reduce deforestation. Carbon dioxide emissions from deforestation exceed those from all the world's cars and trucks. We can use current technologies more efficiently, which will help, but it's not enough. The last choice is to switch to low-carbon emitting technologies. The choices we make for low-carbon technologies will depend largely on our depth of understanding of the science and ecology of how we produce food and distribute it, how we move about and travel, our moral courage and willingness to sacrifice for the good of life on Earth, and our collective political will.

Sectors That Produce the Most Greenhouse Gases

The EPA's 2021 estimates for greenhouse gas emissions by economic sector and electricity end use indicate the following breakdown: agriculture, 11 percent; transportation, 29 percent; industrial, 30 percent; and buildings, 30 percent.[27] The Intergovernmental Panel on Climate Change (IPCC) has compiled greenhouse gas emissions on a global scale. Figures from 2014 indicated this breakdown: electricity and heat, 31 percent; industry, 24 percent; agriculture, forestry, and other land use, 22 percent; and buildings, 6 percent.[28] Significant differences exist between worldwide and USmeasurements as to what sectors produce the most greenhouse gases. Part of this difference is due most likely to different criteria being used for determining what falls under each sector. There appears to be a good deal of overlap as to how and where to categorize energy usage and emissions. In the case of agriculture, I found there is little consensus on how much agriculture contributes. The IPCC indicates that CO_2 accounts for 65 percent of all greenhouse gases. Methane accounts for 16 percent. Nitrous oxide is at 6 percent, and CO_2 from forestry and other land use is at 11 percent. Agriculture contributes a portion to all of these sources. As more is being learned about where the problems are coming from, percentages change. For example, deforestation is a huge driver of CO_2 increases, but it is not easily sorted out in the above percentages unless the percentage of deforestation for new agricultural production is figured into the percentage for agriculture. Another factor is that the manufacture of fertilizers and pesticides and other production inputs are not figured into agriculture's contribution. Only the costs associated with applying or using them are in the lower percentage. The figure for agriculture's contribution to climate change and CO_2 is climbing as time goes on, both because we keep learning more about what actually is happening and because agriculture keeps increasing its contribution. Some analyses show it contributing as much as 40 percent or possibly even higher of total greenhouse gas emissions when everything is factored in. No one seems to agree.

The International Energy Agency (IEA) data from 2021 indicates coal is still the most-used energy source for the emission of greenhouse gases at 44 percent. Oil is second at 32 percent, and natural gas is third at 21 percent. There does appear to be widespread consensus on the three largest fossil fuel usages in the United States and around the world. These three sectors are burning fossil fuels for heat, electricity, and transportation, accounting for 78 percent of all greenhouse gas emissions.[29]

Decrease CO_2 or Sequester Carbon?

One of the most important questions in the United States in terms of agriculture's contribution to climate change deals with monocrop rotations of only corn and soybeans. Long-term agricultural cropping system trials in Wisconsin and Iowa document the value of intensive animal grazing systems and longer crop rotations that are at least three or four years in length and that use small grains such as oats and legume cover and hay crops. The results show more profitability and more positive climate change mitigation effects than do monocropping systems that are so heavily dependent on fertilizers and chemicals.

It is very hard to sequester carbon according to these two long-term studies in the Midwest, where our soils are younger due to the deposition of soil from the glaciation periods over the last 20,000 years, compared with more weathered soils, for example, such as those found in the state of Georgia. Sequestering carbon in already diminished soils with significant erosion and soil degradation due to how it had been farmed for hundreds of years can be accomplished, however. This is because very little carbon is left in soils where organic matter has been so depleted. Bringing these kinds of soils back to a better state can be done slowly over time using the right soil-building cover crops and legumes and less tillage.

Randy Jackson, PhD, agronomist at the University of Wisconsin, and his colleagues have been comparing six different cropping systems used by farmers in Wisconsin for over thirty years. They are (1) continuous corn, (2) minimum-till conventional (with pesticides and fertilizers) corn and soybeans, (3) organic corn/soybeans/wheat with a cover crop, (4) conventional corn/alfalfa/alfalfa/alfalfa, (5) organic corn/oats/alfalfa/alfalfa, and (6) continuous pasture.[30]

To date, results show only the continuous pasture system is positive in being above zero for sequestering carbon in the soil. A key finding is that for only three to four months during the growing season, CO_2 used up in photosynthesis is greater than the CO_2 returned to the system by the growing soil microbial activity. The grazing has to be done in a management-intensive, rotationally grazed manner in order to be positive for sequestering carbon. This could and should have future implications for beef-confinement feedlots with their emphasis on corn for finishing beef. When it comes to economies of scale favoring the large packing plants where many head of beef are processed daily, a step in the right direction could come from grass-based beef or higher percentages of forage operations spread out over the land in appropriate locations all over the country. With our own certified organic beef operation, at least 30 percent of the diet during the grass-growing season (which for us is about May 1 through November 1) is required by the USDA's National Organic Program standards to come from grass or other growing forages.[31] In our case, this is for the last six to seven months of the finishing phase. The finishing phase brings the weight from about 750 pounds to over 1,250 pounds finished weight. The packing plants should not care whether the beef is grain-finished, only that they have sufficient numbers of animals to keep their lines running as efficiently and smoothly as possible. However, when packing plants own so many of the cattle being processed, that statement is probably not true, as carcass value takes on a whole different meaning to the packers.

We currently have ninety-five Red Angus beef mother cows and feed out all of their calves to the finished weight of 1,150 to 1,350 pounds. To be clear, we use some grain (corn, oats, and rye) in our organic beef finishing diets. It is about one-third to one-half less than what a typical feedlot uses. One of the primary reasons for using some grain is that our customers prefer it over 100-percent grass-fed beef. For a number of years, we grass-finished and forage-finished (hay) all of our fall-born calves. We have two beef cow herds: those that have calves born in the spring and those that have calves born in the early fall. We were selling the bulk of the grass-fed animals to a grass-fed beef operation in Minnesota. They paid a premium but not significant or sustainable enough for us to keep doing it without receiving an additional certified organic premium. We were not able to sell very many through our markets. Organic Prairie does not have a program for US-produced certified grass-fed beef.

Grain-finished cattle produce 20 percent less methane than grass-fed beef. Methane is considered the second most significant greenhouse gas leading to global temperature increases. Grass-fed beef do manufacture more methane throughout their lives because eating grass alone produces more methane in the digestive process. They also create more methane because it takes longer to grow a grass-fed animal. However, unless lifecycle analysis is done for all it takes for growing food as we know it, the real truth will escape us. It takes a great deal of fossil fuel energy to grow grain-fed cattle with present intensive fertilizer and chemical cropping systems. When that is factored in, the figures change dramatically.

In terms of greenhouse gases and how they are measured, CO_2 equivalencies also must be considered. Methane is twenty-eight times more powerful than CO_2 as a greenhouse gas, but it has a half-life of only about ten years. CO_2, on the other hand, remains in the atmosphere indefinitely as a constant. According to the current numbers, methane accounts for a range of 4 to 9 percent of all greenhouse gases. A much-quoted United Nations Food and Agriculture Organization (FAO) figure has claimed that ruminant livestock are responsible for 14.5 percent of all human greenhouse gases.[32] Critics point out that they did not use CO_2 equivalency in their report. Unless life cycle analysis is part of every study and is done in an unbiased manner, we will continue to go down a rabbit hole of cherry-picking studies that reinforce particular viewpoints.

Bill Gates and Meat

Bill Gates has proposed wealthy countries eliminate ruminant animal (beef) meats and completely switch to synthetic meats.[33] The two of us would lock horns on that viewpoint. I think we should change how we raise and feed some of our meat animals. We could easily reduce meat consumption in our country by a significant percentage and actually improve the livelihoods of many more farmers, that is, if you had more farmers with livestock and not just a few mega-sized ones. When you look at pork slaughter in the United States, you discover that close to 60 percent of all the hogs butchered are done by the three largest integrated companies: Smithfield, which is owned by Chinese investors; Tyson, which is American-owned; and JBS, out of Brazil.[34] Total annual slaughter is around 121 million head. Seventy million of those hogs are slaughtered by these three companies. Iowa alone contributes one out of every three hogs in the country at over 40 million head annually.

Poultry falls into this same concentration, having played out in the evolution of concentrated ownership and contract production in the eastern and southern parts of the country, which happened thirty to forty years or more before it occurred in pork. The small processing plants went out of business, leaving the small producers with no place to sell their birds and eggs. Another factor was the growing number of chickens and eggs being produced under contract for the large plants.

This says when looking at various aspects of meat production, many issues are occurring in the United States and around the world. One issue is certainly the oligopoly question. Vertical integration where the company owns the animals, slaughters them, and markets the meat has become the name of the game. This brings up all kinds of fairness issues in terms of daily market price discovery to determine how many animals need to be slaughtered on any given day. It has become increasingly difficult for independent producers to market their livestock through a typical large slaughtering plant. The avenue still used by many smaller producers is through an auction barn structure. That option, too, continues to decline as the number of sale barns, as they are referred to, persists in its decline as well.

Bill Gates notes that the production of methane by burping cows is one of the main reasons for eliminating meat. As far as climate change goes, it is true that methane is twenty-eight times more potent a greenhouse gas than CO_2. Methane does not stay in the atmosphere as long, however, which current research shows can be for nine to twelve years, whereas CO_2 stays in the atmosphere for thousands of years. [35]

A discussion about how ruminants produce methane in the first place is needed. How ruminants digest food is specific to the evolution of their four-compartment stomachs, known as the rumen, reticulum, omasum, and abomasum. Ruminant species important to livestock producers include beef and dairy cattle, sheep, and goats. It is in the rumen where the critical digestion process occurs. Digestion in the rumen occurs under anaerobic conditions, meaning very little oxygen is present. The rumen is where bacteria, protozoa, and fungi break down feedstuffs through the process of fermentation. Methanogenic archaea, which evolved before there was oxygen on Earth can then break down these fermentation by-products of CO_2, formic acids, or methylamines into methane (CH_4). Cows must emit this methane by burping or through gas, else they would bloat from the buildup of the gas, causing death by asphyxiation. A good deal of research is examining how to mitigate the methane emissions in ruminants.[36]

Most of the methane comes from grazing animals—more specifically, in cow-calf operations where the cow and the growing calf spend a good part of their lives grazing. It is a remarkable food-producing feat that ruminants have evolved digestive systems that enable them to digest large amounts of roughage to help them grow and produce meat and milk and for the making of butter, cheese, leather, wool, and other products. The roughage is composed of large percentages of cellulose and lignin, which nonruminants cannot digest effectively. This has to be considered when looking at what the diets of humans should or could be.

Ruminants have been able to grow on cheap and abundant sources of feedstocks, namely grass and legume species. It is also noteworthy that many land areas in the world are not suitable for growing food crops for humans. They are either too rocky, too steep, or have soils of very poor quality. But they can support grass- and forage-eating animals. Improvements in the quality of the forage and rotational grazing strategies, where grasses and other plants are consumed at ideal times, are where significant reductions of methane can occur. This also correlates to the time when large amounts of cellulose and lignin are not present, which are harder to break down and tend to increase methane. We need to consider teaching farmers how to better graze on their land before we propose stopping the production of meat. This is especially true in poor and developing countries.

Another factor involves what are called CH_3 and CH_4 grasses. Cool-season grasses and legumes are CH_3 grasses and are more predominant in cool climates. Grasses in the Southern Hemisphere are predominantly CH_4 warm-season species. Corn is a CH_4 species as well as grazing grasses in the South and Tropical South, such as Bermuda grass, elephant grass, and blue fescue. The CH_3 species produce less methane. Alfalfa and clovers and other legumes, especially one called birds foot trefoil, are good methane-reducing species for grazing. We use all these species to complement our cool-season grass pastures.

Are Longer, More Diverse Crop Rotations a Key to Lessening CO_2 in Agriculture?

Matt Liebman, PhD, retired agronomist at ISU, spent twenty years comparing a conventional two-year rotation of corn and soy with two longer rotations.[37] The three-year rotation had a year of oats with a red clover legume with no fertilizer or pesticides used. The four-year rotation comprised corn/soybeans/oats-alfalfa/alfalfa hay with no fertilizer or pesticides used during the two years not in corn or beans. The results clearly show the benefits of reducing the need for fertilizers and pesticides for most, if not all, the critical measurements affecting climate change and soil quality.

Mineral N fertilizer usage is 86 to 91 percent lower in the three- and four-year rotation. Nitrous oxide is a byproduct of nitrogen fertilizer. Its overuse means as much as 50 percent can be wasted annually. Nitrous oxide is 250 to 300 times more powerful a greenhouse gas than carbon dioxide, but until now it has been largely ignored and downplayed as an important contributing factor to the increase in greenhouse gas emissions coming from agriculture. Nitrous oxide is considered to be the third most significant greenhouse gas.

Herbicide use is 96 percent lower when using the three- and four-year rotations that include a small grain and legume. Other benefits of the longer rotations were as follows: Fossil fuel consumption was 60 percent lower. Nitrates were lessened by 57 percent. Soil erosion was reduced by 50 percent. Greenhouse gas and energy usage were both decreased by 64 percent. Net profits and yields, as the rotations were repeated over many years, showed very few significant differences. Liebman's study conducted at one of ISU's field stations called the Marsden farm is quite remarkable.[38] Practitioners of these kinds of cropping practices and sequences are not surprised. Conventional agriculture practitioners find it hard to believe and are convinced it must be flawed research. At worst, the message that ends up being heard by the general public is that these kinds of rotations cannot ultimately feed the world.

The results show a critical need to increase the use of small grains and pulse crops in livestock rations in our present confinement livestock systems as an important first step toward increasing crop diversity and decreasing the detrimental effects of our current food production model.

Pulse crops are dry legumes harvested as seeds versus a "green" legume. Pulse crops include dry beans, dry peas, chickpeas, and lentils. They are valuable sources of protein as an alternative to soybeans, the dominant source of feed protein for swine, chickens, and turkeys. A winter annual grain such as hybrid rye would keep the ground covered from September until the grain is harvested in July of the following year. Then another soil-building and possible carbon-sequestering cover crop could be planted, keeping the ground covered for most of the year. Rye and winter wheat and spring small grains, such as oats and barley, could all be used in swine and poultry diets as a positive step toward diversifying conventional agriculture and becoming more resilient to climate change. I have already described how pulses, small grains, and hybrid rye are used on our own farm.

Unfortunately, just changing from corn and soybeans to small grains does not guarantee that fewer herbicides will be used. In today's world, where there are large acreages of oats in places like North Dakota and southern Canada, Roundup is used to kill anything green when the oat crop is ripening. That way the huge combines can harvest the grain standing versus through windrowing, which lays it in a row on the ground allowing for all of the green material to dry naturally and go through the combine easily.

A clear takeaway from these two important studies by Liebman and Jackson is that, although it may be difficult to sequester carbon depending on where you farm, you can significantly reduce CO_2 emissions through changes in your farming practices. It also shows we are not willing to pay much attention to the most significant practice agriculture could use to decrease greenhouse gases, which is moving away from the monoculture of just corn and soybeans and instead adding longer rotations of third and even fourth crops in our most productive crop-growing regions. Government farm programs simply must change in this regard and provide more incentives for farmers to grow small grains and legumes for food and feed and for improvements in soil quality.

There is plenty of blame to go around for climate change, but the vast majority of it has to be placed on the world's most industrialized and wealthy economies. What are these effects? Some we see in the news every day—increasing wildfires and decreasing water in drought areas; more flooding and extreme rain events, especially in coastal areas; stronger and more frequent storms; and higher temperatures making it harder to keep ourselves reasonably cool. We continue to see feedback loops and forcing mechanisms at play. It is getting hotter, which is causing oceans to get warmer and warmer, which is causing more extreme precipitation events because warmer air can hold more water. The higher temperatures are melting the ice caps, resulting in rising ocean water levels. The frozen tundra is warming, resulting in more biological activity of microorganisms and methane production and plants respiring more CO_2. Where we live can make a difference in our well-being. Ten percent of the world's population live along the coast in low-lying urban centers. Low- and middle-income countries are home to 89 percent of the people exposed to flooding. Who is the first to feel the consequences? It is the poor.

What are we doing about all of this? It appears more former naysayers and doubters are starting to realize the climate has changed because of human activity. I would predict we will make fairly rapid progress on two of the most significant contributors to greenhouse gases, namely vehicles for transportation and electricity production. Electric vehicles and solar and wind energy sources are increasing quickly in order to reduce the growing emergencies we are now facing. That is not to say that progress in this area will be easy. There does appear to be a growing consensus on what constitutes some of the more promising and doable technologies to transform our energy sectors. As modern humans, we seem to be better at finding technological fixes versus changing our behaviors or practices. The latest problem-solving technology is just around the corner in our minds. It has become a god of worship in many ways. I offer a bit of skepticism here:

> I am not so hopeful about agriculture and our food systems around the world, which I base my thought primarily on the ignorance and apathy of the general public about food and farming.

It is perhaps most abundantly true here in America but also worldwide in most urban, and remarkably sad but true, rural areas as well. As people have moved farther and farther from the land, either by choice or obligation, their lack of knowledge and apathy about it has increased. Again, our reliance on technology to solve problems and make agriculture more efficient has blinded us to the other consequences of technology, moving people off the land, rural areas, and communities.

I am continually dismayed at how little attention our media pays to agriculture, to farmers, and to rural America in general. I do not care what kind of media it is either, conservative or liberal, it does not matter. So little attention is given to what goes on in rural America, and when it is, it most often comes with the gibberish and depiction that goes along with the portrayal of the patriotic and hard-working American farmer who is feeding us Americans first and then feeding the world, too. Farmers cannot speak for themselves. They generally have to be represented by an industry or commodity spokesperson. In the organic and sustainable agriculture arena, I see far better farmer participation and representation on the speaker's podium.

Maybe I am coming down too hard on the world's industrial food system. It is easy to be critical when you read about what they consider to be solutions and their contributions to reducing greenhouse gases. There is a good amount of current and potential "greenwashing" on reducing CO_2 and sequestering carbon through the buying of offsets. For instance, it has been reported that two of the largest global food companies, Nestle and Unilever, and one of the largest global financial companies, BlackRock, use the guise of planting monoculture tree plantations that displace poor farmers by taking control of their land so they can use the offsets to keep using fossil fuels and maintain their market share.[39]

Fertilizer companies use the argument that by continually increasing yields through fertilization, they have empowered producers to produce more on less land, thereby cooling the planet and saving the forests from being burned and put into crop production. Real solutions involving agroecology and food sovereignty for poor farmers in poor countries are being blocked. Agricultural innovations like precision fertilization, increased cover cropping, and no-till do improve the soils' carbon and water storage capacity. That is not being denied. However, it is not enough to mitigate the effects of climate change in a larger way. I offer this thought,

> Agriculture emphasizes adaptation over mitigation, and it does not address the broken food system that is controlled by the very few.

Carbon capture has become the buzzword of the fossil fuel, corn ethanol, and industrial food systems. Very few fossil fuel refineries and ethanol plants capture the CO_2 emitted when the fuels are refined into the final products of ethanol, gasoline, and diesel fuels. Another cold hard fact is that some ethanol refineries are still using coal to produce the ethanol. In order to meet the White House and State Department goal of net zero carbon emissions by 2050, many of these plants are proposing to capture the CO_2 at the plant and then pipe it underground to be stored deep in rock formations where it cannot escape. There is no money to be made by any of the fossil fuel or grain ethanol refineries in capturing the CO_2 at the plant and then piping it long distances to be buried underground unless it is going to be used to recover more oil and gas from where it is injected into the ground. Otherwise, only huge amounts of tax credits will make it profitable. Apparently, that is precisely what is intended for three proposed pipelines to run through Iowa, disrupting our precious farmland and lowering crop yields where the pipeline would be laid. At the time of this writing, one of the companies, Summit Agricultural Group, is proposing that its pipeline will collect CO_2 from at least twenty-nine ethanol refineries in Missouri, Iowa, and Minnesota and store it underground in

North Dakota.⁴⁰ If it is approved, it will run directly on the land of many of my neighbors. Eminent domain can be used by the company to force landowners into allowing this to happen if the Iowa Utilities Board approves the project. I think it is a band-aid over a bullet hole at best.

Corn for use as a biofuel has been steeped in controversy ever since it started. It is true that the corn ethanol industry has worked to improve its efficiencies. Yields of corn have increased with better genetics. Ethanol plant efficiencies have improved.

Here are some facts about ethanol at the time of this writing:

- One bushel of corn today yields about 2.86 gallons of ethanol and about 17.5 pounds of distillers dried grains that are fed to cattle in feedlots mostly as a dried product, which takes additional energy to dry because it comes out wet.
- The nation's average corn yield this year was 180 bushels per acre, which yields 486 gallons of ethanol from one acre.
- One gallon of ethanol has the same energy as .66 gallon of gasoline.
- In 2021, 57 percent of Iowa's 1.3 billion bushels of corn went to ethanol production.
- Corn requires a good deal of nitrogen to grow. Nitrogen production at the plant emits six to nine kilograms of nitrous oxide per ton of ammonia produced using the Haber–Bosch process.
- Nitrous oxide represents 6 percent of all greenhouse gases. Three-fourths of that comes from agriculture.
- The current greenhouse gas profile for corn ethanol assesses that it is 39 percent less than gasoline in its CO_2 emissions. This is an improvement over corn ethanol's original emissions of 20 percent. If it is refined at a plant using natural gas, it climbs to 43 percent.⁴¹

The case can be made for ethanol being considerably better than fossil fuels in lowering CO_2 emissions and that it is a renewable fuel. Is that really true? First of all, as a farmer, I want to grow food for people to eat, not necessarily for running their automobiles. That view has intensified with the knowledge of the disastrous effects of climate change building all around us. The simplest and most important argument against using grains for fuel, whether for ethanol or soy biodiesel, is that if corn and soy are used for fuel, then food acres will have to come from someplace else if people continue eating the same amount of food as they are now. We keep hearing we will have 9.9 billion mouths to feed in 2050, up from the 7.8 billion people in 2020. The industrial agriculture system says improvements will continue to be made in fertilizer, genetics, and other best management practices. That flies in the face of the fact soil continues to erode at a faster rate than it can be formed. It flies in the face of the increasing amount of forests being burned and grasslands being plowed up worldwide for the commodities of corn, soy, sugarcane, and palm oil, which release massive amounts of CO_2.

If land is switched out of the USDA's Conservation Reserve Program (CRP) acres and pasture and trees and other waste land to continue to grow more ethanol in the United States, then increased amounts of CO_2 will be released.⁴² If the present acres in CRP are cultivated again in crops, then 750 kg of CO_2eq per year that could have been sequestered carbon will be lost. That exceeds any reduction in CO_2 from ethanol. Given that so much of our country mainly grows two crops, corn and soybeans, another issue is that the expansion of ethanol may decrease soybean production in the United States. If other countries then react by planting more soybeans, it will most likely come either from lower-value crop fields or from virgin grass or forests. If these two scenarios do indeed happen, then the immediate effects of corn ethanol will mean more CO_2 emissions from corn ethanol than what is saved, as compared with fossil fuels. In addition to CO_2 emissions, there are other less mentioned but perhaps more harmful and negative effects of growing corn for ethanol. These include the algae blooms from nutrient runoff, soil erosion, and habitat loss.

In 2022, an eight-year study conducted by the University of Wisconsin–Madison Center for Sustainability and the Global Environment bears out what some have long suspected. Ethanol can actually be worse than fossil fuels when it comes to CO_2 emissions.[43] In fact, it is 22 percent worse when you examine what happens when land is put into corn production that was in grass, trees, prairie, or CRP. This refers to land that has not been in crop production for at least ten years but instead is planted with both cool- and warm-season grasses, legumes, and other native plant species. More CO_2 is lost by the destruction of perennial plants than what is gained by the planting of corn used for ethanol.

Biofuels that come from another starting material, like biomass crops such as *Miscanthus* and prairie grasses grown on land not suitable for other crops would improve the lifecycle analysis of growing plants for fuel. These are perennial crops that require far fewer inputs of chemicals and fertilizers. Second-generation biofuels that use non-food waste stocks coming from agriculture (cellulosic), restaurants, and municipal waste should be more efficient than corn ethanol.

Capturing CO_2 at the plant at the time of refining is certainly better than what is happening right now. But, instead of burying it underground, shouldn't the ethanol plants be working harder toward using it for all the new and emerging uses for captured CO_2? These include more sustainable chemicals that use captured CO_2 to create industrial polymers. Another use that has transformative applications is using carbon dioxide in cement to make it stronger. Present calcium-based cement is responsible for about 7 percent of all global CO_2 emissions. A new technology uses slag, which is a byproduct of the steel industry that contains calcium oxide. The calcium oxide is injected with CO_2 to form calcium carbonate, which replaces the traditional cement portion where primarily limestone is heated to 1480 degrees Celsius to extract calcium oxide. CO_2 makes cement significantly harder. Just as an analogy,

> Question from one police detective to another: Why is it so hard to capture carbon? Answer: Because it just keeps vanishing into thin air!

CO_2 directly captured from the air and combined with hydrogen gets a good deal of press as a breakthrough technology for reducing CO_2. It is technically possible, but at present, it is a very energy-intensive and costly process. It is estimated the price of carbon would have to be in the $90 per ton range to make it economically feasible. Carbon is currently trading at a worldwide average of $3 per ton. Only 4 percent of the world's carbon market is being traded at $40 per ton. That is the same amount of dollars per acre that farmers in a recent US survey said they need to be willing to participate in agricultural carbon sequestration markets. The proposed CO_2 pipelines carrying carbon dioxide to be forced into the ground in North Dakota are purported to earn the companies building them $50 per ton in tax credits from the government, amounting to millions of dollars per year.[44]

Remember, nitrous oxide emissions represent 6 percent of all greenhouse gases. About 62 percent of that comes from agriculture. Nitrous oxide is rarely mentioned for its contributions to increased greenhouse gases. It is nearly 300 times more powerful a greenhouse gas than CO_2, and it stays in the atmosphere for an average of 114 years. Agriculture creates nitrous oxide both directly and indirectly. It does so directly through fertilized soil and livestock manure and represents about 40 percent of the total greenhouse gas emissions coming from nitrous oxide. Around 22 percent comes from the runoff and leaching of fertilizers. New research is bringing to light how important it is to address growing nitrous oxide emissions. For starters, it is the largest remaining cause of ozone depletion. Unless emissions are decreased, the decades-long and mostly successful efforts to lessen the size of the ozone hole will be in jeopardy. The banning of fluorocarbons in refrigeration led to those improvements.

It bears repeating that significant reductions in both methane and nitrous oxide could be made in animal livestock production if the large confinement operations would move away from liquid manure to solid bedding systems using straw as both the source of bedding and the primary carbon source for composting. When manure is placed in windrows and adequately turned and aerated, reductions in both nitrous oxide and methane are obtained. The awful smell of liquid manure could also be significantly reduced. Small grains like oats, barley, and especially hybrid rye could all become more important crops than they are now. For animal agriculture anyway, I have concluded a partial solution to some of the climate issues is to challenge the large, concentrated animal feeding operations and the farmers that grow crops for them to grow and feed something besides corn and soy.

To summarize, I foresee a number of critical steps and changes in behaviors and attitudes that are needed if we are to make more meaningful changes in agriculture. This change is not only for climate resilience but also for other good reasons, including competition, changes in how we grow and distribute food, and how we can bring people and diversified farms back into the equation. Too many people, both urban and rural, do not know what is really going on, or do not care, or feel powerless to do anything about it. When I bring up the subject with my rural friends, too many of them say the powers that be in agriculture and food production kept those possibilities from happening a long time ago.

Some of the most difficult questions to address in fair and unbiased ways revolve around how to increase organic food production and consumption and make it so it is readily available and not cost prohibitive, especially for the poor. It is a real dilemma.

Our food system has become much more concentrated. Walmart now controls over a quarter of the market. Kroger is a distant second with a 10 percent share of the market.[45] Food has gone the same route most other US industries have. That is to say a limited number of corporate entities (three to five) control just about every aspect of our lives in every sector of our economy. This ranges from phones and credit cards to energy and alcohol. For food that includes the processors (many of whom already own the product before processing), and the wholesale and retail sectors. It is justified through the argument of natural efficiencies gained through consolidation. However, the ability to manipulate the market through price collusion and price gouging by those remaining few players also becomes possible and has become more prevalent. The United States has historically had an obsession to lower the share of disposable income spent on food tied to the decline in numbers of farms. In 1920, there were 6.5 million farms in the United States. One hundred years later, in 2020, the number declined to 2 million. In 1929, Americans spent 24 percent of it on food. By 1960, the number dropped to 17 percent. In 1970 it was 13.8 percent. By 1999 it dropped to 10 percent. From 2000 through 2019 it fell to under 10 percent. When the Covid pandemic hit in 2020, that percentage has actually gone up to around 11.2 percent.[46] There are a number of reasons. Some are supply disruptions in the food chain, the war in Ukraine, bird flu in the poultry industry, and substantial increases of 10 percent to 15 percent in corporate profits.

Americans now spend about half of that 10 percent to 11 percent of their disposable income by eating out. Generally speaking, we could spend considerably less if we did more of our own cooking. The raw ingredients, whether organic or conventional, are cheaper than processed foods. They are also more nutrient dense. We know the poor have fewer food choices than people of wealth. Food deserts have become a very real problem in poor neighborhoods. There may not be a grocery store left in that neighborhood. The dependence on fast food coming from both restaurants and convenience stores is on the rise. The price of food in convenience stores is much higher than in a grocery store.

All of this still does not address how to get more healthy and organic foods onto our plates. The first way is to increase disposable income for food through higher paid wages for all of our citizens. After that, a partial list of ways may look something like this:

(1) Increasing urban gardens.

(2) Increasing local food systems through more farmers' markets in poor neighborhoods.

(3) Forming both more for-profit and non-profit local and regional growing, processing, and marketing enterprises. This could involve more civic engagement by city governments to make sure that it happens. It requires a great deal of volume in sales to make it profitable with for-profit entities. Perhaps local option sales taxes could help.

(4) Legislation that would incentivize or even require an adequate number of grocery stores being located in our towns and cities and locations where food deserts occur.

(5) Mandating a certain percentage of food be local and/or organic in school lunch programs, hospitals and nursing homes.

(6) Increasing the number of local food producing and organic farms through incentives to do so. These could be federal, state and local by nature. Lower crop insurance premiums for healthier food systems could be one way to positively influence organic farms and incentivize them to become organic. Lower property taxes may stimulate the creation of many more organic and local farms and food processors.

(7) Improving nutritional quality of food for all people through disincentives for high sugar, high salt and low nutrient density foods

(8) Creating ways to increase conservation stewardship payments and other resource conserving programs through farm bill provisions.

(9) Although it seems impossible to do so at this time, generally incentivize and reward good stewardship practices and disincentivize those practices that harm our health, food, environment and our planet. Another way of putting it is to factor into the final cost the true economic costs of our current food and farming systems.

Another standard response is to have farms the way I describe ours is: it requires people who want to work and nobody (mostly our youth) wants to work anymore. Just how much truth is in that statement? If there is any, it also begs the questions, how and why? Parents first and then schools and workplaces must instill the values of working to support oneself and to teach youth that it is important to learn how to work and have meaningful work that promotes self-worth. We are witnessing too many in our workforce having expectations that do not require enough accountability or consequences for their actions. This appears to be the case whether you are a Gen X person (forty-five to forty-nine years old), a millennial (twenty-eight to forty-three years old), or Gen Z (twelve to twenty-seven years old). Too many people, including some in professional jobs, seem to think they do not have to show up for work but are entitled to start out with salaries beyond their experience or merit. At the same time, wages need to be both fair and reasonable so that people can support themselves. The wage gap between men and women performing comparable work is still unresolved.

In the end, what is so striking is even though we are considered the richest country in the world, we are far down the list in almost every other realm, including health, happiness, and social trust. Could it be because we keep pursuing the wrong things?[47]

Chapter 14
What Can and Will We Do?

The future will be green, or not at all.
—JONATHON PORRITT, BRITISH ENVIRONMENTALIST AND WRITER.[1]

In a *CBS 60 Minutes* interview, Pope Francis spoke about the "globalization of indifference is a very ugly disease."[2] He referred specifically to attitudes toward war, immigration, and the horrific crimes committed against innocent people. When it comes to our changing climate, has this disease of indifference also infiltrated and embodied our society?

An October 19, 2021 article in the *Cornell Chronicle* noted that 99.9 percent of all scientists agree that humans are impacting the climate.[3] Yet, because of the polarization and politicization of these issues, we are not making the kind of progress so badly needed. It only takes a few clicks online to find misinformation and attacks on the credibility of the science of climate change. Powerful vested interests want it that way. The same tactics used to fuel culture wars are being used to demonize environmental and human health concerns over both climate change and modern agricultural practices.

"Big Tech" has promised to lead the way toward renewable energy and climate change mitigation. Currently, Amazon, Google, Microsoft, Apple, and Meta use as much electricity as the entire country of New Zealand, according to an article in the *Financial Times* in 2021.[4] While they have made some progress in adopting renewable energies, much of their supply chain still relies heavily on fossil fuel energy.

A major criticism of "Big Tech" is that it helps to drive emissions by accelerating consumerism. It is also accused of helping to drive division, distraction, and climate disinformation. Four out of five major social media platforms—TikTok, Twitter/X, YouTube (Google), Meta, and Pinterest—do not have a content moderation policy that includes a comprehensive, universal definition of climate misinformation.

The fact remains that we have to lower carbon emissions if we are to lower carbon dioxide emissions and then lower carbon dioxide levels in the atmosphere. There is no getting around that. We can spin it any way we want, but the truth still remains. It is the primary driver of what has occurred since the beginning of the Industrial Age.

Can any of the existing energy sources lower the carbon enough to begin to make a real difference in reducing carbon dioxide emissions? Maybe. Are they the silver bullets we are looking for? Will they make a big difference in how we power much of our electrical, transportation, heating and cooling, and industrial processes? I certainly hope so. The sources so many have their hopes hinged on are solar and wind.

In my view, solar and wind energy have a number of basic advantages:

- They cannot be depleted.
- They both reduce toxic air pollution.
- They both have low operational costs.
- They can both can provide energy to rural areas efficiently.
- They both show more favorable life-cycle analysis for emitting less greenhouse gases than fossil fuel technologies, due in part to not requiring additional energy for production and transport.

Solar and wind have advantages and disadvantages when compared to one another:

- Solar can be used on both new and existing structures in all areas: urban, suburban, and rural.
- Solar panels require less space than wind turbines.
- Solar panels are less conspicuous and make less noise than turbines.
- Wind turbines are more energy efficient. Solar can capture and convert 23 percent of sunlight into energy, while wind can turn roughly 50 percent of captured wind into energy.
- Wind turbines are more susceptible to damage from wind and lightning.
- Current wind energy towers cannot be easily built in water depths greater than two hundred feet.
- Offshore wind power has the potential to create more energy because of higher wind speeds and greater societal support because of a decrease in the negative impact on birds, bats, and aesthetics.
- Wind power has less acceptance due to the increased incidence of annoyance, anxiety, depression, nausea, and vertigo.
- Other complaints include low-frequency hum, infrasound (below the lower limit of audibility), vibration, and shadow flicker of wind power towers.
- Transmission lines need to be built to bring electricity to populated areas for both wind and solar energy.
- Solar and wind can both be unpredictable sources of energy.
- Wind turbines cannot be built in areas with low wind speeds.
- Solar has widespread adaptation capabilities for homeowners as a source of electrical energy to power both their homes and their transportation vehicles.
- If built on a large scale, both wind and solar can negatively impact food production areas in that they take land out of food production. These impacts could be significantly lessened if placed in nonproductive and less populated areas.

- Human physical and emotional health could be enhanced by placing wind farms in areas of low population density with adequate distance from both farm and community residential areas.

At this point in time, I am a bigger proponent of solar energy than I am of wind. Personally, I do not like the noise or the aesthetics of wind generators. Valuable food-producing land is required for wind farms. Solar electricity has fewer moving parts, and the land under solar collectors can be effectively farmed for food and pollinator species if the sites are not too large. Apparently, there is evidence for some climate change effects around wind farms caused by the wind farms themselves. Long-term effects on wildlife are also a concern. An essay in the *World Resources Institute* notes:

> While there is no panacea for energy sources to replace fossil fuels, the fact remains that both solar and wind power have a lower greenhouse gas life cycle than conventional fossil fuel energy sources.[5]

Greenhouse gas (GHG) emissions are produced during the manufacturing and recycling of solar and wind systems, but the generation of energy results in zero GHG emissions. The biggest challenge could be the large increases in the amounts of minerals needed for clean energy sources. Lithium, nickel, cobalt, manganese, and graphite are critical for battery storage and performance. Rare earth elements (REMs) are needed for the magnets in wind turbines. The solar electric grid needs massive amounts of copper and aluminum. Who controls those minerals, how they are mined, and how the workers are treated are all issues that remain unresolved.

Maria and I would like to install solar panels on the roof of one of our barns and perhaps on our farm shop in order to power the whole farm's electrical needs. We will need at least 25-kilowatt (kW) capacity to do so.

We have used geothermal energy for our home since 2005, and it has paid for itself many times over. Three 180-foot-deep wells recycle fifty-five-degree Fahrenheit water through the system. Maria's farm store with display freezers and refrigerators and nine chest freezers for our beef and pork consumes a great deal of electrical energy. Grain bin aeration fans and other electric motors for farm use also consume a significant share. We want to do our part on many different fronts when it comes to agriculture's carbon footprint and GHG emissions. It is no longer a pipedream to be able to plug in an electric or hybrid vehicle at our home, too. Relying so significantly on fossil fuels and waiting to do something different does not seem like a good strategy.

As a society, we have waited far too long already. We could have started to make some of these critical changes forty years ago, but corporations and politicians would not allow it. As has been seen, there are many different sources of energy fighting to keep their seat at the table and new alternative ones trying to get their own seat at the table. Given the inaction and polarization in US Congress and the power of climate deniers, it is really tough to be optimistic about how quickly our energy source transition can significantly reduce the harmful effects of increases in GHGs.

Change must happen everywhere. China burned 50 percent of the world's coal in 2020. Its use of coal for energy production hit a record high in 2023 even though it had pledged steep cuts in emissions by 2030 and to be carbon neutral by 2060. South Africa relies on coal for 71 percent of its energy, China 57 percent, and India 55 percent. The United States reduced its coal usage to 20 percent in 2020.[6]

I first learned about the nonprofit Project Drawdown in 2019.[7] It lists and describes solutions on how GHGs can be reduced. Its core work was published in 2017. Their data were informed and gathered by researchers and scientists all over the world. It ranked one hundred solutions to decrease GHG emissions from 2020 through 2050. It enumerated the potential in gigatons of CO_2 removed from the atmosphere, listing seven areas where this could occur:

- Materials
- Energy
- Food
- Land use
- Transportation
- Buildings and cities
- Women and girls

The results are quite surprising and not what we would at first expect. Refrigeration comes in first for GHGs. The replacement refrigerants for the ozone-depleting compounds of chlorofluorocarbons (CFCs) are known as hydrofluorocarbons (HFCs). While they do less harm to the ozone layer, their capacity for global warming of the atmosphere is one thousand to nine thousand times more powerful than CO_2. Nearly ninety gigatons of CO_2 could be reduced if they were properly disposed of at the end of their useful life. That is when their greatest potential for harm comes due to leaking and other destructive measures that allow the materials to be released into the atmosphere.

As global temperatures continue to rise, so does the demand for air conditioning and cooling. It is imperative that other alternatives be found that do not destroy the ozone but still lower CO_2 levels. It is estimated that 6 percent of the world's energy consumption comes from refrigeration. Space cooling and air conditioning account for 8 percent. It is expected to rise by three times that in 2050 because of hotter temperatures and increased demand. I offer a different perspective:

> I have said that energy and transportation are the largest contributors to GHGs. According to Project Drawdown, I am wrong. It is our food systems around the world.[8]

Project Drawdown notes that, out of the top twenty-five solutions for decreasing CO_2, eleven of them come from the food sector, which was not surprising, but I was surprised to find out what is number one for our food system and number three overall. It is reducing or eliminating food waste. Only wind energy is greater at the number two overall ranking. According to Project Drawdown, the eleven top factors related to food that could contribute to the greatest reduction of GHGs are:

- Reducing food waste #3: 71.5 gigatons
- Plant-rich diet #4: 66 gigatons
- Silvopasture #9: 40 gigatons (permitting or intentionally planting trees and shrubs in grazing pastures to increase CO_2 consumption and carbon sequestration)
- Regenerative agriculture #11: 24 gigatons (using regenerative and organic principles to sequester carbon, not conventional ones using synthetic fertilizers and pesticides)

- Tropical staple food trees #14: 20 gigatons (perennial food trees such as bananas, citrus, avocados, apples, baobab, nuts, mongongo, and many more)
- Conservation agriculture #16: 17 gigatons (no-till and reducing soil erosion)
- Tree intercropping #17: 17 gigatons (planting trees and shrubs in alternate strips with crops—also known as alley cropping)
- Management-intensive grazing #19: 16 gigatons (rotational grazing of pasture strips throughout the grazing season)
- Clean cookstoves #21: 16 gigatons (eliminating black carbon, incomplete burning, carbon monoxide, and premature deaths of 4.3 million people annually)
- Farmland restoration #23: 14 gigatons (restoring fertility to abandoned and worn-out soils)
- Improved rice cultivation #24: 11 gigatons (transplanting single seedlings versus mass planting of rice seedlings, watering versus continual flooding of rice paddies—traditional rice paddy flooding greatly increases methane from methane producing microbes).[9]

These suggested improvements account for over *300* gigatons of CO_2 that could be removed by the year 2050. The next seven greatest areas of reducing GHGs come from the energy sector:

- Wind turbines #2: 85 gigatons (onshore)
- Solar farms #8: 37 gigatons
- Rooftop solar #10: 25 gigatons
- Geothermal #18: 17 gigatons
- Nuclear #20: 16 gigatons
- Offshore wind turbines #22: 15 gigatons
- Concentrated solar #25: 11 gigatons (using mirrors to optimize solar power).[10]

These seven energy areas combine to lower GHGs by over *200* gigatons of CO_2. The next four out of the top twenty-five ways to reduce GHGs revolve around our world's forests and peatlands:

- Tropical forests #5: 61 gigatons (preventing loss)
- Temperate forests #12: 23 gigatons (preventing loss)
- Peatlands #12: 22 gigatons (preventing drainage, burning, fuel, and horticultural uses)
- Afforestation #15: 18 gigatons (reforestation of cut down, burned, degraded, and destroyed forests).[11]

These four areas alone account for over *123* gigatons of GHGs. The last two in the list of the top twenty-five ways to reduce emissions involve women and girls. They are Family Planning and Education of Girls. Why are they so important? Most often, education and family planning go hand in hand. Educated girls tend to have smaller but healthier families. Both a decrease in the number of children they give birth to and the improved health of the ones they have would have a significant impact on the amount of global emissions based on the sheer numbers involved. Another notable factor, however, is that women play a large role in food and in income production around the world. Their roles in political institutions and in societal functions are becoming more critical as time progresses. The more educational opportunities they have, the greater their potential influence on how the land, families, and social institutions are cared for. Educating girls is #6 and Family Planning is #7 on the list, both at 59.6 gigatons of CO_2.[12]

In developing countries, 43 percent of the agricultural labor force are women. We need more women farmers in our own country. They have been a major driver for the increased numbers of small farmers in the United States. The number of women entering the field of agricultural studies has been steadily increasing. The number of women studying in Iowa State University's (ISU) Department of Animal Science in 2022 was 765 female students, compared to 171 male students. Women make up 80 percent of the enrollment in veterinary medicine in the United States and Canada. And as to the future of farms, I offer this thought:

> What kind of livable future the Earth will be able to provide is not dependent only on striving to mitigate the harmful effects of our changing climate and rising GHGs.
>
> I want to make it as clear as I can by stating, again, that we need *more* farmers around the world, not fewer.

A more livable future for the world addresses climate change and builds more just and equitable societies comprising more diverse and vibrant rural communities and more just and equitable food systems around the world. We rarely hear talk about needing additional farmers because of agrifood corporations' continued defense of the economic status quo of industrialized systems for food and farming, believing that the inevitable outcome is wanting and needing fewer farms and farmers. That is what I have been trying to articulate throughout much of this book. Why do we have to keep going down this path? I reiterate: *It is because we do not understand nature, our place in it, and what we must do in order to live in harmony with it*. We continue to make the mistake of assuming that larger farms are more efficient. We do that without putting the full cost in the equation. What are the public health and human capital costs associated with energy-intensive inputs in food production systems (for example, nitrogen-based fertilizers and pesticides)? What are the ecological costs? What are the air, water, and soil health costs? What are the economic costs of the decline of rural communities? What will be the cost derived from the displacement of millions of people from their homes and homelands due to the rise in sea level? I suggest that,

> Growing our own food and owning the land where it is grown gives a deep meaning and purpose for living that we are losing.

For example, if small farmers in Central America continue to be forced from their land due to the changing climate and the economics of a powerful industrial food system that chews people up and spits them out, they will continue to migrate. Even then, if there are no decent jobs, or even any job, waiting for them in crowded cities, there's a strong risk they will turn to alternatives such as crime or drugs. This holds true for much of the world where foreign countries, aided by industrial food, chemical, and seed companies, continue to grab land from farmers and industrialize their food economies. Moving forward, consider that,

> For most of us, tending and caring for the "Garden of Eden" seems further and further away from our memory, our consciousness, and our being. That is the core of what we are up against.

We could do something about it — if we were willing to invest in educating farmers and industry in our own country and around the world in the principles of green, biological, regenerative, and practical farming and by more emphasis on putting resources into food processing and distribution systems that are just and fair and local and regional in nature.

It could be a new kind of Cooperative Agricultural Extension or a new kind of Peace Corps that would train literally millions of educators and industries, which in turn would work with millions of farmers to create the change so many desperately want and hope for. A core challenge, however, is that only farmers would be making money from such an endeavor. Agribusiness would find that hard to accept. If we would allow Mother Nature to do most of the "heavy lifting," which she does mostly for free with her multiple ecosystem services, then we could make meaningful progress. In order to do that, we would have to change how campaign contributions and powerful lobbyists influence the political and economic spheres.

We need to build on the positive models of local and regional food hubs, for some obvious reasons and some not so obvious. An apparent component for reducing the energy and carbon footprints is how far food travels before it is consumed. Perhaps a not-so-obvious reason is to bring new life back to rural areas and communities. For instance, there could be literally hundreds of farmers farming and living in my county growing food for the greater Omaha area. We are hardly fifty miles from the growing urban centers that make up Omaha, Nebraska, and Council Bluffs, Iowa, where nearly a million people now live. Shelby County has a population of 11,000 people.

Although it would appear that the "Titanic" is sinking, there is reason to be hopeful. Human suffering has historically brought about change and redemption for many of us. It can again. The stakes are higher now, but we can do it. Finding ways to increase land ownership by those working the land and to increase the sheer number of farmers around the world are two key goals that we need to keep in sight. They should be right at the top.

However, there will be resistance to this. Corporations' market share and bottom line always come first. If industrialized food giants were to have the goals I previously proposed and worked to make them happen, then just maybe I would have a different opinion of them, which echoes what I said in this book's introduction. I still do not see it happening in a socially just and equitable way. But I want to be proven wrong.

The word "sustainable" has been so co-opted that I can hardly recognize it or agree with its definitions. Most every commodity group — ranging from those like soy and corn growers, the egg industry, and pork producers — says what they are producing is sustainably raised. To be fair, certain parts that claim may be true or at least partially so. However, it comes from a narrow definition of sustainability. For example, today I see advertising that says more of whatever is being raised or grown is being done on less land. It is an argument that is often used to negate the possibility of more organically farmed acreage in the future. It does not just come from commodity groups. It comes from climate experts who say more agricultural land being put into food production automatically means more GHGs because of how much carbon is lost in the process of transition and how much more will come from annual food production thereafter. This does not account for obvious questions such as what will be grown. If it is 10,00 acres for palm oil production with all its energy and fertilizer- and chemical-intensive inputs, that is one thing. It is quite another if it is a small farmer in Honduras planting a diverse perennial and annual crop and vegetable mix on five acres comprising three acres of perennials like mandarin oranges, coffee, plantain, grapefruit, cacao, and leguminous trees to contribute nitrogen to all of those perennials. The annual crop production may consist of an acre of white corn for tortillas and a half acre each of carrots and casava. How many people are being fed has to do with what they are eating. How much CO_2 is being emitted into the atmosphere has much to do with how the food is raised, processed, and transported.

The solutions I have presented for the regeneration of our rural communities, our farms, our environment, and climate change could hardly be perceived as radical in nature. They could not be viewed as being socialist, communist, or fascist. They actually represent what is best about a just and fair capitalistic economy that includes widespread farm and land ownership, just prices up and down the food and business chains, and responsible free enterprise. Being rewarded for doing the right thing in terms of our environment and our natural resources and treating our fellow human beings and all living things with the respect and dignity they all deserve is not something out of the realm of possibility.

With that said, there has to be some ranking of my own, given what I think our priorities for a livable planet need to be.

Climate Change Must Be at the Top of the List

In the year 2000, catastrophic storms, wildfires, floods, and droughts seemed far into the distant future. I feel a sense of urgency now. Two of the biggest questions we face are, will we work to mitigate and to lessen the effects of climate change or will we only learn how to adapt? Many of the subsequent topics on the list could be used to lessen the harmful effects of climate change. All of these topics are important in their own right, climate change or not. The reality of what climate change is bringing only makes them that much more urgent.

> The decline of species appears to be directly correlated to increases in temperatures caused by climate change.[13]

One-third of all plant and animal species could face extinction by 2070 due to the effects of climate change. Researchers at the University of Arizona have found that the maximum annual temperature (the highest daily highs in summer) is the key variable best predicting whether a species will go extinct. They found that about 50 percent of the species will have local extinctions if the temperature increases by more than 0.9 degrees Fahrenheit. The decline will increase to 95 percent if temperatures increase by more than 5.2 degrees.[14] It appears that we will be hard pressed to keep the temperature under 2.7 degrees Fahrenheit, which is what the Paris Climate Accord of 2016 declared as its legally binding goal.[15] The mating behavior and breeding cycles of insects are very sensitive to increases in temperatures. Degraded habitats and increased rates of disease are also leading to their decline.

Maintaining Soil Quality and Preventing Soil Loss Must Come Next

> So it is not one whit mysterious that we poison the water and air and topsoil, and construct ever more doomsday devices, both industrial and military.
>
> —*KURT* **VONNEGUT**, *TIMEQUAKE*[16]

With or without the negative effects of climate change, numerous vibrant rural societies throughout human history have declined or disappeared due to not taking care of their soil resources, including the ancient Egyptians and Greeks, citizens of the Roman Empire, and the Mayans of Central America.

The world's food system has an addiction to fertilizers, pesticides, and big machines. Negative results come to mind immediately for these three addictions: increased nutrient loads and waste of fertilizer, increased health problems and pest resistance, and soil compaction and decreased soil quality. Research at the University of Illinois over the past few years has found that two-thirds of the annual yields in corn productivity should come from activity in the soil, meaning the life and health of the soil itself.[17] Only one-third should be coming from direct supplemental fertilization. The long-standing paradigm has been just the opposite. It has stated that soil can supply only one-third of the needed soil fertility for any given year in corn production and that two-thirds has to come from the addition of fertilizer. There is a soil test, also developed at the University of Illinois, that can accurately measure whether any fertilizer should be added for optimum corn yields. Unfortunately, this test has not been adequately tested out in the field but only in self-contained growing containers in a closed system. The test has not proved reliable in testing done by Iowa State University. However, the question of more usable nitrogen being able to come from soil itself seems to me to be a legitimate one for further study.

Diversity

Diversity includes a wide variety of cropping systems, crop rotations, and maintaining species heterogeneity, which will become even more difficult with temperature rises. More diverse cropping systems and crop rotations help to lower harmful emissions.

Water

Water is perhaps the sleeping giant that will have far greater negative impacts than we can even begin to anticipate.

We are running out of water in so many places, drying up aquifers, and irrigating where we probably should not. The low quality of our water is due to inadequate crop, fertilizer, and pesticide management. Desertification is increasing in already arid regions because of overgrazing, increased water demand, deforestation, and, most importantly, climate change. Different water-saving strategies need to be considered, including restoring wetlands and creating ponds and water-holding structures on farms and water stops or shutoffs on tile drainage systems that effectively push the water back up a slope for subirrigation of crops. Other strategies include the slowing down of water through the use of riparian areas, wetlands, buffer strips, and grass waterways.[18]

Perennials

Perennials as staple crops will have to take on more importance if we are to curb the harmful effects of what food production and climate change have wrought. These already include long-lived trees, vines, shrubs, and herbs. One staple crop that the Land Institute in Kansas has been working on is now coming to fruition. It is a form of perennial wheat known as Kernza.[19] China has been working on perennial rice. Permanent pastures for grazing of ruminants is also an important perennial staple crop. Some pastures have been in intensively managed rotational grazing for beef animals for nearly thirty years on our farm. The soil has not been disturbed by any form of tillage for that amount of time. Perennials for biomass fuels may also play a role in meeting our alternative energy goals. These might include crops such as miscanthus and switch grass or even some plants that are now considered as weeds.

It is hard to know where to begin and end with the list of what should be our priorities for a more livable planet. Below, I list some of the most important ones, in my estimation. (*Disclaimer*: The following list is not comprehensive and may not include other equally important items.)

- Vibrant rural communities
- Gardening, cooking, and growing our own food.
- Recognizing food sovereignty: respecting the right of other countries to grow the bulk of their own food themselves.
- The United States does not have the right or the moral authority to make other countries subservient to it for food imports.
- Increasing competition and decreasing concentration in the food industry.
- Creating laws and preventing foreign governments in poor countries from "land grabbing."
- Getting livestock back on the land and largely out of huge confinement operations.
- Respecting the rights of indigenous peoples.
- Developing a just and fair immigration policy.
- Creating regional and local food hubs as the primary food production, processing, distribution, and consumption modes of operation around the world.
- Employing the "cooperative business model."
- Making products that last.
- Making recycling mandatory.
- Creating opportunities for beginning farmers, minority farmers, and women farmers.
- Creating a new extension service around the world that will train and work with farmers in becoming more climate-resilient.
- Providing more opportunities for farm ownership around the world.
- Creating more regenerative and organic farms.
- Creating agricultural policies, incentives, and insurance programs that reward farmers for farming in climate- and ecology-resilient ways.
- Tempering and lowering materialism, consumptive, and wasteful behaviors.
- Planting trees, shrubs, prairie plants, perennials, herbs, and other beneficial plants.
- Improving the quality and nutrition of our food.
- Reinvigorating public plant and animal breeding.
- Making land-grant schools and colleges of agriculture less dependent on and beholden to the corporate food and agriculture industries.

- Putting more emphasis on conducting on-farm research trials and investing in sustainable, organic, and climate-resilient research at public land-grant universities.
- Providing healthier diets and more organic options for school lunch programs and for people in institutional care, such as nursing homes and hospitals.
- Including food and agriculture, as well as growing and cooking our own food as curriculums in secondary and postsecondary programs of study.

The year 2024 marked the forty-first year since our farm stopped using pesticides and commercial fertilizers. Often, I have to ask myself why the way we farm has not been adopted by more of our neighbors. If what we are doing is considered so great, why aren't more people farming in ways similar to what we are doing? It is a valid question. I suppose part of the answer could be simply that it is too much work, or at least it appears to be that way to others. I think it is fair to say that it is a bit more work, but to me the most relevant questions would require farmers to ask themselves whether they are willing to take on more management; have a more creative mindset, one that is willing to think outside the box, and are willing and able to resist the incredible peer pressure that exists when they start doing things differently on their farms.

It is also fair to say that many farmers respond positively to what certain entities tell them to do, based on the entities' priorities. One such entity is the government farm program. Another is the supply and input side of farming, which has historically meant purchasing inputs as substitutes for a sense of what nature is telling us to do. As farming has become more and more mechanized and more technology-oriented, we are losing the indigenous knowledge and skills that farmers traditionally had. Many farmers no longer perform once-required daily tasks. Many do not have daily chores because they no longer have animals to feed and care for. There's no need for fences to keep livestock out of neighbors' fields. The old statement that "good fences make good neighbors" is no longer relevant. Very few of today's farmers would know how to make a good fence anyway. Many do not plant, apply their own fertilizer and weed control measures, harvest their own crops, or even market their own crops, but hire it done by others, including the fertilizer, chemical, and seed dealers that dominate that marketplace. Many farmers have become capital managers more than anything.

When it comes to talking or dealing with the general public, farmers have all too often permitted someone else to do the talking for them. I still hear the phrase: "I'm just a dumb farmer" or "I'm just a dirt farmer." Maybe that is too often what the general public really believes, too.

The Practical Farmers of Iowa and other sustainable, organic, or regenerative agricultural groups do not fit the stereotypes that I just presented. That is because they are made up of like-minded farmers who have learned the value of thinking not only for themselves but also to have the self-confidence and humility to ask questions of each other and to learn from one another. I have noted numerous times throughout this book the words of my mentor, Dick Thompson: "I don't have all the answers, I am just trying to ask the right questions."

The Role of Policy and Politics

First of all, I would argue that the 2010 US Supreme Court ruling in *Citizens United v. Federal Elections Commission* needs to be overturned.[20] The campaign finance laws prior to this ruling were in existence for more than one hundred years. The court's majority opinion was that political spending limits by corporations and other groups was a violation of the First Amendment right to free speech. This ruling has led to ever-greater amounts of "dark money" being poured into every election cycle. Historically, wealthy individuals, corporations, and special interest groups have had a significantly outsized role in influencing elections, but not to the extent they do now. I would also argue that this ruling has done more than any other Supreme Court ruling in the history of our country to further partisanship and polarized politics. I do not see this changing in the foreseeable future.

When it comes to agricultural policy, asking the right questions is not easy. Most people, whether rural or urban, do not really have a clue about what agricultural policies exist today nor how those policies came to be shaped historically. Most do not care unless the cost of food in grocery stores goes up. Everyone has become used to the idea that food should be relatively low in price in the scheme of things. People have become pretty used to the variety and abundance of food, too. However, a growing number of people now are beginning to understand that the corporate industrial food system controls much of what happens in farm country, whether we are Republican or Democrat. Both parties are influenced significantly by our system of industrial, corporate agriculture.

The last forty years have also shown a steady decline in the election of Democrats to office in rural areas, along with a steady decline of progressive Republicans in rural areas. This first became evident under President Ronald Reagan, whose administration sought the foreclosure of as many farms as possible. Reagan changed the demographics of the Democratic Party in the South by appealing to previous Democratic voters who felt betrayed by Democratic support for the civil rights of Black people. Rural Catholics, who had mostly formed a strong Democratic bloc in my county, for instance, left the Democratic Party over the issue of abortion.

The chasm between rural and urban people is now being cast within the framework of conservative values versus liberal values like never before. I believe it is being expressed by increasingly more visible cracks in our rule of law. Benjamin Franklin famously stated that "we have a republic, if we can keep it."[21] The United States can be described as being both a "representative democracy" and a "democratic republic," which means we elect people to represent us and make laws in a fully democratic process. In a republic, there is an official set of fundamental laws. As Americans, it is our US Constitution and Bill of Rights. The Constitution states that we are to have three branches of government: the Executive (Presidential), Legislative (Senators and Representatives), and Judicial (Supreme Court). There is to be a separation of power among these three branches where the power of any one branch is kept in check by the other two branches. We see ourselves in the increasingly tenuous position where one or more branches are testing the boundaries of this separation by usurping the rule of law. That is to say that any one branch is above the rule of law originally established by all three branches. In a republic, laws are made by representatives chosen by the people and must comply with a constitution that protects the rights of the minority from the will of the majority. We are not supposed to be a monarchy nor a dictatorship. It does not say we are to be conservative or liberal if constructs such as those are even valid in their definitions.

Unemployment, at the time I am writing this, is very low. In reality, there are not many people who do not want to work. However, it is true that they no longer want to work at the jobs that my generation has in rural areas, such as packing houses, roofing, road construction, farm labor, or more generally in jobs requiring physical work. These are the very jobs that immigrant workers are willing to take. Families, schools, policymakers, companies, and businesses have put little value on hard physical work, both monetarily and socially. These jobs and what they paid did not keep pace with the rest of the jobs in our society. It was always a race to the bottom in these kinds of jobs. The fertilizer, chemical and seed, machinery and equipment manufacturers, and meat processors and pork, chicken, and dairy confinements have greatly increased their influence at the expense of small and medium-sized businesses, whether related to farming or other sectors.

What generations of children witnessed as they grew up is now mostly in decline. Every day there are reminders of decline in opportunities for young people choosing to remain in rural areas, decline in opportunities for young people to go into farming, decline in good-paying jobs and meaningful employment, loss of belonging and the feeling of community, and, at its worst, the hollowing out and unsettling of America that Wendell Berry so aptly stated in his book *The Unsettling of America: Culture and Agriculture*.[22]

I think most of us living in rural areas would agree with that statement regardless of how we vote. I have tried to refrain from talking much about politics. I am sure that most of us want the same things and would agree on much more than we disagree on. That is, as long as we are all willing to listen and talk with one another. Communicating only with those who agree with us has become poisonous and destructive. Endless grievances harden hearts.

For the past forty years, we have generally been fed a steady diet that most politicians are crooked and are only looking to be reelected, especially those sent to Washington, DC, which requires massive amounts of money for campaigning. This behavior has steadily grown in state and local political races, too. I think the unlimited amount of money allowed to be spent helps to reinforce the notion that most, if not all, politicians cannot be trusted. Most politicians feel like they have to take the money because, generally speaking, the more money they get, the better their chances. I think corporations may have more power than the government does.

Many people may say I am too anti-corporation and too pro-socialism. I am not anti-corporation in principle, only when it comes to practices that allow them to have undue monopoly, oligopoly, and power, and undue influence over policy. I am for a blend of socialism and capitalism where restraints are placed on the excesses of both. Social Security and Medicare provide senior citizens with some retirement and medical care stability. Private medical insurance should continue to be the option for those who can afford it. The public option could provide the basics for those who cannot afford it. Farmers have had no qualms about accepting subsidies from the government for disasters, low prices, and things out of their control. But maybe our consumers and citizens also have the responsibility to help pay for investing in conserving and protecting our most valuable food, soil, water, and air quality ecosystems, too. Pharmaceutical companies should be able to be profitable but not to set such ridiculously high prices that no one can afford to get sick. There is a role for government to step in and do something about excesses and price gouging. Some say that corporations are "people too," but then others say that corporations "have no soul."

Some people living in rural areas do not seem to have any more knowledge about agriculture than their urban counterparts. That is due in large part to the loss of the diversified crop and livestock family farms that were at the heart of what the term "heartland" really means, as far as I am concerned. Many of the rural kids who lived in town had summer jobs baling hay or walking beans or cleaning out hog barns on the farms that existed when I was growing up. Maria had a summer job detasseling corn for a seed corn company. "Urban" seemed much closer to "rural" then.

Why Are We So Afraid?

I now not only fear for our country and for our kids and grandchildren in terms of climate change and the environment; I also fear for our country in terms of our democracy and democratic institutions and our intolerance for one another and disrespect for differences in opinion, race, religion, sexual identity, and the so-called values that are now are being used to separate and tear us apart. These very behaviors are exhibited in dictator-led and tightly controlled countries that have historically had less freedom than us, but that now could happen to us.

One only has to look at what Russia has attempted to do with the invasion of Ukraine to see the dangers we face. Who would have thought that one person, Vladimir Putin, could wreak so much havoc over the entire world? The power of nationalism and propaganda has become so powerful in Russia that its citizens do not dare to question or protest what is happening. As time has gone on, it appears that most people in Russia cannot do anything but buy into the legitimacy of Putin's war. Is it easier to ignore or start believing the lies than it is to be arrested or worse for not believing it? That seems like an easy choice.

Many were lulled into thinking that this type of conflict could never happen again after the Second World War. Yet, history seems to have a way of repeating itself over and over again. Ideas of empire, whether historically distorted or not, seem to never go away.

Now, it, too, can be seen how prevailing fossil fuels have been and continue to be. It remains to be seen how firm European countries' resolve to wean themselves off fossil fuels will be, to help avoid the unthinkable atrocities and thus genocide of innocent lives, and whether and how others will follow in the near future. As of right now, such an existential crisis does not seem to make a difference.

I cannot understand why we choose not to believe the science behind climate change but have no qualms respecting the science of other things that hold so much influence over our lives, whether positive, negative, or somewhere in between. Is it what we listen to and where we get our information that largely determines what we think or choose to believe?

One of the effects of climate change here where I live that makes me very sad is the apparent loss of the integrity of our four seasons. There used to be real differences in the makeup of our four seasons. The changing of those seasons has been a very real part of what has made living here so appealing. Historically, because of where we are located on the planet, you could count on having four seasons. Winter was winter and summer was summer. That is no longer the case. Now you do not know if the ground will even freeze through the winter. This, of course, is leading to more insect and disease problems for the following spring and summer crop season. Very cold temperatures used to freeze the ground to depths of greater than three or four feet some winters. That was the norm. That was the case less than twenty years ago; 2024 was a very strange and extreme year weather-wise. January through the end of March reinforced the four-year drought with way below normal moisture. April and May brought nearly sixteen inches of moisture and violent storms. That is one-half of our average rainfall for the year. The summer from June through September went back to being very dry. September and October were two of the driest months on record. A hard killing frost usually happens sometime in October but that did not occur in 2024. Now you might have to wait until Thanksgiving or later for that to occur. That is why it is so imperative for farmers, such as us, to find ways to be more resilient and ready for the inevitable unknowns that our weather and climate are creating.

None of us knows for sure what all of the effects will be due to our changing climate, but we are already witnessing more intense storms, rising sea levels, higher temperatures, and displacement of people from coastal areas. Regardless of climate change, rural areas, food and farming, and the health and well-being of people in our country and countries around the world could benefit from some of the ideas that have been presented in this book. That is my hope. The historical context in the book could provide at least some insight, experience, and practical solutions for how we can create a more livable planet.

In the scheme of things, our lives on Earth are very short. The stories of what happened after white Europeans discovered the Missouri River, how my ancestors came to farm and form a thriving community in the tiny town of Westphalia, Iowa, and now the challenges of what we face with climate change and the decline of all things rural are but a mere blink of an eye. I want to believe that my story of the intrinsic goodness of our family life of farming and rural living and care and love of all creation means more than a blink of an eye, that the practical knowledge and lessons learned are timeless. I end this book with a closing thought to ponder:

I am your Mother Earth. Your Father and I have formed you in the palm of our hands. We have given you everything needed for life. We have nourished you and cared for you. When you were hungry, we gave you food. Naked, we gave you clothes. We have given you the great gifts of the living ecosystems of the world to sustain you and the generations that follow you. We have given you the great gifts of intellect and reason. We have given you the greatest gift of all, love. You must now grow up and realize that so much of me is dying because of selfishness, indifference, and neglect. What will be your response to my pleas?

Notes

Introduction

1   For one of the many times Pope Francis has discussed this topic, see his Encyclical Letter, *Laudato Si'*, given in Rome at Saint Peter's on May 24, 2015, https://www.vatican.va/content/francesco/en/encyclicals/documents/papa-francesco_20150524_enciclica-laudato-si.html (accessed January 10, 2025).

Chapter 1

1   A quote from Ponca Chief White Eagle, "Top 10 Native American Quotes," *Xavier University*, n.d., https://www.xavier.edu/jesuitresource/online-resources/quote-archive1/native-american1 (accessed January 10, 2025).
2   This is a thought envisioned by the author. From time to time, the author will offer a fictionalized scene set out in block quotation format.
3   Mark Awakuni-Swetland, "Omaha," Encyclopedia of World Cultures Supplement, *Encyclopedia.com*, January 8, 2025, https://www.encyclopedia.com/history/united-states-and-canada/north-american-indigenous-peoples/omaha-indians (accessed January 18, 2025).
4   *New World Encyclopedia*, "Omaha (tribe)," last updated November 18, 2022, https://www.newworldencyclopedia.org/p/index.php?title=Omaha_(tribe)&oldid=1087914 (accessed January 18, 2025); Entrepreneurship in Omaha, "History and Culture of the Omaha Tribe," May 31, 2023, https://www.omahaimc.org/history-and-culture-of-the-omaha-tribe/ (accessed January 18, 2025).
5   Judith A. Boughter, *Betraying the Omaha Nation, 1790-1916* (Norman: University of Oklahoma Press, 1998).
6   Mari Sandoz, *Old Jules* (Boston: Little, Brown, 1935).
7   Harold A. Innis, *The Fur Trade in Canada* (Toronto: University of Toronto Press, 1930; 2001), 34.
8   On the building of Fort Orleans and the activities of Ettience de Veniard, sieur de Bourgmont, see Louis Houck, *A History of Missouri: From the Earliest Explorations and Settlements until the Admission of the State into the Union*, in 2 vols. (Chicago: R. R. Donnelly, 1908), vol. 1, 273, 281.
9   Hinmaton-Yalatit, commonly known as Chief Joseph, said in a speech with a US commission, led by Lt. Col. H. Clay Wood, in Lapwai, Idaho, on November 13, 1876, quoted in Chester Anders Fee, *Chief Joseph: The Biography of a Great Indian* (New York: Wilson-Erickson, 1936), 88, https://babel.hathitrust.org/cgi/pt?id=inu.32000000324311&seq=114&q1=.+I+claim+a+right+to+live+on+my+land (accessed January 11, 2025).
10 Wendell Berry, "Private Property and the Common Wealth" (essay), posted February 10, 2008, last updated December 19, 2008, http://tipiglen.co.uk/berryprivate.html (accessed January 11, 2025). Also see Berry, *A Continuous Harmony: Essays Cultural & Agricultural* (Boston: Houghton Mifflin Harcourt, 1979).

11   Tim Flannery, *The Eternal Frontier: An Ecological History of North America and Its Peoples* (Melbourne: Text Publishing; New York: Grove Publishing, 2001).

12   Jacob Van der Zee, "Episodes in the Early History of the Western Iowa Country," *Iowa Journal of History and Politics* 11, no. 3 (July 1913): 323–63, at 328, https://babel.hathitrust.org/cgi/pt?id=wu.89067281931&seq=340 (accessed January 11, 2025).

13   Treaty of Prairie du Chien, July 15, 1830, see the State of Oklahoma, "Treaty with the Missouria, Fox, Sioux, Menominee, Otoe, Iowa, Sauk, Winnebago, and Omaha, 1830," https://treaties.okstate.edu/treaties/treaty-with-the-missouria-fox-sioux-menominee-ottoe-iowa-sauk-winnebago-and-omaha-1830-22923 (accessed January 13, 2025); For the Treaty of Chicago, September 26, 1833, see Oklahoma, "Treaty with the Chippewa, etc., 1833," https://treaties.okstate.edu/treaties/treaty-with-the-chippewa-etc-1833-0402 (accessed January 13, 2025). A list of all treaties in the area can be found at Oklahoma, "Treaties, Agreements, and Documents," 1722–1954, https://treaties.okstate.edu/treaties/.

14   See US National Park Service, "Blackbird Hill," last updated December 30, 2021, https://www.nps.gov/places/blackbird-hill.htm (accessed January 13, 2025); and Ponca Tribe of Nebraska, "Office Locations," 2025, https://www.poncatribe-ne.gov/contact/office-locations/ (accessed January 13, 2025).

15   This phrase became popular with the publication of the book *Critical Race Theory: An Introduction*, ed. Richard Delgado and Jean Stefancic (New York: New York University Press, 2001). The book is now available in its fourth edition and continues to be both praised and criticized. Critical race theory (CRT) is an academic field that grew from the post-civil rights era in the 1960s.

16   Wendell Berry, *Standing by Words* (Washington, DC: Shoemaker, 2005; Berkeley, CA: Counterpoint Press, 2011), 14.

17   Catherine Locks, Sara K. Mergel, Pamela Thomas Roseman, and Tamara Spike, "Jacksonian America (1815–840)," in *History in the Making: A History of the People of the United States of America to 1877* (Dahlonega: University of North Georgia Press, 2013), chap. 12, 525–82, https://ung.edu/university-press/_uploads/files/us-history/US-History-I-Chapter-12.pdf (accessed January 13, 2025).

18   See John O'Sullivan, "Annexation (1845)," *United States Magazine and Democratic Review* 17, no. 1 (July–August 1845): 5–10, https://pdcrodas.webs.ull.es/anglo/OSullivanAnnexation.pdf.

19   William M. Denevan, ed., *The Native Population of the Americas in 1492* (Madison: University of Wisconsin Press, 1976); 2nd ed., 1992.

20   While this quote cannot be verified, attributions to Sheridan can be found in Dee Brown's 1970 book, *Bury My Heart at Wounded Knee: An Indian History of the American West* (New York: Holt, Rinehart & Winston), 147.

21   For more detail, see Community Environmental Legal Defense Fund (CELDF), "The Enclosure Movement," last updated March 20, 2021, https://celdf.org/the-enclosure-movement/ (accessed January 14, 2025).

22   See an essay by Eleonora Montuschi, "Order of Man, Order of Nature: Francis Bacon's Idea of a 'Dominion' Over Nature," paper presented at the London School of Economics, October 27–8, 2010, https://iris.unive.it/retrieve/handle/10278/24867/23441/MontuschiBacon.pdf (accessed January 14, 2025).

23   To review the journals with interpretation by the editor, see Meriwether Lewis and William Clark, *The Journals of Lewis and Clark*, ed. Bernard DeVoto (New York: Harper Perennial, 1997).

24   This estimate was given in a radio broadcast by KCUR Public Radio, Kansas City, Missouri, "This American Life," a talk show interview featuring historian Paula Rose, Steamboat Arabia Presentation, July 14, 2015.

25   Col. George Croghan, *Army Life on the Western Frontier: Selections from the Official Reports Made Between 1826 and 1845 by Colonel George Croghan*, ed. Francis Paul Prucha (Norman: University of Oklahoma Press, 1958; 2014).

Chapter 2

1. George Washington, "From George Washington to Joshua Holmes," Papers of George Washington, *Founders Online*, December 2, 1783, https://founders.archives.gov/documents/Washington/99-01-02-12127.
2. See US National Archives, "Treaty Between the United States of America and the French Republic," last updated May 1, 2022, https://www.archives.gov/milestone-documents/louisiana-purchase-treaty#:~:text=In%20this%20transaction%20with%20France,size%2C%20expanding%20the%20nation%20westward (accessed January 14, 2025).
3. For an early overview of Iowa's population and economy, see Dorothy Schwieder, "History of Iowa," *Iowa Official Register*, November 15, 1999, https://publications.iowa.gov/135/1/history/7-1.html (accessed January 14, 2025).
4. U.S. Congress, *Union Pacific Railroad Act of 1862*, July 1, 1862, 12 STAT 489, 37th Cong., 2nd sess., https://www.archives.gov/milestone-documents/pacific-railway-act.
5. For more on railroads in Iowa, see Iowa PBS, "Iowa Pathways: Railroads," December 2021, https://www.iowapbs.org/iowapathways/mypath/2536/railroads (accessed January 14, 2025).
6. U.S. Congress, *Homestead Act of 1862*, May 20, 1862, 12 STAT 392, 37th Cong., 2nd sess., https://www.archives.gov/milestone-documents/homestead-act.
7. For more on the *United States Constitution*, amendments, and the Bill of Rights, see https://constitutioncenter.org/the-constitution.
8. For the principles and almanac of the American Party, see "The Know Nothing Party, 1856," *Digital Public Library of America* (exhibitions), https://dp.la/exhibitions/outsiders-president-elections/anti-outsider-platforms/know-nothing-party-1856 (accessed January 15, 2025).
9. Iowa Board of Immigration, *Iowa: The Home of Immigrants* (Des Moines: Mills, 1870).
10. Wendell Berry, *What I Stand for Is What I Stand On* (New York: Penguin Classics, 2021).
11. Wendell Berry, "Private Property and the Common Wealth" (essay), posted February 10, 2008, last updated December 19, 2008, http://tipiglen.co.uk/berryprivate.html (accessed January 11, 2025).
12. For more information on the reorganized church (RLDS), see Paul M. Edwards, *Our Legacy of Faith: A Brief History of the Reorganized Church of Jesus Christ of Latter-Day Saints* (Independence, MO: Herald House, 1991).
13. Emil Flusche, quoted in Ron Rosmann and Maria Vakulskas Rosmann, *Preserving Our Past, Ensuring Our Future, 1872–997, Westphalia, Iowa* (Audubon, IA: Audubon Media, 1997).
14. Adam Schmitz, quoted in Rosmann and Rosmann, *Preserving Our Past*.
15. Antonia Sasse, quoted in Rosmann and Rosmann, *Preserving Our Past*.

Chapter 3

1. While this quote has often been attributed to Margaret Mead, there is no actual source specifically. She often spoke and wrote about hard work; therefore, this quote continues.
2. For more on the Spanish Flu epidemic, see John M. Barry, *The Great Influenza: The Story of the Deadliest Pandemic in History* (New York: Penguin, 2005). Keep in mind that this was published prior to COVID-19, but is well-researched and documents the event.
3. Pope Leo XIII, "Rerum Novarum – Encyclical Letter of Pope Leo XIII on the Conditions of Labor" (1891), Historical Catholic and Dominican Documents. 13, https://digitalcommons.providence.edu/cgi/viewcontent.cgi?article=1014&context=catholic_documents (accessed January 17, 2025).

4. Pope Pius XI, "Rerum Novarum – Quadragesimo Anno (In the 40th Year), Reconstruction of the Social Order" (1931), https://www.papalencyclicals.net/pius11/p11quadr.htm (accessed January 17, 2025).

5. US Congress, *Emergency Banking Relief Act of 1933*, March 9, 1933, 48 STAT 1, 73rd cong., 1st sess., https://govtrackus.s3.amazonaws.com/legislink/pdf/stat/48/STATUTE-48-Pg1.pdf.

6. See updates at the World Health Organization (WHO), "Household Air Pollution," last updated October 16, 2024, https://www.who.int/news-room/fact-sheets/detail/household-air-pollution-and-health (accessed January 17, 2025).

7. For a history of cooperative farming in the United Kingdom, see Notes form the UK, "Early British Consumer Co-ops," January 10, 2020, https://notesfromtheuk.com/2020/01/10/co-ops-in-britain-yet-another-bit-of-history/ (accessed January 17, 2025).

8. For a recent history of cooperatives, see Lynn Pitman's report, "History of Cooperatives in the United States: An Overview," University of Wisconsin-Madison, Center for Cooperatives, December 2018, https://resources.uwcc.wisc.edu/History_of_Cooperatives.pdf (accessed January 17, 2025).

9. US Congress, *Homestead Act of 1862*, May 20, 1862, 12 STAT 392, 37th Cong., 2nd sess., https://www.archives.gov/milestone-documents/homestead-act.

10. Fr. Hubert Duren, quote by Dorothy Day, "On Pilgrimage," *Catholic Worker Movement*, June 1, 1947, https://catholicworker.org/454-html/ (accessed January 17, 2025).

Chapter 4

1. Epictetus was a Greek philosopher who lived *c.* 50 to 135 AD.

2. Story retold by the daughter of Ann Schwarte, quoted in Ron Rosmann and Maria Vakulskas Rosmann, *Preserving Our Past, Ensuring Our Future, 1872–997, Westphalia, Iowa* (Audubon, IA: Audubon Media, 1997).

3. W. H. Gemmill, quoted in Rosmann and Rosmann, *Preserving Our Past.*

4. Matt Jancer, of Vice Media and Broadcasting group, writes "Millions of Americans Have a Parasite and Don't Realize It," *Vice*, December 19, 2018, https://www.vice.com/en/article/millions-of-americans-have-a-parasite-and-dont-realize-it/ (accessed February 5, 2025).

5. For more information on suicide and depression among farmers, see Megan Thompson and Melanie Staltzman's report, "How Rural Communities are Tackling a Suicide and Depression Crisis Among Farmers," *PBS News* (video), January 14, 2024, https://www.pbs.org/newshour/show/how-rural-communities-are-tackling-a-suicide-and-depression-crisis-among-farmers (accessed January 18, 2025).

6. US Congress, *Civil Rights Act of 1964*, July 2, 1964, PL 88–352, 88th Cong., 2nd sess., https://www.archives.gov/milestone-documents/civil-rights-act.

7. US Congress, *Morrill Act of 1862*. July 2, 1862, PL 37–108, 37th Cong., 2nd sess., https://www.archives.gov/milestone-documents/morrill-act.

8. See United States. Treaty with the Sauk and Foxes, 1842, October 11, 1842, 7 Stat. 596, March 23, 1843, https://treaties.okstate.edu/treaties/treaty-with-the-sauk-and-foxes-1842-0546.

9. US Congress, *Morrill Act of 1890*, August 30, 1890, PL 111–122, 51st Cong., 1st sess., https://www.nifa.usda.gov/sites/default/files/asset/document/First%20and%20Second%20Morrill%20Act.pdf.

10. For more on Black farmers, see Daniel Aminetzah et al., "Black Farmers in the US: The Opportunity for Addressing Racial Disparities in Farming," *McKinsey*, November 10, 2021, https://www.mckinsey.com/industries/agriculture/our-insights/black-farmers-in-the-us-the-opportunity-for-addressing-racial-disparities-in-farming (accessed January 18, 2025).

11. US Congress, *Morrill Act of 1862*.

12 US Congress, *GI Bill of Rights of 1944*, June 22, 1944, PL. 346-268, 78th Cong., 2nd sess., https://www.archives.gov/milestone-documents/servicemembers-readjustment-act.

13 For more information, see Margaret Mead, "Coming of Age in Samoa," *World History Commons*, n.d., https://worldhistorycommons.org/margaret-mead-coming-age-samoa (accessed February 5, 2025).

Chapter 5

1   George Washington, although this quote has often been attributed to him, the source is unclear. It may have been a spoken response to a newspaper criticism of his presidency, but not confirmed.

2   For more on the Swing Riots and mechanization see, Bruno Caprettini and Hans-Joachim Voth, "Rage against the Machines: Labor-Saving Technology and Unrest in Industrializing England," *American Economic Review: Insights* 2, no. 3 (2020): 305–20, https://pubs.aeaweb.org/doi/pdfplus/10.1257/aeri.20190385.

3   Community Environmental Legal Defense Fund (CELDF), "The Enclosure Movement," last updated March 20, 2021, https://celdf.org/the-enclosure-movement/ (accessed January 14, 2025).

4   While this is quoted in many works as Humphrey's words, there is no source cited. This quote comes from "Hubert H. Humphrey Quotes," *BrainyQuote*, BrainyMedia, 2025, https://www.brainyquote.com/quotes/hubert_h_humphrey_152591 (accessed February 5, 2025).

5   Originally written in 1974 by the author.

6   D. H. Lawrence, *Apocalypse and the Writings on Revelation*, ed. Mara Kalnins (Cambridge: Cambridge University Press, 2002), 101.

7   Kris De Decker, "Bring Back the Horses," *Low-Tech Magazine*, April 18, 2008, https://solar.lowtechmagazine.com/2008/04/bring-back-the-horses/ (accessed January 20, 2025).

8   M. H. Benda, "An Economic Comparison of Traditional and Conventional Agricultural Systems at a County Level," *American Journal of Alternative Agriculture* 16, no. 1 (2001): 2–15, http://www.jstor.org/stable/44503173.

9   Jennifer E. Ifft and Youwei Yang, "Horses vs. Tractors? Old Order Amish Population Growth and New York Farmland Markets," (paper), Agricultural and Applied Economics Association, July 26–8, 2020, Kansas City, MO, https://econpapers.repec.org/paper/agsaaea20/304565.htm (accessed January 20, 2025).

10 De Decker, "Bring Back the Horses."

11 Ray Hunt, quoted in Gretel Ehrlich, "Ray Hunt: The Cowboy Sage," *Shambhala Sun*, July 1998.

Chapter 6

1   John Muir, *Yosemite* (New York: The Century Company, 1912).

2   Curtis K. Stadtfeld, *From the Land and Back: What Life Was Like on a Family Farm and How Technology Changed It* (New York: Charles Scribner's Sons, 1972).

3   For more on Iowa's more than 400 species of birds, see Iowa Ornithologists' Union, "Birding in Iowa," 2025, at https://iowabirds.org/birds/ (accessed January 20, 2025).

4   For up-to-date drought conditions in Iowa, see Drought.gov, "Drought by Location, Iowa," 2025, https://www.drought.gov/states/iowa. (accessed January 20, 2025).

5   The abbreviation, 4-H, stands for Head, Heart, Hands, and Health, and its emblem is a 4-leaf clover. It was founding in 1902 and continues today. 4-H was a federal program administered through the

land-grant university system, under the National Institute of Food and Agriculture (NIFA), https://4-h.org/about/.

Chapter 7

1. Bob Dylan, "The Times They Are a-Changin," *The Times They Are a-Changin* (album) (New York: Columbia, 1965).
2. See an essay by Peter Shapiro, "Freshman Deferments End as Nixon Signs New Draft Legislation," *Harvard Crimson*, September 29, 1971, https://www.thecrimson.com/article/1971/9/29/freshman-deferments-end-as-nixon-signs/ (accessed January 21, 2025).
3. For similar sentiments, see New York Times, "61% in Poll Assert Entry Into the War Was a U.S. 'Mistake,'" *New York Times*, June 6, 1971, https://nyti.ms/42Idk8Y (accessed January 21, 2025).

Chapter 8

1. See Johns Hopkins Center for a Livable Future, 2014; 2025, https://clf.jhsph.edu/.
2. George Orwell, *Animal Farm: A Fairy Story* (New York: Secker and Warburg, 1945), 47.
3. Earl Butz, quoted in IowaPBS, "The 1970s See Good Times in Agriculture," (essay; YouTube), July 1, 2013, https://www.iowapbs.org/shows/farmcrisis/clip/5310/1970s-see-good-times-agriculture (accessed January 22, 2025).
4. Maurice J. Dingman, Bishop, Des Moines Catholic Diocese, Des Moines, Iowa, quoted this article from a speech presented at the Conference on Religious Ethics and Technological Change, Iowa State University, February 21, 1986.
5. See comments made in 1945 by the US Chamber of Commerce, calling the elimination of family farms, US Congress, "Long Range Farm Program," in *Hearings before the Committee on Agriculture*, July 25, August 4, 6, 1953, 83rd Cong., 1st sess. (Washington, DC: GPO, 1953), 1244–5, https://www.google.com/books/edition/Long_Range_Farm_Program/zRDiyAZsXY0C?hl=en&gbpv=1&bsq=family%20farm (accessed January 22, 2025).
6. US Congress, "Long Range Farm Program," 1244.
7. Senate Bill 4808, McNary-Haugen Farm Relief Bill, was passed by Congress but vetoed by the president. For the history of this bill, see Calvin Coolidge, "Message to the Senate Returning Without Approval S. 4808," *American Presidency Project*, February 25, 1927, https://www.presidency.ucsb.edu/documents/message-the-senate-returning-without-approval-s-4808-the-mcnary-haugen-farm-relief-bill (accessed January 22, 2025).
8. US Congress, *Agricultural Adjustment Act (AAA) of 1933*, May 12, 1933, 48 Stat. 31, 73rd Cong., 1st sess., https://nationalaglawcenter.org/wp-content/uploads/assets/farmbills/1933.pdf (accessed January 22, 2025).
9. Still based in Lyons, NE, see Center for Rural Affairs, "Home," 2025, https://www.cfra.org/.
10. Don McCabe, "Amendment May Be Dead, Unless Supreme Court Hears Appeal," *Farm Progress*, January 26, 2007, https://www.farmprogress.com/farm-business/nebraska-can-no-longer-enforce-i-300 (accessed January 22, 2025). The US Supreme Court declined to take up the appeal.
11. Murphy Farms, quoted by Nancy L. Thompson, "Center for Rural Affairs and South Dakota Family Farm Coalition," *P2InfoHouse.org*, n.d., https://p2infohouse.org/ref/21/20027.htm (accessed January 22, 2025).

12 Iowa Office of National Agricultural Statistics Service may be viewed at: USDA National Agricultural Statistics Service, Iowa Field Office, https://www.nass.usda.gov/Statistics_by_State/Iowa/index.php (accessed January 22, 2025).

13 Orwell, *Animal Farm*.

Chapter 9

1 James Earl Jones is an actor and a 1949 graduate of Dickson Agricultural School, Brethren, Michigan. This quote is from an interview for his American Academy of Achievement Award, Class of 1996, "James Earl Jones on Preparation," December 5, 2016, https://youtu.be/hnze3oquDpI (accessed January 23, 2025).

2 The Vietnam War officially began on November 1, 1955 and ended April 20, 1975, with an estimated 58,281 US military casualties recorded.

3 US Senator Tom Harkin, "The Tiger Cages of Con Son: A Congressional Investigating Committee Finds Secret Cells in a Concentration Camp for South Vietnamese Political Prisoners" (photograph), *Life Magazine*, July 17, 1970, 26, https://www.originallifemagazines.com/product/life-magazine-july-17-1970/ (accessed January 23, 2025).

4 Written by Joe Rosmann, eulogy for the funeral of his father, Raymond J. Rosmann, on May 13, 1980. https://iagenweb.org/shelby/obit/bumancoll/1970-1986/r-surnames-18xx-1986/ROSMANN-RAYMOND-JOSEPH-1907-1980-2.pdf (accessed January 13, 2025).

5 Research and Policy Committee of the Committee for Economic Development, *An Adaptive Program for Agriculture* (New York: Committee for Economic Development, 1962), https://babel.hathitrust.org/cgi/pt?id=mdp.39015008782123&seq=7 (accessed January 23, 2025).

Chapter 10

1 Wendell Berry, *The Unsettling of America: Culture and Agriculture* (San Francisco: Sierra Club Books, 1977; 2nd ed., Berkeley, CA: Counterpoint Press, 1996), 90.

2 For issues of the magazine, see Rodale Institute, *New Farm* Archive, https://rodaleinstitute.org/education/resources/new-farm/ (accessed January 23, 2025).

3 USDA Study Team on Organic Farming, "Report and Recommendations on Organic Farming," July 1980, available online at *Center for Inquiry*, https://centerforinquiry.org/wp-content/uploads/sites/33/quackwatch/usda_organic_1980.pdf (accessed January 23, 2025).

4 USDA Study Team on Organic Farming, "Report," 93.

5 Earl Butz from a statement made in 1971, quoted in John Reganold, "Can We Feed 10 Billion People on Organic Farming Alone?" *Guardian*, August 14, 2016, https://www.theguardian.com/sustainable-business/2016/aug/14/organic-farming-agriculture-world-hunger (accessed January 23, 2025).

6 Iowa Corn, "Corn Facts and Fun: Use for Corn," 2023, https://www.iowacorn.org/corn-facts-faq/ (accessed January 23, 2025).

7 Sir Albert Howard, *Farming and Gardening for Health or Disease* (Oxford: Oxford City Press, 2011; 2019), https://gutenberg.net.au/ebooks02/0200311.txt# (accessed January 23, 2025).

8 Charles M. Benbrook, "Impacts of Genetically Engineered Crops on Pesticide Use in the US—the First Sixteen Years," *Environmental Sciences Europe* 24, no. 24 (2012), https://doi.org/10.1186/2190-4715-24-24.

9. Vicky C. Chang et al., "Glyphosate Exposure and Urinary Oxidative Stress Biomarkers in the Agricultural Health Study," *Journal of the National Cancer Institute* 115, no. 4 (April 2023): 394–404, https://doi.org/10.1093/jnci/djac242.

10. See an essay by Carey Gillam, "CDC Finds Weed Killer tied to Cancer in Over 80 pct of US Urine Samples," *New Lede*, July 9, 2022, https://www.thenewlede.org/2022/07/cdc-finds-weed-killer-tied-to-cancer-in-over-80-pct-of-us-urine-samples/ (accessed January 24, 2025).

11. Chang et al., "Glyphosate Exposure."

12. On the current status of Bayer and Monsanto and recent research, see an essay by Corporate Europe Observatory, "Bayer and Monsanto Merger: What Role Did Revolving Doors Play?" February 26, 2024, https://corporateeurope.org/en/2024/02/bayer-and-monsanto-merger-what-role-did-revolving-doors-play (accessed January 24, 2025).

13. For general information on the pesticide dicamba, see National Pesticide Information Center, "Dicamba: General Fact Sheet," February 2012, https://npic.orst.edu/factsheets/dicamba_gen.html (accessed January 24, 2025).

14. Iowa State University, Extension and Outreach, "Soybean Cupping 2021: Iowa Update," July 22, 2021, https://crops.extension.iastate.edu/blog/prashant-jha/soybean-cupping-2021-iowa-update (accessed January 24, 2025).

15. Claudine Samanic et al., "Cancer Incidence Among Pesticide Applicators Exposed to Dicamba in the Agricultural Health Study," *Environmental Health Perspectives* 114, no. 10 (2006): 1521–6, https://doi.org/10.1289/ehp.9204.

16. U.S. Department of Environmental Protection (EPA), "Registration of Dicamba for Use on Dicamba-Tolerant Crops," February 14, 2024, https://www.epa.gov/ingredients-used-pesticide-products/registration-dicamba-use-dicamba-tolerant-crops# (accessed January 25, 2025).

17. American Farm Bureau Foundation for Agriculture, "Does Organic Production Use Pesticides or synthetic Fertilizers?" 2025, https://www.agfoundation.org/questions/does-organic-production-use-pesticides-or-synthetic-fertilizers# (accessed January 25, 2025).

18. Steffen Schneider, "The Haber-Brosch Process," *Institute for Mindful Agriculture*, March 24, 2016, https://www.instituteformindfulagriculture.org/writings-1/2016/3/24/the-haber-bosch-process-1# (accessed January 25, 2025).

19. Winston Churchill, "Parliamentary Debate, April 25, 1918," in *Churchill by Himself: The Definitive Collection of Quotations*, ed. Richard Langworth (New York: Public Affairs; London: Ebury Press, 2008), 469.

20. Iowa Corn, "Corn Facts and Fun: Use for Corn."

21. US Department of Commerce, National Oceanic and Atmospheric Administration (NOAA), "Gulf of Mexico 'Dead Zone' Larger than Average, Scientists Find," August 1, 2024, https://www.noaa.gov/news-release/gulf-of-mexico-dead-zone-larger-than-average-scientists-find# (accessed January 25, 2025).

22. For National Organic Program list, see US National Archives, Code of Federal Regulations, "Nonsynthetic Substances Prohibited for Use in Organic Crop Production," Title 7, Sec. 205.602, https://www.ecfr.gov/current/title-7/subtitle-B/chapter-I/subchapter-M/part-205/subpart-G/subject-group-ECFR0ebc5d139b750cd/section-205.602 (accessed January 25, 2025).

23. For more on Alfred Blackmer test, see an essay by Rod Swoboda, "Sampling Cornfields for Soil Nitrate," *Wallaces Farmer, Urbandale* 123, no. 8 (May 1998): 14, https://www.proquest.com/trade-journals/sampling-cornfields-soil-nitrate/docview/219323336/se-2 (accessed January 25, 2025).

24. For more on Blackmer's influence, see Dan Jayne's references to water testing using Blackmer's test at: Wallaces Farmer, "Long-Term Partnership between USDA and Iowa State University Researchers Produces Innovative Results," *FarmProgress*, September 16, 2020, https://www.farmprogress.com/farm-life/partnership-works-to-improve-water-quality (accessed January 25, 2025).

25. Practical Farmers of Iowa (PFI), farmer led research publication, with Ron Rosmann and Maria Vakulskas Rosmann, and Iowa Future Farmers of America, "Corn Population Trials," 1988–1999, https://practicalfarmers.org/research/9810/ (accessed February 6, 2025). For more information email info@practicalfarmers.org.
26. Alejandra Borunda, "U.S. Corn Production Is Booming—But Not for the Reasons Scientists Hoped," *National Geographic*, January 24, 2022, https://www.nationalgeographic.com/environment/article/us-corn-production-booming-but-its-future-isnt-clear (accessed January 25, 2025).
27. Deepranjan Sarkar et al., "Low Input Sustainable Agriculture: A Viable Climate-Smart Option for boosting Food Production in a Warming World," *Ecological Indicators* 115 (August 2020): Article 106412, https://doi.org/10.1016/j.ecolind.2020.106412.
28. See previously cited, US National Archives, Code of Federal Regulations, "Nonsynthetic Substances Prohibited for Use in Organic Crop Production," Title 7, Sec. 205.602.
29. On categorization of livestock by the USDA, see Samantha Capaldo, "USDA Publishes Final Rule Amending Organic Livestock and Poultry Standards," *National Agricultural Law Center*, November 2, 2023, https://nationalaglawcenter.org/usda-publishes-final-rule-amending-organic-livestock-and-poultry-standards/ (accessed January 25, 2025).
30. For more on our farm and products, see Ron Rosmann and Marie Rosmann, "Farm Sweet Farm," 2025, https://www.farmsweetfarm.org/ (accessed January 25, 2025).

Chapter 11

1. Friedrich Schiller, *Kallias Letters*, 2.2 "On Grace and Dignity," quoted from Moland, Lydia L., "Friedrich Schiller," *The Stanford Encyclopedia of Philosophy* (Summer 2025 Edition), Edward N. Zalta and Uri Nodelman (eds.), https://plato.stanford.edu/archives/sum2025/entries/schiller/ (accessed May 16, 2025).
2. Sustainable Agricultural Research and Education (SARE), "Farmer-Driven Innovations in Agriculture that Improve Profitability, Stewardship, and Quality of Life," 2025, https://www.sare.org/ (accessed January 26, 2025).
3. National Sustainable Agriculture Coalition (NSAC), "Home," 2025, https://sustainableagriculture.net/ (accessed January 26, 2025).
4. Organic Farming Research Foundation (OFRF), "Why Organic?" 2025, https://ofrf.org/ (accessed January 26, 2025).
5. Friends of the Earth, "Friends of the Earth Is a Bold Voice for Justice and the Planet," 2025, https://foe.org/ (accessed January 26, 2025).
6. California Certified Organic Farmers (CCOF), "Help Make Organic the Norm!" 2025, https://www.ccof.org/ (accessed January 26, 2025).
7. See also Rodale Institute, *New Farm* Archive, https://rodaleinstitute.org/education/resources/new-farm/ (accessed January 23, 2025).
8. For the history of the *California Organic Food Act of 1979* and subsequent legislation, see California Department of Food and Agriculture. "California's State Organic Program," 2017, https://www.cdfa.ca.gov/is/i_&_c/pdfs/CalOrganicPrgrmFactSheet.pdf (accessed January 26, 2025).
9. On GMO, see Charles M. Benbrook, "Impacts of Genetically Engineered Crops on Pesticide Use in the U.S.—the First Sixteen Years," *Environmental Sciences Europe* 24, no. 24 (2012), https://doi.org/10.1186/2190-4715-24-24.
10. Organic Research and Education Initiative (OREI), "Organic Research, Education, and Extension Programs," 2025, https://www.usda.gov/farming-and-ranching/organic-farming/organic-research-education-and-extension-programs (accessed January 26, 2025).

11 For a recently published study linking prostrate cancer and pesticides, see Simon John Christoph Soerensen et al., "Pesticides and Prostate Cancer Incidence and Mortality: An Environment-Wide Association Study," *Cancer* 131, no. 1 (January 2024): e35572, https://doi.org/10.1002/cncr.35572.

Chapter 12

1 This sentiment of hope and faith is echoed in many religions throughout the world. In the Christian faith, it mirrors the verse from the Holy Bible, 1 Corinthians 13:13, among many others.

2 US Census Bureau, "2020 Census, Westphalia, Iowa," https://data.census.gov/profile?q=Westphalia%20city,%20Iowa%20Education (accessed January 26, 2025), notes the population at 126. The World Population Review notes a population of 135 in 2024, https://worldpopulationreview.com/us-cities/iowa/westphalia (accessed January 26, 2025).

3 Martha Luther King Jr., *Strength to Love* (Minneapolis, MN: Fortress Press, 1963), 71.

4 For more on this topic, see Trent Horn, "Thomas Aquinas's Five Proofs for God Revisited," *Christian Research Journal* 41, no. 2 (2018): JAF3412, https://www.equip.org/articles/thomas-aquinas-five-proofs-for-god-revisited/ (accessed January 26, 2025).

5 For an overview on carbon dating, see Steve Koppes and Louise Lerner, "Carbon-14 Dating, Explained," *UChicago News*, April 27, 2023; last updated January 16, 2025, https://news.uchicago.edu/explainer/what-is-carbon-14-dating (accessed January 26, 2025).

6 For more on the Big Bang Theory and Planck time, see a lecture by Dale E. Gary, "Cosmology and the Beginning of Time" (lecture), *New Jersey Institute of Technology (NJIT)*, n.d., https://web.njit.edu/~gary/202/Lecture26.html (accessed January 26, 2025).

7 Joseph Bernardin, quoted in Pedro Gabriel, "The Seamless Garment Is the Catholic Position," *Catholic Outlook*, February 9, 2020, https://catholicoutlook.org/the-seamless-garment-is-the-catholic-position (accessed January 27, 2025). The essay notes that the phrase was originally reported to have been written in 1971 by Eileen Egan, in reference to John 19:23.

8 Francis (Pope), Encyclical Letter, *Laudato Si'* (Praise be to you), Rome, Italy, at Saint Peter's, May 24, 2015, para. 91, https://www.vatican.va/content/francesco/en/encyclicals/documents/papa-francesco_20150524_enciclica-laudato-si.html (accessed January 10, 2025).

9 Leviticus 25:23, New American Bible.

10 Luke 4:18, New American Bible.

11 Iowa Catholic Conference, "STRANGERS AND GUESTS: Toward Community in the Heartland, a Regional Catholic Bishops' Statement on Land Issues," May 1, 1980, https://iowacatholicconference.org/wp-content/uploads/2021/05/Strangers-and-Guests-1980.pdf (accessed January 27, 2025).

12 For a biography on Rev. Edwin Vincent O'Hara and his work, see Lucinda A. Nolan, "Edwin Vincent O'Hara," *Biola University*, 2009, https://www.biola.edu/talbot/ce20/database/edwin-vincent-ohara (accessed January 27, 2025).

13 For a biography on Monsignor Ligutti, see David S. Bovee, "Ligutti, Luigi Gino," *Biographical Dictionary of Iowa*, University of Iowa, n.d., https://uipress.lib.uiowa.edu/bdi/DetailsPage.aspx?id=232 (accessed January 27, 2025).

14 On the resettlement of Iowa coal miners and farmers, see USDA National Agricultural Library, "Granger Homesteads," May 15, 1935, https://www.nal.usda.gov/exhibits/ipd/small/exhibits/show/subsistence/item/52 (accessed January 27, 2025).

15 John Paul II, "Letter to Reverend George V. Coyne, S.J." *Vatican*, June 1, 1988, https://www.vatican.va/content/john-paul-ii/en/letters/1988/documents/hf_jp-ii_let_19880601_padre-coyne.html (accessed January 27, 2025).

16 See an essay on the case by Monica Miller, "US Farmers Win Right to Repair John Deere Equipment," *BBC News* (Singapore), January 8, 2023, https://www.bbc.com/news/business-64206913 (accessed January 27, 2025).

17 National Young Farmer's Coalition, "Release 2022: National Young Farmer Survey Reveals a Generation Defying the Odds," September 1, 2022, https://www.youngfarmers.org/2022/09/643050/ (accessed February 6, 2025).

18 Sam Becker, "US Farms Are Making an Urgent Push into AI. It Could Help Feed the World," *BBC*, March 27, 2024, https://www.bbc.com/worklife/article/20240325-artificial-intelligence-ai-us-agriculture-farming (accessed February 6, 2025).

19 Francis (Pope), Encyclical Letter, *Laudato Si'* (Praise be to You), 2015.

20 Richard Conniff, "What the Luddites Really Fought Against," *Smithsonian Magazine*, March 2011, https://www.smithsonianmag.com/history/what-the-luddites-really-fought-against-264412/ (accessed January 27, 2025).

21 C. S. Lewis, *The Problem of Pain* (1940) (New York: HarperCollins, 1996), 153.

22 See Bernard J. F. Lonergan, *Insight: A Study in Human Understanding* (New York: Philosophical Library; London: Darton Longman and Todd, 1957); and his *Method in Theology* (London: Darton Longman and Todd, 1972); and Daniel A. Helminiak, *The Human Core of Spirituality: Mind as Psyche and Spirit* (New York: University of New York, 1996), xii–xiii; and his *Meditation Without Myth* (New York: Crossroad, 2005) 14.

23 C. S. Lewis, *A Joyful Christian* (New York: Macmillan, 1977), 138.

Chapter 13

1 Pope Francis, Encyclical Letter, *Laudato Si'*, presented in Rome at Saint Peter's on May 24, 2015, https://www.vatican.va/content/francesco/en/encyclicals/documents/papa-francesco_20150524_enciclica-laudato-si.html (accessed January 10, 2025).

2 US Department of Commerce, National Oceanic and Atmospheric Administration (NOAA), "Larger-than-Average Gulf of Mexico 'Dead Zone' Measured," August 3, 2021, https://www.noaa.gov/news-release/larger-than-average-gulf-of-mexico-dead-zone-measured (accessed January 27, 2025).

3 Waypoint Team, "Essential for Optimum Plant Growth," *Waypoint*, February 20, 2024, https://www.waypointcommodities.com/news/5-vital-reasons-why-nitrogen-is-essential-for-optimum-plant-growth (accessed January 27, 2025).

4 Waypoint Team, "Essential for Optimum Plant Growth."

5 Steffen Schneider, "The Haber-Bosch Process," *Institute for Mindful Agriculture*, March 24, 2016, https://www.instituteformindfulagriculture.org/writings-1/2016/3/24/the-haber-bosch-process-1# (accessed January 25, 2025).

6 For the video version of this slogan, see Iowa Corn, "Corn Grows Iowa," (YouTube), February 19, 2020, https://www.youtube.com/watch?v=n2oDt391GVk&list=PLxepamPolX9LEtdWlqZuDKLpebKwYxPGi&index=2 (accessed January 27, 2025).

7 Iowa Corn, "Corn Facts and Fun: Use for Corn," 2023, https://www.iowacorn.org/corn-facts-faq/ (accessed January 23, 2025).

8 US Department of Commerce (NOAA), "Larger-than-Average Gulf of Mexico 'Dead Zone' Measured"; and also by NOAA, "Gulf of Mexico 'Dead Zone' Larger than Average, Scientists Find," August 1, 2024. https://www.noaa.gov/news-release/gulf-of-mexico-dead-zone-larger-than-average-scientists-find# (accessed January 25, 2025).

9   Union of Concerned Scientists, "Rural Iowans Bear Brunt of Water Treatment Costs for Nitrate Pollution from Farms and CAFOs," January 14, 2021, https://www.ucsusa.org/about/news/rural-iowans-bear-brunt-water-treatment-costs-nitrate-pollution-farms-and-cafos (accessed January 27, 2025).

10 Union of Concerned Scientists, "Rural Iowans Bear Brunt of Water Treatment Costs."

11 One of many reports can be viewed at the Minnesota Department of Health, "Nitrate in Drinking Water," last updated January 27, 2025, https://www.health.state.mn.us/communities/environment/water/contaminants/nitrate.html (accessed January 27, 2025).

12 Iowa Committee on Ways and Means, *Iowa Groundwater Protection Act of 1987*, House File 631, June 9, 1987, https://www.legis.iowa.gov/docs/publications/LGI/72/HF631.pdf (accessed January 27, 2025).

13 Leopold Center for Sustainable Agriculture, Iowa State University, 2025, https://www.leopold.iastate.edu/leopold-center-sustainable-agriculture (accessed January 27, 2025).

14 Leopold Center for Sustainable Agriculture, Iowa State University, "Statement from the Leopold Center Regarding the Decision of the Iowa Legislature and the Governor," May 15, 2017, https://www.leopold.iastate.edu/statement-leopold-center-regarding-decision-iowa-legislature-and-governor (accessed January 27, 2025).

15 U.S. Department of Environmental Protection, "Gulf Hypoxia Action Plan 2008," last updated May 31, 2024, https://www.epa.gov/ms-htf/gulf-hypoxia-action-plan-2008 (accessed January 27, 2025).

16 Iowa Environmental Council, "The Slow Reality of the Nutrient Reduction Strategy," July 16, 2019, https://www.iaenvironment.org/webres/file/slow%20reality%20of%20the%20nrs_final_7_16_19.pdf (accessed January 28, 2025).

17 See essay by Erin Jordan, "Iowa Would Spend $1 Million to Revise Iowa's Nitrogen Fertilizer Recommendations," *Gazette*, May 26, 2022, https://www.thegazette.com/state-government/iowa-would-spend-1-million-to-revise-iowas-nitrogen-fertilizer-recommendations/ (accessed January 28, 2025); and Iowa State University, Extension and Outreach, "Nitrogen Use in Iowa Corn Production," March 2018, https://store.extension.iastate.edu/Product/14281 (accessed January 28, 2025).

18 Iowa Environmental Council, "The Iowa Nutrient Reduction Strategy: Ten Years and No Progress," 2022, https://www.iaenvironment.org/webres/File/NRS%20Report%20and%20Recommendations%202022.pdf (accessed January 28, 2025).

19 USDA Natural Resources Conservation Service, "Conservation Stewardship Program," 2025, https://www.nrcs.usda.gov/programs-initiatives/csp-conservation-stewardship-program (accessed January 28, 2025).

20 David Lawrence, "What Hath Man Wrought!" *U.S. News & World Report*, August 17, 1945, https://www.usnews.com/news/special-reports/the-manhattan-project/articles/2015/09/28/editorial-from-1945-what-hath-man-wrought (accessed January 28, 2025).

21 For example, see an essay by Shannon Hall, "Exxon Knew about Climate Change almost 40 Years Ago," *Scientific American*, October 26, 2015, https://www.scientificamerican.com/article/exxon-knew-about-climate-change-almost-40-years-ago/ (accessed January 28, 2025).

22 Kevin Krajick, "James Hansen's Climate Warning, 30 Years Later," Columbia Climate School, *State of the Planet*, June 26, 2018, https://news.climate.columbia.edu/2018/06/26/james-hansens-climate-warning-30-years-later/ (accessed January 28, 2025).

23 For a history of where the Paris Climate Agreement stands today, see United Nations, Climate Change, "The Paris Agreement," 2025, https://unfccc.int/process-and-meetings/the-paris-agreement (accessed January 28, 2025).

24 For an explanation and why it is important as it relates to climate change, see an essay by Devens Gust, "What Is Photosynthesis?" Arizona State University, Center for Bioenergy and Photosynthesis, 1996, https://bioenergy.asu.edu/why-study-photosynthesis (accessed January 28, 2025).

25 U.S. Department of Environmental Protection (EPA), "Climate Change Indicators: Atmospheric Concentrations of Greenhouse Gases," last updated January 15, 2025, https://www.epa.gov/climate

-indicators/climate-change-indicators-atmospheric-concentrations-greenhouse-gases (accessed February 6, 2025).

26. Andrew Krosofsky, "Scientists Believe the Earth Will Eventually Run Out of Oxygen—But When?" *Green Matters*, April 28, 2021, https://www.greenmatters.com/p/will-earth-run-out-of-oxygen (accessed February 6, 2025).

27. For today's percentages, see U.S. Department of Environmental Protection, "Sources of Greenhouse Gas Emissions," 2025, https://www.epa.gov/ghgemissions/sources-greenhouse-gas-emissions (accessed January 28, 2025).

28. U.S. Department of Environmental Protection, Intergovernmental Panel on Climate Change (IPCC), "Global Greenhouse Gas Overview," 2014; 2022, https://www.epa.gov/ghgemissions/global-greenhouse-gas-overview (accessed January 28, 2025).

29. International Energy Agency (IEA), "Global Energy Review: CO2 Emissions in 2021," Paris, March 2022, https://www.iea.org/reports/global-energy-review-co2-emissions-in-2021-2 (accessed January 28, 2025).

30. Clarissa L. Dietz, Randall D. Jackson, Matthew D. Ruark, and Gregg R. Sanford, "Soil Carbon Maintained by Perennial Grasslands Over 30 Years in Field Crop System in a Temperate Mollisol," *Communications Earth & Environment* 5 (2024): article 360, https://doi.org/10.1038/s43247-024-01500-w.

31. USDA National Organic Program, "Home," 2025, https://www.ams.usda.gov/about-ams/programs-offices/national-organic-program (accessed January 28, 2025).

32. United Nations, Food and Agriculture Organization, "Livestock Solutions for Climate Change," 2017, https://openknowledge.fao.org/items/2985e4e2-3c37-4e7c-aa7c-3655de93d53c (accessed January 28, 2025).

33. Bill Gates, "Rich Countries Should Only Eat Synthetic Beef," *Sky News*, February 16, 2021, https://news.sky.com/story/rich-countries-should-eat-synthetic-beef-says-bill-gates-12219763 (accessed January 28, 2025).

34. See an essay about the largest producers, by Julie Creswell, "How Many Hogs Can Be Slaughtered Per Hour? Pork Industry Wants More," *New York Times*, August 19, 2019, https://www.nytimes.com/2019/08/09/business/pork-factory-regulations.html (accessed January 28, 2025).

35. Gates, "Rich Countries."

36. A recent essay on ruminants by more than 20 scientists can be found at Claudia Arndt et al., "Full Adoption of the Most Effective Strategies to Mitigate Methane Emissions by Ruminants …," *Proceedings of the National Academy of Sciences (PNAS)* 10, no. 119 (2022), https://doi.org/10.1073/pnas.2111294119.

37. For an article on Dr. Lieberman and links to his research, see Jean Caspers-Simmet, "Matt Liebman, More Diverse Rotations Can Create Profitable, Sustainable Farms," *Post Bulletin*, July 12, 2015, https://www.postbulletin.com/news/more-diverse-rotations-can-create-profitable-sustainable-farms (accessed January 28, 2025).

38. Matt Liebman, "Marsden Long-Term Study," Iowa State University, 2001–2025, https://www.cals.iastate.edu/inrc/marsden-long-term-rotation-study (accessed January 28, 2025).

39. A discussion of these companies, and others, along with fuel, plantations, and monoculture trees can be found in this report, Agrifood Atlas, "Facts and Figures about the Corporations that Control What We Eat," 2017, https://eu.boell.org/sites/default/files/agrifoodatlas2017_facts-and-figures-about-the-corporations-that-control-what-we-eat.pdf (accessed January 28, 2025).

40. For an overview, see this recent article by the Associated Press, "North Dakota Regulators Consider Underground Carbon Dioxide Storage Permits for Midwest Pipeline," *MPR News*, December 12, 2024, https://www.mprnews.org/story/2024/12/12/north-dakota-regulators-underground-carbon-dioxide-storage (accessed January 28, 2025).

41 Iowa Renewable Fuels Association, "Distillers Grains Facts," December 2, 2014, last updated July 10, 2015, https://iowarfa.org/ethanol-center/ethanol-co-products/distillers-grains-facts/ (accessed January 28, 2025).

42 For more on the program, see USDA Farm Service Agency, "Conservation Reserve Program (CRP)," 2024, https://www.fsa.usda.gov/resources/programs/conservation-reserve-program (accessed January 28, 2025).

43 See an article about this study and Dr. Tyler Lark of University of Wisconsin-Madison, by Leah Douglas, "U.S. Corn-Based Ethanol Worse for the Climate than Gasoline Study Finds," *Reuters*, February 14, 2022, https://www.reuters.com/business/environment/us-corn-based-ethanol-worse-climate-than-gasoline-study-finds-2022-02-14/ (accessed January 28, 2025).

44 Associated Press, "North Dakota Regulators."

45 For percentages of all top grocery market share holders, see Zachary Russell, "Walmart Led Grocery Dollar Share in 2023, followed by …" *Chain Store Age (CSA)*, March 8, 2024, https://chainstoreage.com/walmart-led-grocery-dollar-share-2023-followed (accessed January 28, 2025).

46 See report by Seth Berkman, "How Farming Has Changed in Every State the Last 100 Years," *Hulton Archive*, *News on 6*, March 13, 2020, https://www.news9.com/story/5e7e3e72700eaf6f16c5bc7f/how-farming-has-changed-in-every-state-the-last-100-years (accessed January 28, 2025).

47 Douglas Harris, quoted in Keith Bannon, "Bipartisan State of the Nation Report Where US is Excelling—and Falling Behind," *Tulane University*, February 3, 2025, https://news.tulane.edu/pr/bipartisan-state-nation-report-reveals-where-us-excelling-and-falling-behind (accessed February 6, 2025).

Chapter 14

1 While this quote has also been attributed to others, he was quoted in an article by the Green Party: Jonathon Porritt, "Green Party Has Been 'Right All Along'," March 13, 2009, https://www.greenparty.org.uk/news/2009-03-13-porritt.html (last retrieved June 22, 2017).

2 Pope Francis, "Pope Francis Tells *60 Minutes* in Rare Interview," *CBS News*, December 29, 2024, https://www.cbsnews.com/news/pope-francis-interview-60-minutes-transcript/ (accessed January 28, 2025).

3 Krishna Ramanujan, "More Than 99.9% of Studies Agree: Humans Caused Climate Change," *Cornell Chronicle*, October 19, 2021, https://news.cornell.edu/stories/2021/10/more-999-studies-agree-humans-caused-climate-change (accessed January 28, 2025).

4 Leslie Hook and Dave Lee, "How Tech Went Big on Green Energy," *Financial Times*, February 10, 2021, https://www.ft.com/content/0c69d4a4-2626-418d-813c-7337b8d5110d (accessed January 28, 2025).

5 Susan Tierney and Lori Bird, "Setting the Record Straight About Renewable Energy," *World Resources Institute*, March 12, 2020, https://www.wri.org/insights/setting-record-straight-about-renewable-energy (accessed January 28, 2025).

6 To review coal usage on a global scale from 2020 through mid-2024, see the International Energy Agency (IEA), "Coal Mid-Year Update – July 2024, Demand," https://www.iea.org/reports/coal-mid-year-update-july-2024/demand (accessed January 25, 2028).

7 See climate essays at: Project Drawdown, 2025, https://drawdown.org/.

8 Project Drawdown, 2025, https://drawdown.org/.

9 Project Drawdown, 2025, https://drawdown.org/.

10 Project Drawdown, 2025, https://drawdown.org/.

11 Project Drawdown, 2025, https://drawdown.org/.

12 Project Drawdown, 2025, https://drawdown.org/.

13. International Fund for Animal Welfare (IFAW), "Which Animals Are Most Impacted by Climate Change?" August 9, 2023, https://www.ifaw.org/journal/animals-most-impacted-climate-change (accessed January 29, 2025).
14. Cristian Román-Palacios and John J. Wiens, "Recent Responses to Climate Change Reveal the Drivers of Species Extinction and Survival," *Proceedings of the National Academy of Sciences (PNAS)* 117, no. 8 (2020): 4711–17, https://doi.org/10.1073/pnas.1913007117.
15. United Nations, Climate Change, "The Paris Agreement," 2025, https://unfccc.int/process-and-meetings/the-paris-agreement (accessed January 28, 2025).
16. Kurt Vonnegut, *Timequake* (New York: Berkley Books; Putnam, 1997), 2.
17. Ryan Sriver and David Lafferty, "Researchers Uncovered Long-Term Shortcomings in Predicting Corn Yields," University of Illinois, November 12, 2021, https://las.illinois.edu/news/2021-11-12/researchers-uncover-long-term-shortcomings-predicting-corn-yields (accessed January 29, 2025).
18. See previously discussed Iowa legislation: Iowa Committee on Ways and Means, *Iowa Groundwater Protection Act of 1987*, House File 631, June 9, 1987, https://www.legis.iowa.gov/docs/publications/LGI/72/HF631.pdf (accessed January 27, 2025).
19. Land Institute, "Kernza Grain," October 21, 2024, https://landinstitute.org/our-work/perennial-crops/kernza/ (accessed January 29, 2025).
20. U.S. Supreme Court, *Citizens United v. Federal Elections Commission*, 558 U.S. 310, January 21, 2010, https://supreme.justia.com/cases/federal/us/558/310/.
21. Quoted at the Constitutional Convention of 1787, Benjamin Franklin, "Essay: Republican Government," *Bill of Rights Institute*, November 5, 2020, https://billofrightsinstitute.org/essays/republican-government (accessed January 29, 2025).
22. Wendell Berry, *The Unsettling of America: Culture and Agriculture* (San Francisco: Sierra Club Books, 1977; 2nd ed., Berkeley, CA: Counterpoint Press, 1996).

Bibliography

Agrifood Atlas. "Facts and Figures about the Corporations that Control What We Eat." 2017. https://eu.boell.org/sites/default/files/agrifoodatlas2017_facts-and-figures-about-the-corporations-that-control-what-we-eat.pdf (accessed January 28, 2025).

American Farm Bureau Foundation for Agriculture. "Does Organic Production Use Pesticides or Synthetic Fertilizers?" 2025. https://www.agfoundation.org/questions/does-organic-production-use-pesticides-or-synthetic-fertilizers# (accessed January 25, 2025).

Aminetzah, Daniel, Jane Brennan, Wesley Davis, Bekinwari Idoniboye, Nick Noel, Jake Pawlowski, and Shelley Stewart. "Black Farmers in the US: The Opportunity for Addressing Racial Disparities in Farming." *McKinsey,* November 10, 2021. https://www.mckinsey.com/industries/agriculture/our-insights/black-farmers-in-the-us-the-opportunity-for-addressing-racial-disparities-in-farming (accessed January 18, 2025).

Arndt Claudia et al. "Full Adoption of the Most Effective Strategies to Mitigate Methane Emissions by Ruminants..." *Proceedings of the National Academy of Sciences (PNAS)* 10, no. 119 (2022): https://doi.org/10.1073/pnas.2111294119.

Associated Press. "North Dakota Regulators Consider Underground Carbon Dioxide Storge Permits for Midwest Pipeline." *MPR News,* December 12, 2024. https://www.mprnews.org/story/2024/12/12/north-dakota-regulators-underground-carbon-dioxide-storage (accessed January 28, 2025).

Awakuni-Swetland, Mark. "Omaha." Encyclopedia of World Cultures Supplement. *Encyclopedia.com,* January 8, 2025. https://www.encyclopedia.com/history/united-states-and-canada/north-american-indigenous-peoples/omaha-indians (accessed January 18, 2025).

Bannon, Keith. "Bipartisan State of the Nation Report Where US is Excelling—and Falling Behind." *Tulane University,* February 3, 2025. https://news.tulane.edu/pr/bipartisan-state-nation-report-reveals-where-us-excelling-and-falling-behind (accessed February 6, 2025).

Barry, John M. *The Great Influenza: The Story of the Deadliest Pandemic in History*. New York: Penguin, 2005.

Becker, Sam. "US Farms Are Making an Urgent Push into AI. It Could Help Feed the World." *BBC,* March 27, 2024. https://www.bbc.com/worklife/article/20240325-artificial-intelligence-ai-us-agriculture-farming (accessed February 6, 2025).

Benbrook, Charles M. "Impacts of Genetically Engineered Crops on Pesticide Use in the U.S.—the First Sixteen Years." *Environmental Sciences Europe* 24, no. 24 (2012). https://doi.org/10.1186/2190-4715-24-24.

Bender, M. H. "An Economic Comparison of Traditional and Conventional Agricultural Systems at a County Level." *American Journal of Alternative Agriculture* 16, no. 1 (2001): 2–15. http://www.jstor.org/stable/44503173.

Berkman, Seth. "How Farming Has Changed in Every State the Last 100 Years." *Hulton Archive, News on 6,* March 13, 2020. https://www.news9.com/story/5e7e3e72700eaf6f16c5bc7f/how-farming-has-changed-in-every-state-the-last-100-years (accessed January 28, 2025).

Berry, Wendell. *A Continuous Harmony: Essays Cultural & Agricultural*. Boston: Houghton Mifflin Harcourt, 1979.

Berry, Wendell. "Private Property and the Common Wealth" (essay). Posted February 10, 2008, last updated December 19, 2008. http://tipiglen.co.uk/berryprivate.html (accessed January 11, 2025).

Berry, Wendell. *Standing by Words*. Washington, DC: Shoemaker, 2005; Berkeley, CA: Counterpoint Press, 2011.

Berry, Wendell. *The Unsettling of America: Culture and Agriculture*. San Francisco: Sierra Club Books, 1977; 2nd ed., Berkeley, CA: Counterpoint Press, 1996.

Berry, Wendell. *What I Stand for Is What I Stand On*. New York: Penguin Classics, 2021.

Borunda, Alejandra. "U.S. Corn Production Is Booming—But Not for the Reasons Scientists Hoped." *National Geographic,* January 24, 2022. https://www.nationalgeographic.com/environment/article/us-corn-production-booming-but-its-future-isnt-clear (accessed January 25, 2025).

Boughter, Judith A. *Betraying the Omaha Nation, 1790-1916*. Norman: University of Oklahoma Press, 1998.

Bovee, David S. "Lugutti, Luigi Gino." *Biographical Dictionary of Iowa*. University of Iowa, n.d. https://uipress.lib.uiowa.edu/bdi/DetailsPage.aspx?id=232 (accessed January 27, 2025).

Brown, Dee. *Bury My Heart at Wounded Knee: An Indian History of the American West*. New York: Holt, Rinehart & Winston, 1970.

California Certified Organic Farmers (CCOF). "Help Make Organic the Norm!" 2025. https://www.ccof.org/ (accessed January 26, 2025).

California Department of Food and Agriculture. "California's State Organic Program." 2017. https://www.cdfa.ca.gov/is/i_&_c/pdfs/CalOrganicPrgrmFactSheet.pdf (accessed January 26, 2025).

Capaldo, Samantha. "USDA Publishes Final Rule Amending Organic Livestock and Poultry Standards." *National Agricultural Law Center,* November 2, 2023. https://nationalaglawcenter.org/usda-publishes-final-rule-amending-organic-livestock-and-poultry-standards/ (accessed January 25, 2025).

Caprettini, Bruno, and Hans-Joachim Voth. "Rage against the Machines: Labor-Saving Technology and Unrest in Industrializing England." *American Economic Review: Insights* 2, no. 3 (2020): 305–20. https://pubs.aeaweb.org/doi/pdfplus/10.1257/aeri.20190385.

Caspers-Simmet, Jean. "Matt Liebman, More Diverse Rotations Can Create Profitable, Sustainable Farms." *Post Bulletin,* July 12, 2015, https://www.postbulletin.com/news/more-diverse-rotations-can-create-profitable-sustainable-farms (accessed January 28, 2025).

Center for Rural Affairs. "Home." 2025. https://www.cfra.org/.

Chang Vicky C., Gabriella Andreotti, Maria Ospina, Christine G. Parks, Danping Liu, Joseph J. Shearer, Nathaniel Rothman, Debra T. Silverman, Dale P. Sandler, Antonia M. Calafat, Laura E. Beane Freeman, and Jonathan N. Hofmann. "Glyphosate Exposure and Urinary Oxidative Stress Biomarkers in the Agricultural Health Study." *Journal of the National Cancer Institute* 115, no. 4 (April 2023): 394–404. https://doi.org/10.1093/jnci/djac242.

Churchill, Winston. "Parliamentary Debate, April 25, 1918." In *Churchill by Himself: The Definitive Collection of Quotations*, edited by Richard Langworth. New York: Public Affairs; London: Ebury Press, 2008.

Community Environmental Legal Defense Fund (CELDF). "The Enclosure Movement." Last updated March 20, 2021, https://celdf.org/the-enclosure-movement/ (accessed January 14, 2025).

Conniff, Richard. "What the Luddites Really Fought Against." *Smithsonian Magazine,* March 2011. https://www.smithsonianmag.com/history/what-the-luddites-really-fought-against-264412/ (accessed January 27, 2025).

Coolidge, Calvin. "Message to the Senate Returning Without Approval S. 4808." *American Presidency Project,* February 25, 1927. https://www.presidency.ucsb.edu/documents/message-the-senate-returning-without-approval-s-4808-the-mcnary-haugen-farm-relief-bill (accessed January 22, 2025).

Corporate Europe Observatory. "Bayer and Monsanto Merger: What Role Did Revolving Doors Play?" February 26, 2024. https://corporateeurope.org/en/2024/02/bayer-and-monsanto-merger-what-role-did-revolving-doors-play (accessed January 24, 2025).

Creswell, Julie. "How Many Hogs Can Be Slaughtered Per Hour? Pork Industry Wants More." *New York Times,* August 19, 2019. https://www.nytimes.com/2019/08/09/business/pork-factory-regulations.html (accessed January 28, 2025).

Croghan, George, Col. *Army Life on the Western Frontier: Selections from the Official Reports Made Between 1826 and 1845 by Colonel George Croghan*. Edited by Francis Paul Prucha. Norman: University of Oklahoma Press, 1958; 2014.

Day, Dorothy. "On Pilgrimage." *Catholic Worker Movement,* June 1, 1947. https://catholicworker.org/454-html/ (accessed January 17, 2025).

De Decker, Kris. "Bring Back the Horses." *Low-Tech Magazine,* April 18, 2008. https://solar.lowtechmagazine.com/2008/04/bring-back-the-horses/ (accessed January 20, 2025).

Deepranjan Sarkar, Saswat Kumar Kar, Arghya Chattopadhyay, Shikha, Amitava Rakshit, Vinod Kumar Tripathi, Pradeep Kumar Dubey, and Purushothaman Chirakkuzhyil Abhilash. "Low Input Sustainable Agriculture: A Viable Climate-Smart Option for Boosting Food Production in a Warming World." *Ecological Indicators* 115 (August 2020): Article 106412. https://doi.org/10.1016/j.ecolind.2020.106412.

Delgado, Richard, and Jean Stefancic, eds. *Critical Race Theory: An Introduction.* New York: New York University Press, 2001.

Denevan, William M., ed. *The Native Population of the Americas in 1492.* Madison: University of Wisconsin Press, 1976; 2nd ed., 1992.

Dietz, Clarissa L., Randall D. Jackson, Matthew D. Ruark, and Gregg R. Sanford. "Soil Carbon Maintained by Perennial Grasslands Over 30 Years in Field Crop System in a Temperate Mollisol." *Communications Earth & Environment* 5 (2024): article 360. https://doi.org/10.1038/s43247-024-01500-w.

Douglas, Leah. "U.S. Corn-Based Ethanol Worse for the Climate than Gasoline Study Finds." *Reuters,* February 14, 2022. https://www.reuters.com/business/environment/us-corn-based-ethanol-worse-climate-than-gasoline-study-finds-2022-02-14/ (accessed January 28, 2025).

Drought.gov. "Drought by Location, Iowa." 2025. https://www.drought.gov/states/iowa (accessed January 20, 2025).

Dylan, Bob. "The Times They Are a-changin." *The Times They Are a-Changin* (album). New York: Columbia, 1965.

Edwards, Paul M. *Our Legacy of Faith: A Brief History of the Reorganized Church of Jesus Christ of Latter Day Saints.* Independence, MO: Herald House, 1991.

Ehrlich, Gretel. "Ray Hunt: The Cowboy Sage." *Shambhala Sun,* July 1998.

Entrepreneurship in Omaha. "History and Culture of the Omaha Tribe." May 31, 2023. https://www.omahaimc.org/history-and-culture-of-the-omaha-tribe/ (accessed January 18, 2025).

Farm Table Delivery. "We Sell Local Food; Delivery." 2025. https://farmtabledelivery.localfoodmarketplace.com/ (accessed January 26, 2025).

Fee, Chester Anders. *Chief Joseph: The Biography of a Great Indian.* New York: Wilson-Erickson, 1936. https://babel.hathitrust.org/cgi/pt?id=inu.32000000324311 (accessed January 11, 2025).

Flannery, Tim. *The Eternal Frontier-An Ecological History of North America and Its Peoples.* Melbourne: Text Publishing; New York: Grove Publishing, 2001.

4-H.org. "4-H, About." 2025. https://4-h.org/about/.

Francis (Pope). Encyclical Letter, *Laudato Si'* (Praise be to you). Rome, Italy, at Saint Peter's, May 24, 2015. https://www.vatican.va/content/francesco/en/encyclicals/documents/papa-francesco_20150524_enciclica-laudato-si.html (accessed January 10, 2025).

Francis (Pope). "Pope Francis Tells *60 Minutes* in Rare Interview." *CBS News,* December 29, 2024. https://www.cbsnews.com/news/pope-francis-interview-60-minutes-transcript/ (accessed January 28, 2025).

Franklin, Benjamin. "Essay: Republican Government." *Bill of Rights Institute,* November 5, 2020. https://billofrightsinstitute.org/essays/republican-government (accessed January 29, 2025)

Friends of the Earth. "Friends of the Earth Is a Bold Voice for Justice and the Planet." 2025. https://foe.org/ (accessed January 26, 2025).

Gabriel, Pedro. "The Seamless Garment Is the Catholic Position." *Catholic Outlook,* February 9, 2020. https://catholicoutlook.org/the-seamless-garment-is-the-catholic-position (accessed January 27, 2025).

Gary, Dale E. "Cosmology and the Beginning of Time" (lecture). *New Jersey Institute of Technology (NJIT)*, n.d. https://web.njit.edu/~gary/202/Lecture26.html (accessed January 26, 2025).

Gates, Bill. "Rich Countries Should Only East Synthetic Beef." *Sky News,* February 16, 2021. https://news.sky.com/story/rich-countries-should-eat-synthetic-beef-says-bill-gates-12219763 (accessed January 28, 2025).

Gillam, Carey. "CDC Finds Weed Killer tied to Cancer in Over 80 pct of US Urine Samples." *New Lede,* July 9, 2022. https://www.thenewlede.org/2022/07/cdc-finds-weed-killer-tied-to-cancer-in-over-80-pct-of-us-urine-samples/ (accessed January 24, 2025).

Gust, Devens. "What Is Photosynthesis?" Arizona State University. Center for Bioenergy and Photosynthesis, 1996. https://bioenergy.asu.edu/why-study-photosynthesis (accessed January 28, 2025).

Hall, Shannon. "Exxon Knew about Climate Change almost 40 Years Ago." *Scientific America*, October 26, 2015. https://www.scientificamerican.com/article/exxon-knew-about-climate-change-almost-40-years-ago/ (accessed January 28, 2025).

Harkin, Tom. "The Tiger Cages of Con Son: A Congressional Investigating Committee Finds Secret Cells in a Concentration Camp for South Vietnamese Political Prisoners" (photograph). *Life Magazine*, July 17, 1970, 26, https://www.originallifemagazines.com/product/life-magazine-july-17-1970/ (accessed January 23, 2025).

Helminiak, Daniel A. *The Human Core of Spirituality: Mind as Psyche and Spirit*. New York: University of New York, 1996.

Helminiak, Daniel A. *Meditation Without Myth*. New York: Crossroad, 2005.

Hook, Leslie, and Dave Lee. "How Tech Went Big on Green Energy." *Financial Times,* February 10, 2021. https://www.ft.com/content/0c69d4a4-2626-418d-813c-7337b8d5110d (accessed January 28, 2025).

Horn, Trent. "Thomas Aquinas's Five Proofs for God Revisited." *Christian Research Journal* 41, no. 2 (2018): JAF3412. https://www.equip.org/articles/thomas-aquinass-five-proofs-for-god-revisited/ (accessed January 26, 2025).

Houck, Louis. *A History of Missouri: From the Earliest Explorations and Settlements until the Admission of the State into the Union*. 2 vols. Chicago: R. R. Donnelly, 1908.

Howard, Sir Albert. *Farming and Gardening for Health or Disease*. Oxford: Oxford City Press, 2011; 2019. https://gutenberg.net.au/ebooks02/0200311.txt# (accessed January 23, 2025).

Humphrey, Hubert H. "Hubert H. Humphrey Quotes." *BrainyQuote*, BrainyMedia, 2025. https://www.brainyquote.com/quotes/hubert_h_humphrey_152591 (accessed February 5, 2025)

Ifft, Jennifer E., and Youwei Yang. "Horses vs. Tractors? Old Order Amish Population Growth and New York Farmland Markets" (paper). *Agricultural and Applied Economics Association,* July 26-28, 2020. Kansas City, MO. https://econpapers.repec.org/paper/agsaaea20/304565.htm (accessed January 20, 2025).

Innis, Harold A. *The Fur Trade in Canada*. Toronto: University of Toronto Press, 1930; 2001.

International Energy Agency (IEA). "Coal Mid-Year Update – July 2024, Demand." https://www.iea.org/reports/coal-mid-year-update-july-2024/demand (accessed January 25, 2028).

International Energy Agency (IEA). "Global Energy Review: CO2 Emissions in 2021." Paris, March 2022. https://www.iea.org/reports/global-energy-review-co2-emissions-in-2021-2 (accessed January 28, 2025).

International Fund for Animal Welfare (IFAW). "Which Animals Are Most Impacted by Climate Change?" August 9, 2023. https://www.ifaw.org/journal/animals-most-impacted-climate-change (accessed January 29, 2025).

Iowa Board of Immigration. *Iowa: The Home of Immigrants*. Des Moines: Mills, 1870.

Iowa Catholic Conference. "STRANGERS AND GUESTS: Toward Community in the Heartland, a Regional Catholic Bishops' Statement on Land Issues." May 1, 1980. https://iowacatholicconference.org/wp-content/uploads/2021/05/Strangers-and-Guests-1980.pdf (accessed January 27, 2025).

Iowa Committee on Ways and Means. *Iowa Groundwater Protection Act of 1987*. House File 631. June 9, 1987. https://www.legis.iowa.gov/docs/publications/LGI/72/HF631.pdf (accessed January 27, 2025).

Iowa Corn. "Corn Facts and Fun: Use for Corn." 2023. https://www.iowacorn.org/corn-facts-faq/ (accessed January 23, 2025).

Iowa Corn. "Corn Grows Iowa." (YouTube), February 19, 2020. https://www.youtube.com/watch?v=n2oDt391GVk&list=PLxepamPolX9LEtdWIqZuDKLpebKwYxPGi&index=2 (accessed January 27, 2025).

Iowa Environmental Council. "The Iowa Nutrient Reduction Strategy: Ten Years and No Progress." 2022. https://www.iaenvironment.org/webres/File/NRS%20Report%20and%20Recommendations%202022.pdf (accessed January 28, 2025).

Iowa Environmental Council. "The Slow Reality of the Nutrient Reduction Strategy." July 16, 2019. https://www.iaenvironment.org/webres/file/slow%20reality%20of%20the%20nrs_final_7_16_19.pdf (accessed January 28, 2025).

Iowa Ornithologists' Union. "Birding in Iowa." 2025. https://iowabirds.org/birds/ (accessed January 20, 2025).

IowaPBS. "IowaPathways: Railroads." December 2021. https://www.iowapbs.org/iowapathways/mypath/2536/railroads (accessed January 14, 2025).

IowaPBS. "The 1970s See Good Times in Agriculture." (essay; YouTube), July 1, 2013. https://www.iowapbs.org/shows/farmcrisis/clip/5310/1970s-see-good-times-agriculture (accessed January 22, 2025).

Iowa Renewable Fuels Association. "Distillers Grains Facts." December 2, 2014. Last updated July 10, 2015. https://iowarfa.org/ethanol-center/ethanol-co-products/distillers-grains-facts/ (accessed January 28, 2025).

Iowa State University. "Historical Hog and Lamb Prices." *Ag Decision Maker,* File B2-10. https://www.extension.iastate.edu/agdm/livestock/html/b2-10.html (accessed February 7, 2025).

Iowa State University. "Historical Iowa Farmland Values Survey by County." *Ag Decision Maker,* File C2-72. https://www.extension.iastate.edu/agdm/wholefarm/html/c2-72.html (accessed February 7, 2025).

Iowa State University. "Iowa Cash Corn and Soybean Prices (USDA NASS)." *Ag Decision Maker,* File A2-11. https://www.extension.iastate.edu/agdm/crops/pdf/a2-11.pdf (accessed February 7, 2025).

Iowa State University. Extension and Outreach. "Nitrogen Use in Iowa Corn Production." March 2018. https://store.extension.iastate.edu/Product/14281 (accessed January 28, 2025).

Iowa State University. Extension and Outreach. "Soybean Cupping 2021: Iowa Update." July 22, 2021. https://crops.extension.iastate.edu/blog/prashant-jha/soybean-cupping-2021-iowa-update (accessed January 24, 2025).

Jancer, Matt. "Millions of Americans Have a Parasite and Don't Realize It." *Vice,* December 19, 2018. https://www.vice.com/en/article/millions-of-americans-have-a-parasite-and-dont-realize-it/ (accessed February 5, 2025).

John Paul II (Pope). "Letter to Reverend George V. Coyne, S.J." *Vatican,* June 1, 1988. https://www.vatican.va/content/john-paul-ii/en/letters/1988/documents/hf_jp-ii_let_19880601_padre-coyne.html (accessed January 27, 2025).

Johns Hopkins Center for a Livable Future, 2014; 2025. https://clf.jhsph.edu/.

Jones, James Earl. "James Earl Jones on Preparation" (interview; YouTube). *American Academy of Achievement,* December 5, 2016. https://youtu.be/hnze3oquDpl (accessed January 23, 2025).

Jordan, Erin. "Iowa Would Spend $1 Million to Revise Iowa's Nitrogen Fertilizer Recommendations." *Gazette,* May 26, 2022. https://www.thegazette.com/state-government/iowa-would-spend-1-million-to-revise-iowas-nitrogen-fertilizer-recommendations/ (accessed January 28, 2025)

King, Martin Luther, Jr. *Strength to Love*. Minneapolis, MN: Fortress Press, 1963.

"The Know Nothing Party, 1856." *Digital Public Library of America* (exhibitions). https://dp.la/exhibitions/outsiders-president-elections/anti-outsider-platforms/know-nothing-party-1856 (accessed January 15, 2025).

Koppes, Steve, and Louise Lerner. "Carbon-14 Dating, Explained." *UChicago News*, April 27, 2023. Last updated January 16, 2025. https://news.uchicago.edu/explainer/what-is-carbon-14-dating (accessed January 26, 2025).

Krajick, Kevin. "James Hansen's Climate Warning, 30 Years Later." Columbia Climate School. *State of the Planet,* June 26, 2018. https://news.climate.columbia.edu/2018/06/26/james-hansens-climate-warning-30-years-later/ (accessed January 28, 2025)

Krosofsky, Andrew. "Scientists Believe the Earth Will Eventually Run Out of Oxygen—But When?" *Green Matters,* April 28, 2021. https://www.greenmatters.com/p/will-earth-run-out-of-oxygen (accessed February 6, 2025).

Land Institute. "Kernza Grain." October 21, 2024. https://landinstitute.org/our-work/perennial-crops/kernza/ (accessed January 29, 2025)

Lawrence, D. H. *Apocalypse and the Writings on Revelation*. Edited by Mara Kalnins. Cambridge: Cambridge University Press, 2002.

Lawrence, David. "What Hath Man Wrought!" *U.S. News & World Report*, August 17, 1945. https://www.usnews.com/news/special-reports/the-manhattan-project/articles/2015/09/28/editorial-from-1945-what-hath-man-wrought (accessed January 28, 2025).

Leo XIII (Pope). "Rerum Novarum – Encyclical Letter of Pope Leo XIII on the Conditions of Labor." 1891. Historical Catholic and Dominican Documents. 13. https://digitalcommons.providence.edu/cgi/viewcontent.cgi?article=1014&context=catholic_documents (accessed January 17, 2025).

Leopold Center for Sustainable Agriculture. Iowa State University. 2025. https://www.leopold.iastate.edu/leopold-center-sustainable-agriculture (accessed January 27, 2025).

Leopold Center for Sustainable Agriculture. . "Statement from the Leopold Center Regarding the Decision of the Iowa Legislature and the Governor." May 15, 2017. https://www.leopold.iastate.edu/statement-leopold-center-regarding-decision-iowa-legislature-and-governor (accessed January 27, 2025).

Lewis, C. S. *A Joyful Christian.* New York: Macmillan, 1977.

Lewis, C. S. *The Problem of Pain* (1940). New York: HarperCollins, 1996.

Lewis, Meriwether, and William Clark. *The Journals of Lewis and Clark.* Edited by Bernard DeVoto. New York: Harper Perennial, 1997.

Liebman, Matt. "Marsden Long-Term Study." *Iowa State University*, 2001–2025. https://www.cals.iastate.edu/inrc/marsden-long-term-rotation-study (accessed January 28, 2025).

Locks, Catherine, Sara K. Mergel, Pamela Thomas Roseman, and Tamara Spike. "Jacksonian America (1815–1840)." In *History in the Making: A History of the People of the United States of America to 1877*, chap. 12, 525–82. Dahlonega: University of North Georgia Press, 2013. https://ung.edu/university-press/_uploads/files/us-history/US-History-I-Chapter-12.pdf (accessed January 13, 2025).

Lonergan, Bernard J. F. *Insight: A Study in Human Understandings.* New York: Philosophical Library; London: Darton Longman and Todd, 1957.

Lonergan, Bernard J. F. *Method in Theology.* London: Darton Longman and Todd, 1972.

McCabe, Don. "Amendment May Be Dead, Unless Supreme Court Hears Appeal." *FarmProgress,* January 26, 2007. https://www.farmprogress.com/farm-business/nebraska-can-no-longer-enforce-i-300 (accessed January 22, 2025).

Mead, Margaret. "Coming of Age in Samoa." *World History Commons,* n.d. https,//worldhistorycommons.org/margaret-mead-coming-age-samoa (accessed February 5, 2025).

Minnesota Department of Health. "Nitrate in Drinking Water." Last updated January 27, 2025. https://www.health.state.mn.us/communities/environment/water/contaminants/nitrate.html (accessed January 27, 2025).

Montuschi, Eleonora. "Order of Man, Order of Nature: Francis Bacon's Idea of a 'Dominion' Over Nature." Paper presented at the London School of Economics, October 27–28, 2010. https://iris.unive.it/retrieve/handle/10278/24867/23441/MontuschiBacon.pdf (accessed January 14, 2025).

Muir, John. *Yosemite.* New York: The Century Company, 1912.

National Pesticide Information Center. "Dicamba: General Fact Sheet." February 2012. https://npic.orst.edu/factsheets/dicamba_gen.html (accessed January 24, 2025).

National Sustainable Agriculture Coalition (NSAC). "Home." 2025. https://sustainableagriculture.net/ (accessed January 26, 2025).

National Young Farmer's Coalition. "Release 2022: National Young Farmer Survey Reveals a Generation Defying the Odds." September 1, 2022. https://www.youngfarmers.org/2022/09/643050/ (accessed February 6, 2025).

New World Encyclopedia. "Omaha (tribe)." Last updated November 18, 2022. https://www.newworldencyclopedia.org/p/index.php?title=Omaha_(tribe)&oldid=1087914 (accessed January 18, 2025).

New York Times, "61% in Poll Assert Entry Into the War Was a U.S. 'Mistake'." *New York Times*, June 6, 1971. https://nyti.ms/42Idk8Y (accessed January 21, 2025).

Nolan, Lucinda A. "Edwin Vincent O'Hara." *Biola University,* 2009. https://www.biola.edu/talbot/ce20/database/edwin-vincent-ohara (accessed January 27, 2025).

Notes From the UK. "Early British Consumer Co-ops." January 10, 2020. https://notesfromtheuk.com/2020/01/10/co-ops-in-britain-yet-another-bit-of-history/ (accessed January 17, 2025).

Oklahoma [State of]. "Treaties, Agreements, and Documents." 1722–1954. https://treaties.okstate.edu/treaties.

Oklahoma [State of]. "Treaty with the Chippewa, etc., 1833." https://treaties.okstate.edu/treaties/treaty-with-the-chippewa-etc-1833-0402 (accessed January 13, 2025).

Oklahoma [State of]. "Treaty with the Missouria, Fox, Sioux, Menominee, Otoe, Iowa, Sauk, Winnebago, and Omaha, 1830." https://treaties.okstate.edu/treaties/treaty-with-the-missouria-fox-sioux-menominee-ottoe-iowa-sauk-winnebago-and-omaha-1830-22923 (accessed January 13, 2025).

Organic Farming Research Foundation (OFRF). "Why Organic?" 2025. https://ofrf.org/ (accessed January 26, 2025).

Organic Research and Education Initiative (OREI). "Organic Research, Education, and Extension Programs." 2025. https://www.usda.gov/farming-and-ranching/organic-farming/organic-research-education-and-extension-programs (accessed January 26, 2025).

Orwell, George. *Animal Farm: A Fairy Story*. London: Secker and Warburg, 1945.

O'Sullivan, John. "Annexation (1845)." *United States Magazine and Democratic Review* 17, no. 1 (July–August 1845): 5–10. https://pdcrodas.webs.ull.es/anglo/OSullivanAnnexation.pdf.

Pearson, April. "Milk & Honey: Harlan's Sweet Spot for Brunch." January 29, 2024. https://www.iowafoodandfamily.com/blog/milk--honey-harlan's-sweet-spot-for-brunch (accessed January 26, 2025).

Petersen, William J., ed. "Iowa Land Values, 1803–1967." In *The Palimpsest* (Iowa City: State Historical Society of Iowa, 1967). https://farmland.card.iastate.edu/files/inline-files/Murray-1967-Palimpsest-Iowa-Land-Values-1803-1967_0.pdf (accessed February 7, 2025).

Pitman, Lynn. "History of Cooperatives in the United States: An Overview." December 2018. University of Wisconsin-Madison, Center for Cooperatives. https://resources.uwcc.wisc.edu/History_of_Cooperatives.pdf (accessed January 17, 2025).

Pius XI (Pope). "Rerum Novarum – Quadragesimo Anno (In the 40th Year), Reconstruction of the Social Order." 1931. https://www.papalencyclicals.net/pius11/p11quadr.htm (accessed January 17, 2025).

Ponca Tribe of Nebraska. "Office Locations." 2025. https://www.poncatribe-ne.gov/contact/office-locations/ (accessed January 13, 2025).

Porritt, Jonathon. "Green Party Has Been 'Right All Along'." March 13, 2009. https://www.greenparty.org.uk/news/2009-03-13-porritt.html (last retrieved June 22, 2017).

Project Drawdown. 2025. https://drawdown.org/.

Ramanujan, Krishna. "More Than 99.9% of Studies Agree: Humans Caused Climate Change." *Cornell Chronical*, October 19, 2021. https://news.cornell.edu/stories/2021/10/more-999-studies-agree-humans-caused-climate-change (accessed January 28, 2025).

Reganold, John. "Can We Feed 10 Billion People on Organic Farming Alone?" *Guardian*, August 14, 2016. https://www.theguardian.com/sustainable-business/2016/aug/14/organic-farming-agriculture-world-hunger (accessed January 23, 2025).

Research and Policy Committee of the Committee for Economic Development. *An Adaptive Program for Agriculture*. New York: Committee for Economic Development, 1962. https://babel.hathitrust.org/cgi/pt?id=mdp.39015008782123&seq=7 (accessed January 23, 2025).

Rodale Institute. *New Farm* Archive. https://rodaleinstitute.org/education/resources/new-farm/ (accessed January 23, 2025).

Román-Palacios, Cristian, and John J. Wiens. "Recent Responses to Climate Change Reveal the Drivers of Species Extinction and Survival." *Proceedings of the National Academy of Sciences (PNAS)* 117, no. 8 (2020): 4711–17. https://doi.org/10.1073/pnas.1913007117.

Rosmann, Ron, and Maria Vakulskas Rosmann. "Farm Sweet Farm." 2025. https://www.farmsweetfarm.org/ (accessed January 25, 2025).

Rosmann, Ron, and Maria Vakulskas Rosmann. *Preserving Our Past, Ensuring Our Future, 1872-1997, Westphalia, Iowa*. Audubon, IA: Audubon Media, 1997.

Rosmann, Ron, and Maria Vakulskas Rosmann, with Practical Farmers of Iowa and Iowa Future Farmers of America. "Corn Population Trials." 1988–1999, https://practicalfarmers.org/research/9810/ (accessed February 6, 2025).

Russell, Zachary. "Walmart Led Grocery Dollar Share in 2023, followed by ..." *Chain Store Age (CSA)*, March 8, 2024. https://chainstoreage.com/walmart-led-grocery-dollar-share-2023-followed (accessed January 28, 2025).

Samanic, Claudine, Jennifer Rusiecki, Mustafa Dosemeci, Lifang Hou, Jane A. Hoppin, Dale P. Sandler, Jay Lubin, Aaron Blair, and Michael C. R. Alavanja. "Cancer Incidence Among Pesticide Applicators Exposed to Dicamba in the Agricultural Health Study." *Environmental Health Perspectives* 114, no. 10 (2006): 1521–6. https://doi.org/10.1289/ehp.9204.

Sandoz, Mari. *Old Jules*. Boston: Little, Brown, 1935.

Schiller, Friedrich. *Kallias Letters,* 2.2 "On Grace and Dignity." Quoted from Moland, Lydia L. "Friedrich Schiller." *The Stanford Encyclopedia of Philosophy* (Summer 2025 Edition). Edward N. Zalta and Uri Nodelman (eds.). https://plato.stanford.edu/archives/sum2025/entries/schiller/ (accessed May 16, 2025).

Schneider, Steffen. "The Haber-Brosch Process." *Institute for Mindful Agriculture*, March 24, 2016. https://www.instituteformindfulagriculture.org/writings-1/2016/3/24/the-haber-bosch-process-1# (accessed January 25, 2025).

Schwieder, Dorothy. "History of Iowa." *Iowa Official Register*, November 15, 1999. https://publications.iowa.gov/135/1/history/7-1.html (accessed January 14, 2025).

Shapiro, Peter. "Freshman Deferments End as Nixon Signs New Draft Legislation." *Harvard Crimson*, September 29, 1971. https://www.thecrimson.com/article/1971/9/29/freshman-deferments-end-as-nixon-signs/ (accessed January 21, 2025).

Soerensen, Simon John Christoph, David S. Lim, Maria E. Montez-Rath, Glenn M. Chertow, Benjamin I. Chung, David H. Rehkopf, and John T. Leppert. "Pesticides and Prostate Cancer Incidence and Mortality: An Environment-Wide Association Study." *Cancer* 131, no. 1 (January 2024): e35572. https://doi.org/10.1002/cncr.35572.

Sriver Ryan, and David Lafferty. "Researchers Uncovered Long-Term Shortcomings in Predicting Corn Yields." University of Illinois, November 12, 2021. https://las.illinois.edu/news/2021-11-12/researchers-uncover-long-term-shortcomings-predicting-corn-yields (accessed January 29, 2025).

Stadtfeld, Curtis K. *From the Land and Back: What Life Was Like on a Family Farm and How Technology Changed It*. New York: Charles Scribner's Sons, 1972.

Sustainable Agricultural Research and Education (SARE). "Farmer-Driven Innovations in Agriculture that Improve Profitability, Stewardship and Quality of Life." 2025. https://www.sare.org/ (accessed January 26, 2025).

Swoboda, Rod. "Sampling Cornfields for Soil Nitrate." *Wallaces Farmer, Urbandale* 123, no. 8 (May 1998): 14. https://www.proquest.com/trade-journals/sampling-cornfields-soil-nitrate/docview/219323336/se-2 (accessed January 25, 2025).

Thompson, Megan, and Melanie Staltzman. "How Rural Communities are Tackling a Suicide and Depression Crisis Among Farmers." *PBS News* (video). January 14, 2024. https://www.pbs.org/newshour/show/how-rural-communities-are-tackling-a-suicide-and-depression-crisis-among-farmers (accessed January 18, 2025).

Thompson, Nancy L. "Center for Rural Affairs and South Dakota Family Farm Coalition." *P2InfoHouse.org*, n.d. https://p2infohouse.org/ref/21/20027.htm (accessed January 22, 2025).

Tierney, Susan, and Lori Bird. "Setting the Record Straight About Renewable Energy." *World Resources Institute,* March 12, 2020. https://www.wri.org/insights/setting-record-straight-about-renewable-energy (accessed January 28, 2025).

Union of Concerned Scientists. "Rural Iowans Bear Brunt of Water Treatment Costs for Nitrate Pollution from Farms and CAFOs." January 14, 2021. https://www.ucsusa.org/about/news/rural-iowans-bear-brunt-water-treatment-costs-nitrate-pollution-farms-and-cafos (accessed January 27, 2025).

United Nations. Climate Change. "The Paris Agreement." 2025. https://unfccc.int/process-and-meetings/the-paris-agreement (accessed January 28, 2025).

United Nations. Food and Agriculture Organization (FAO). "Livestock Solutions for Climate Change." 2017. https://openknowledge.fao.org/items/2985e4e2-3c37-4e7c-aa7c-3655de93d53c (accessed January 28, 2025).

United States. Treaty with the Sauk and Foxes, 1842. October 11, 1842. 7 Stat., 596. March 23, 1843. https://treaties.okstate.edu/treaties/treaty-with-the-sauk-and-foxes-1842-0546.

United States. *United States Constitution.* Amendments, and the Bill of Rights. https://constitutioncenter.org/the-constitution.

U.S. Census Bureau. "2020 Census, Westphalia, Iowa." https://data.census.gov/profile?q=Westphalia%20city,%20Iowa%20Education (accessed January 26, 2025).

U.S. Congress. *Agricultural Adjustment Act (AAA) of 1933*. May 12, 1933. 48 Stat. 31. 73rd Cong., 1st sess. https://nationalaglawcenter.org/wp-content/uploads/assets/farmbills/1933.pdf (accessed January 22, 2025).

U.S. Congress. *Civil Rights Act of 1964*. July 2, 1964. PL 88-352. 88th Cong., 2nd sess. https://www.archives.gov/milestone-documents/civil-rights-act.

U.S. Congress. *Emergency Banking Relief Act of 1933*. March 9, 1933. 48 STAT 1. 73rd cong., 1st sess. https://govtrackus.s3.amazonaws.com/legislink/pdf/stat/48/STATUTE-48-Pg1.pdf.

U.S. Congress. *GI Bill of Rights of 1944*. June 22, 1944. PL. 346-268. 78th Cong., 2nd sess. https://www.archives.gov/milestone-documents/servicemens-readjustment-act.

U.S. Congress. *Homestead Act of 1862*. May 20, 1862. 12 STAT 392. 37th cong., 2nd sess. https://www.archives.gov/milestone-documents/homestead-act.

U.S. Congress. "Long Range Farm Program." In *Hearings before the Committee on Agriculture*. July 25, August 4, 6, 1953. 83rd Cong., 1st sess. Washington, DC: GPO, 1953. https://www.google.com/books/edition/Long_Range_Farm_Program/zRDiyAZsXY0C?hl=en&gbpv=1&bsq=family%20farm (accessed January 22, 2025).

U.S. Congress. *Morrill Act of 1862*. July 2, 1862. PL 37-108. 37th Cong., 2nd sess. https://www.archives.gov/milestone-documents/morrill-act.

U.S. Congress. *Morrill Act of 1890*. August 30, 1890. PL 111-122. 51st Cong., 1st sess. https://www.nifa.usda.gov/sites/default/files/asset/document/First%20and%20Second%20Morrill%20Act.pdf.

U.S. Congress. *Union Pacific Railroad Act of 1862*. July 1, 1862. 12 STAT 489. 37th cong., 2nd sess. https://www.archives.gov/milestone-documents/pacific-railway-act.

USDA Farm Service Agency. "Conservation Reserve Program (CRP)." 2024. https://www.fsa.usda.gov/resources/programs/conservation-reserve-program (accessed January 28, 2025).

USDA National Agricultural Library. "Granger Homesteads." May 15, 1935. https://www.nal.usda.gov/exhibits/ipd/small/exhibits/show/subsistence/item/52 (accessed January 27, 2025).

USDA National Agricultural Statistics Service. Iowa Field Office. https://www.nass.usda.gov/Statistics_by_State/Iowa/index.php (accessed January 22, 2025).

USDA National Organic Program. "Home." 2025. https://www.ams.usda.gov/about-ams/programs-offices/national-organic-program (accessed January 28, 2025).

USDA Natural Resources Conservation Service. "Conservation Stewardship Program." 2025. https://www.nrcs.usda.gov/programs-initiatives/csp-conservation-stewardship-program (accessed January 28, 2025).

USDA Study Team on Organic Farming. "Report and Recommendations on Organic Farming." July 1980. Available online at *Center for Inquiry*. https://centerforinquiry.org/wp-content/uploads/sites/33/quackwatch/usda_organic_1980.pdf (accessed January 23, 2025).

U.S. Department of Commerce. National Oceanic and Atmospheric Administration (NOAA). "Gulf of Mexico 'Dead Zone' Larger than Average, Scientists Find." August 1, 2024. https://www.noaa.gov/news-release/gulf-of-mexico-dead-zone-larger-than-average-scientists-find# (accessed January 25, 2025).

U.S. Department of Commerce. National Oceanic and Atmospheric Administration (NOAA). "Larger-than-Average Gulf of Mexico 'Dead Zone' Measured." August 3, 2021. https://www.noaa.gov/news-release/larger-than-average-gulf-of-mexico-dead-zone-measured (accessed January 27, 2025).

U.S. Department of Environmental Protection (EPA). "Climate Change Indicators: Atmospheric Concentrations of Greenhouse Gases." Last updated January 15, 2025. https://www.epa.gov/climate-indicators/climate-change-indicators-atmospheric-concentrations-greenhouse-gases (accessed February 6, 2025).

U.S. Department of Environmental Protection (EPA). "Gulf Hypoxia Action Plan 2008." Last updated May 31, 2024. https://www.epa.gov/ms-htf/gulf-hypoxia-action-plan-2008 (accessed January 27, 2025).

U.S. Department of Environmental Protection (EPA). Intergovernmental Panel on Climate Change (IPCC). "Global Greenhouse Gas Overview." 2014; 2022. https://www.epa.gov/ghgemissions/global-greenhouse-gas-overview (accessed January 28, 2025).

U.S. Department of Environmental Protection (EPA). "Registration of Dicamba for Use on Dicamba-Tolerant Crops." February 14, 2024. https://www.epa.gov/ingredients-used-pesticide-products/registration-dicamba-use-dicamba-tolerant-crops# (accessed January 25, 2025).

U.S. Department of Environmental Protection (EPA). "Sources of Greenhouse Gas Emissions." 2025. https://www.epa.gov/ghgemissions/sources-greenhouse-gas-emissions (accessed January 28, 2025).

U.S. National Archives. "Treaty Between the United States of America and the French Republic." Last updated May 1, 2022. https://www.archives.gov/milestone-documents/louisiana-purchase-treaty#:~:text=In%20this%20transaction%20with%20France,size%2C%20expanding%20the%20nation%20westward (accessed January 14, 2025).

U.S. National Archives. Code of Federal Regulations. "Nonsynthetic Substances Prohibited for Use in Organic Crop Production." Title 7, Sec. 205.602. https://www.ecfr.gov/current/title-7/subtitle-B/chapter-I

/subchapter-M/part-205/subpart-G/subject-group-ECFR0ebc5d139b750cd/section-205.602 (accessed January 25, 2025).

U.S. National Park Service. "Blackbird Hill." Last updated December 30, 2021, https://www.nps.gov/places/blackbird-hill.htm (accessed January 13, 2025).

U.S. Supreme Court. *Citizens United v. Federal Elections Commission*. 558 U.S. 310. January 21, 2010. https://supreme.justia.com/cases/federal/us/558/310/.

Van der Zee, Jacob. "Episodes in the Early History of the Western Iowa Country." *Iowa Journal of History and Politics* 11, no. 3 (July 1913): 323–63. https://babel.hathitrust.org/cgi/pt?id=wu.89067281931 (accessed January 11, 2025).

Vonnegut, Kurt. *Timequake*. New York: Berkley Books; Putnam, 1997.

Wallaces Farmer. "Long-Term Partnership Between USDA and Iowa State University Researchers Produces Innovative Results." *FarmProgress,* September 16, 2020. https://www.farmprogress.com/farm-life/partnership-works-to-improve-water-quality (accessed January 25, 2025).

Washington, George (President). "From George Washington to Joshua Holmes." Papers of George Washington. *Founders Online*, December 2, 1783. https://founders.archives.gov/documents/Washington/99-01-02-12127.

Waypoint Team. "Essential for Optimum Plant Growth." *Waypoint*, February 20, 2024. https://www.waypointcommodities.com/news/5-vital-reasons-why-nitrogen-is-essential-for-optimum-plant-growth (accessed January 27, 2025).

White Eagle. Quoted in "Top 10 Native American Quotes." *Xavier University*, n.d. https://www.xavier.edu/jesuitresource/online-resources/quote-archive1/native-american1 (accessed January 10, 2025).

World Health Organization (WHO). "Household Air Pollution." Last updated October 16, 2024. https://www.who.int/news-room/fact-sheets/detail/household-air-pollution-and-health (accessed January 17, 2025).

World Population Review. "Westphalia, Iowa, 2024." https://worldpopulationreview.com/us-cities/iowa/westphalia (accessed January 26, 2025).

Index

adaptation   196, 204
Age of Enlightenment   14, 16
Agrarian Revolt   41, 42
Agricultural Adjustment Act   106, 224
Agricultural Industrialization   103
Agricultural Mechanization   42, 60, 91
Alar   160
allelopathy   142
ammonia   120, 143, 145, 146, 182, 186, 197
anhydrous ammonia   77, 131, 145, 148
Animal Farm   104, 108, 115
Appleby, John   60
artificial intelligence (AI)   177
Atkinson, Colonel Henry   18, 19

Bacon, Francis   17
Baltimore Catechism   169
base acres   106, 107
baseball   44, 45, 77, 81, 163
Bearpaw Sea   11
beaver   8, 9, 12, 18
Belgian draft horse   34, 35, 64, 88
Berkshire pork   153
Bernardin, Cardinal Joseph   172
Berry, Wendell   11, 14, 27, 129, 215
big bang theory   170
Big Tech   203
Birds   74
Black gold   24
Black holes   170
Blackbird, Chief   8, 9
Blackmer, Alfred   148, 149
Blizzards   67, 68
Blue Baby Syndrome   183
Blum, Ben   48
Book of Genesis   169, 170, 172
book of Leviticus   85, 172
Boom and bust cycles   41–3, 105
Boyer River   12, 23
Broadleaf weeds   139
buffalo   16, 84

Buffalo farm equipment   129, 134, 138, 140
Buffalo ridge-till   129, 138–40, 149
Buffer strips   186, 211
Butz, Earl   104, 107, 131

C:N rates   143
California   16, 24, 76, 160–2
California Certified Organic Farmers (CCOF)   160
Camp Dodge   36
Cancer   136–8, 157, 164, 165, 183
carbon
  Capture   196
  dating   169
  pipeline   196, 198
  sequestration   72, 198, 206
Catholic immigration   25
Catholic Worker Movement   44
Center for Rural Affairs   110, 119
Cheap food policy   105, 124
Chief Joseph   10
Church attendance   175
Civil War   16, 22–4, 41–2, 56
Climate change   72, 74, 76, 98, 134, 142, 150, 154–5, 187–99, 203–13, 216–17
CO_2 188–98, 204
coal   16, 17, 34, 40, 130, 145, 175, 190, 196, 205
College of Agriculture   95, 97
Columbus, Christopher   14, 16
Commodity crops   16, 105–8, 146, 209
Commons   17
Complete Life Program   38–9, 45
composting   129, 133–4, 142–4, 148
Concentrated Animal Feeding Operations (CAFO's)   110, 144
Conservation Reserve Program (CRP)   197–8
consumerism   203
Cook stoves   54
Coolidge, Calvin   105
cooperatives   40–2, 175

corn hybridization   44, 105
corn yields   143, 148, 150, 211
corporate agriculture   105, 214
cotton   16, 42, 105–6, 138, 146
Council Bluffs   7, 12–13, 18, 22–3, 28, 108, 168, 209
Council of Nicaea   178
cover crops   140–2, 150–2, 191, 194
Covid   37, 83, 167, 199
cream Separator   63
creation   14, 169–75
Critical Race Theory   13
Croghan, Colonel George   19
crop rotation   78, 103, 106–7, 128, 133–4, 140–8, 152, 191, 194, 211

Darwin, Charles   169–70
Day, Dorothy   44
Dead Zone   2, 182, 183, 185
defiance   13, 30, 37, 53, 54, 84, 93, 94, 168
Delate, Kathleen   149
democracy   159, 214, 216
democratic Party   16, 104, 118, 214
Des Moines   10, 36, 48, 117, 118, 123, 153, 174, 175
desertification   211
DeWitt, Jerry   158
Dicamba   137–8
Dingman, Bishop Maurice J   123–4
diversity   2, 16, 72, 79, 105, 107, 140, 184, 194, 211
down syndrome   46, 92
draft horses   34–5, 59, 64, 67, 88
draft number   36, 55, 91, 96, 117, 125
Dubuque   22, 26, 27, 29
Duren, Rev. Hubert   38–40, 44–5, 48, 50–1, 54, 85, 175
Duren, Ryne   45
Dust Bowl   106

ear corn   84, 85
earling   30, 37, 54, 163
early spring annuals   141
ecology   97–8, 173, 190
economic constraints   11, 16
enclosure movement   17, 60
England   8, 14, 16–17, 21–2, 25–6, 40–1, 60, 178
ergot   142
Ertmer, Anna   26
ethanol   2, 133, 146, 182, 196–8
ethics, land, spiritual   123, 169, 187
Ettiene de Veniard Bourgmont   9
evolution   169, 171, 193

Exner, Rick   157–8

farm consolidation   111, 115, 131, 193, 199
farm operations curriculum   95
farm ownership   209–12
farm processing   161, 193, 199–200, 209, 212
farm programs   105–6, 159, 195, 213
Farmall Super M tractor   60
Farmer's Holiday   40
federalism   16
Feral pigs   115
First frost   84, 216
First Fruits   85, 168
Floyd, George   13
Fluorocarbons   189, 199
Flusche Brothers   28, 30, 48
food, deserts   200
food, local   153, 200, 212
food, prices   146, 200
food, processing   209
food, production   146, 155, 194, 199, 204, 208–9, 212
food, sovereignty   177, 196, 212
Foreign Agricultural Service (FAS)   164
Forestry   71, 81, 190
Fossil fuels   17, 40, 65, 188–91, 196–8, 205, 216
France   8, 14, 21–2, 24, 36
Fresh *vs.* frozen meat   153–4
frostbite   68
futures trading   107–8

G.I. Bill of Rights   56
gag orders   115
Gemmill, Witt   48
genetically modified organisms   41, 136–8, 153, 160
geothermal energy   205, 207
Germany   1, 25–7, 29–30, 37, 50–1, 141, 146
Glyphosate   129, 136–8
Gold   18, 24, 26, 42
Gophers, pocket   75–6
Grass waterways   74, 181, 186–7, 211
Great Grain Robbery   107
Great Plains   8, 18, 42, 106
Greenhouse gases   40, 142, 149, 188–99, 204–5
Groundwater Protection Act   183–4

Haber/Bosch process   145
Hardin, Clifford   104
Harkin, Senator Thomas   118
Harlan   10, 28, 36, 39, 46, 54, 67, 84, 94, 107, 109, 114, 117–18, 162–4

Harpoon fork   61
Hassebrook, Chuck   110
Headlands   72, 80, 186
Herbicides   77–78, 108, 129–30, 134–40, 145, 150, 177, 194–5
Heer, Fridolin   29
high interest rates   113, 124
high mass   35
high nitrates   183–5
Hodapp, Rose   33
Hodapp, Wendell   33
Hoefner, Ferd   159–60
Homestead Act   23, 42
Honduras   164, 209
Horses   29, 33–5, 59–60, 64–7, 87–8
Howard, Sir Albert   134
Hudson Bay   9
Humphrey, Hubert   118
hunting   12, 19, 79, 83–6
Hybrid Rye   141–2, 194–5, 199
Hypoxia   181, 185

immigration   24–7, 125, 203, 212
Indian Summer   84
Industrial Revolution   14, 16–17, 22, 24, 38, 41
initiative 300   110
Intergovernmental panel on climate change (IPCC)   190
International Harvester   60
Investment Tax Credit   110–11
Iowa Environmental Council   185
Iowa Groundwater Protection Act   183–4
Iowa Institute of Cooperatives   132
Iowa Journal of History and Politics   12
Iowa Nutrient Reduction Strategy   185
Iowa State University   1, 55, 65, 94–5, 97, 104, 109, 118, 130, 157, 208, 211
Ioway   12, 55
Irradiation   160

Jackson, Andrew   15
Jackson, Randy   191, 195
John Deere   42, 59, 61, 176
Jubilee year   172–3

Kallem, Larry   132
Kanesville   28
Keeney, Dennis   184
Kettler, A. H.   28
Kirschbaum, Katie   36
Knecht Ruprecht   50
Know Nothings   25

land-grant university   23, 55–6, 158

land prices   43–4, 108, 113–14, 174
LaSalle, Robert   22
Laudato Si   172, 178, 181
legumes   75, 78, 105–6, 133, 140–2, 144–6, 148, 152, 182, 191, 193–5, 198
Leopold Center   183–4
Lewis and Clark   8, 10, 18–19, 21, 23, 53
Lewis, C. S.   178–9
Liebman, Matt   194–5
Life Magazine   104, 118
Ligutti, Msgr. Luigi   174–5
Lincoln, Abraham   23, 33, 55–6, 131
Lipson, Mark   161
Liquid manure   109, 114, 120, 142, 144, 199
Livestock antibiotics   130, 144, 151–3
Livestock confinement   109–10, 114–15, 120, 144, 152–3, 191, 194, 199, 212, 215
Louisiana Purchase   18, 21–2, 26

McCormick Cyrus   42, 60
McCormick Deering   60
Mckenzie, Alexander   18
Manifest Destiny   15–16, 56
Marquette and Joliet   9
Maximum Return to Nitrogen Calculator   185
Mertens, Henry, Katie, Ellen   36–7
Methane   120, 142–4, 188, 190, 192–5, 199, 207
Methanogenic archaea   193
Milk and Honey   164
Mississippi River   2, 9–10, 12, 15, 21–3, 26–7, 55, 147, 181–2, 185
Missouri River   1, 7–13, 18–19, 22–3, 53, 97, 147, 161, 181–2, 187, 217
Morrill Act   55–6
Mortgage Lifters   109

Napoleon Bonaparte   22
National Register of Historic Places   168
National Sustainable Agriculture Coalition (NSAC)   159–61
Native Americans
  Fox   12, 55
  Mandan   10
  Omaha   8–9, 12–13
  Otoe   12–13, 18
  Sac   12, 55
  Sioux   8–9, 12
  Ponca   8, 13
  Potawatomi   13
Natural Pork   153
New Holland square baler   62
Niman Ranch pork   153
Nishnabotna River   12–13, 28, 75

Nitrates   146, 182–3, 185, 194
Nitrogen Cycle   2, 181–2
Nitrogen trials   149
nitrous oxide   149, 189–90, 194, 197–9
Nuns   49, 52, 54

oats   35, 42, 59–61, 64, 74, 78–9, 85, 87, 105, 131, 133, 140–5, 148–9, 152, 191–5
oligopolies   112, 193, 215
organic certification, transition, principles   135, 145, 151–5
Organic Farming Research Foundation (OFRF)   160–2
Organic Prairie   41, 192
Organic Research and Education Initiative (OREI)   161
Organic Valley   41, 114, 144, 154
Orwell, George   94, 104, 115
O'Sullivan, John   15
Overproduction   105, 174
Ozone   189, 199, 206

Panama   30, 37, 53, 163, 169
Paris Climate Accord   188, 210
parity   105–6
Partners of the Americas   164
Perennials   133, 146, 186, 198, 207, 209, 211–12
pesticide drift   108, 137–8
phosphorous   2, 145, 147, 181, 183, 185
photosynthesis   2, 142, 150, 170, 182, 188–91
polarization   203, 205, 214
politics   213–15
popcorn   45, 153
Pope Francis   2, 93, 172, 178, 181, 203
Pork consolidation   111–15
pork prices   111–14
Portsmouth   30, 37, 53, 169
potassium   11–12, 133, 145–8
Practical Farmers of Iowa (PFI)   132, 135, 138–9, 148–51, 157–8, 161, 164
Project Drawdown   206
Public Service and Administration in Agriculture   164

railroads   21–3, 28, 30, 38, 41–2, 91
reaper   42, 60
Red Angus beef   144, 192
Regenerative farming   131, 177, 206, 209, 212–13
religion   14, 24, 30, 38, 167–79, 216
Reno, Milo   40
Reorganized Church of Jesus Christ of Latter-Day Saints   28

Republican Party   16, 25, 105, 118, 214
Rerum Novarum   39
Robotics   177–8
Rochdale Society   40–1
Rodale Research Initiative   131, 160
Rodale, Robert   131
romantic nationalism   15
Rosmann, Adeline   36
Rosmann, Daniel   33, 82, 125, 162–5
Rosmann, David   12, 72, 124–5, 132, 162–5
Rosmann, Ellen   27, 35–7, 45–6, 49, 52, 60, 63–4, 67, 71, 79–81, 87, 92, 95–6, 108, 118, 123
Rosmann, Ellen Walsh   164
Rosmann, George   26, 29
Rosmann, Joe   33, 36, 60
Rosmann, John   26, 29
Rosmann, Joseph   45, 59, 63, 66, 76, 92, 121
Rosmann, Ken   52, 151, 153
Rosmann, Larry   46, 63, 92, 125
Rosmann, Leonard   36–7, 52, 60–3
Rosmann, Mark   125, 162–4
Rosmann, Mike   63, 66, 76, 78, 92–3, 97, 113
Rosmann, Ray   12, 26–7, 33–8, 40, 45–6, 48–9, 51–2, 59–67, 71, 77–8, 81–2, 85, 87–8, 92, 94–8, 107–9, 118, 120–1, 123, 163–5
Rosmann, Rebecka Thompkins   72, 164
Rosmann, Louis (Sr.)   45
Rosmann, Virginia Lehner   164
Ruminants   133, 192–3, 211
Rural Electrification Cooperative (REC)   40
Russia   107, 113, 131, 216

Saint Boniface Church   1, 27, 29–30, 38, 85, 167–8
Saint Boniface School   25, 37, 39, 47–54, 93–4
Saltpeter   146
Sandoz, Jules   9
Sandoz, Mari   9
Sasse, Antonia   30–1
Sawyer, John   185
School Sisters of St. Francis   45, 48–9, 92
Schwarte, Ann   47
Scowcroft, Bob   160–1
Second Summer   84
Self-propelled combine   84
Sewage sludge   160
Sexuality   93
Shelby County   1, 10, 12, 23, 26–8, 43–4, 54–5, 66, 84, 92, 108, 176, 209
Sheridan, Lieutenant General Phil   16
Sisal   59–62
Sister Bartholomew Marie   53

Sister Jarento   52
Sister Romaine   50
slavery   16, 23–5, 56
small farm energy project   119
small grains   79, 85, 107, 141–5, 148, 152, 186, 191, 194–5, 199
smallpox   8, 16
sodium nitrate   146
soil conservation   106, 189
soil erosion   2, 105, 129, 186–7, 189, 191, 194, 198, 207
soil quality   16, 98, 107, 133, 142, 144, 146, 150–1, 173, 177, 189, 194–5, 210–12
solar energy   65, 119, 189, 195, 204–5, 207
Solar Farrowing house   113, 119
soul   171, 178–9
Spain   8, 14, 21–2
Spanish Flu   36
spirit   178
St. Nicholas   50–1
St. Paul High School   53
Stadtfeld, Curtis   73
Steam power   16, 60–1
Steamboats   18
Steel moldboard plow   42, 74–5, 80
Stewardship   10, 73, 125, 169, 173, 200
Strangers and Guests   172, 174
Subsistence farming   17, 42, 175
Succotash   141, 143–4
summer annuals   141
Summit Ag. CO2 pipeline   196
Sustainable Agriculture Research and Education (SARE)   149, 159
Swing Riots   60

tall grass prairie   7, 11–12, 22, 28, 30, 71–4, 79–80, 186, 198, 212
technology, appropriate   2, 176–7
technology, costs   2, 176–7
Testament, New   171–3
Testament, Old   85, 171–2
Thompson, Dick and Sharon   129–34, 140, 143, 150, 158, 213
Thresher   42, 59–61
Tiger cages   118
tobacco   16, 34, 158
Tofu Soy   151
trypsin inhibitor   141

Ukraine   199, 216
United States Constitution   25, 106, 214
United States Peace Corps   164, 209

Vietnam War   91–2, 94–6, 117–18
vocations   1, 53, 55, 91, 174

Westphalia Community Credit Union   39
Westphalia Consumer's Cooperative Association   39
Westphalia, Iowa   1, 10, 21, 25–30, 33–4, 36–40, 44–5, 47–55, 61, 80–4, 91, 93–4, 97, 103, 107, 117–18, 163, 167–8, 175–6, 217
Westphalia, Kansas, Missouri, Texas, Michigan, Indiana   27, 30
wetlands   23, 181, 185–6, 211
Whigs   15–16
White Eagle, Chief   9
wildlife   71–89, 181, 205
wind energy   195, 204, 206
women religious   175
World War One   24, 27, 33, 37, 42, 105, 145–6
World War Two   40, 42–4, 54–6, 60, 73, 106, 125, 146, 175, 216

Yellowstone   18–19

About the Author

Ron and Maria Vakulskas Rosmann, along with two of their sons, David and Daniel, and families, operate Rosmann Family Farms near Harlan in western Iowa. Their third son, Mark, is an agricultural attaché with the USDA. Ron is a fourth-generation farmer. Ron stopped the use of pesticides in 1983 on their 700-acre farm, and the farm has been certified organic since 1994.

Rosmann Family Farms is a diversified organic crop and livestock farm which produces corn, oats, soy, hay, pasture, barley, wheat, field peas, and multiple annual forages and cover crops. The farm has a cow-calf operation, a farrow-to-finish hog operation, and egg layers. The Rosmann farm features a farm store called "Farm Sweet Farm."

Ron helped launch Practical Farmers of Iowa (PFI) in 1986 and has completed over fifty on-farm research trials with PFI and the Iowa State University Organic Program. Ron was on the board of the Organic Farming Research Foundation and National Catholic Rural Life.

Ron has been a policy advocate for more sustainable, diversified family-farm and food systems for his entire career.

www.ingramcontent.com/pod-product-compliance
Lightning Source LLC
Chambersburg PA
CBHW081211230426
43666CB00015B/2718